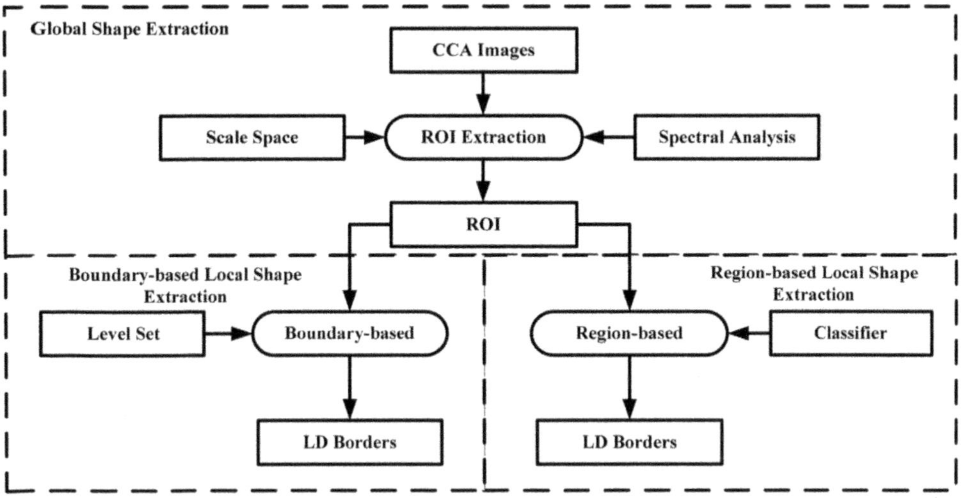

Figure 8.6. Region and boundary-based model for CCA segmentation.

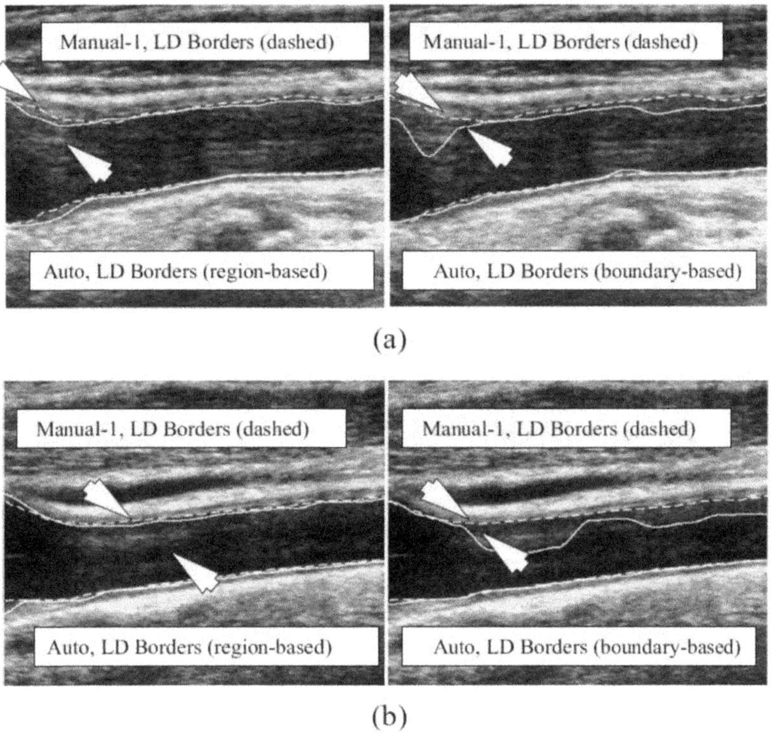

Figure 8.7. Results from the region- and boundary-based model (image source: [29]).

segmentation. CNNs use convolution filters in multiple layers to extract feature maps from the original image. It also uses pooling layers for dimensionality reduction. FCNs on the other hand use a series of upsample layers and a skipping operation for segmenting the image.

Suri and his team [35] used a combination of CNNs and FCNs for lumen segmentation from CCA images. The model used consisted of three phases: image pre-processing, DL and performance evaluation as shown in figure 8.8. The pre-processing multiresolution framework consists of four stages: cropping, binarization, reduction and downsampling. In the cropping stage all background information is removed. In the binarization stage ground truth (GT) information in the form of manually delineated lumen border coordinates from GT images are collected for training. During the US probe usage, the best information is collected from the centre of the probe due to full contact with the neck region. However, the image information collected at the sides of probe become blurry due to partial contact. Therefore, 10% reduction is done from the sides of the CCA images to retrieve only the crisp information. Further, 50% downsampling is done for speedy computation. The images are input to the DL phase. The DL phase consists of two stages: an encoder for feature extraction and a decoder for segmentation. The encoder constitutes the first 13 layers (13 convolution layers with intermediate five max pooling layers) of the VGG16 [36] CNN. The features extracted from the encoder are fed into the decoder. The decoder consists of the three upsample layers of the FCN. It uses two skipping operations to merge outputs from the last two max pool layers with outputs of the first two

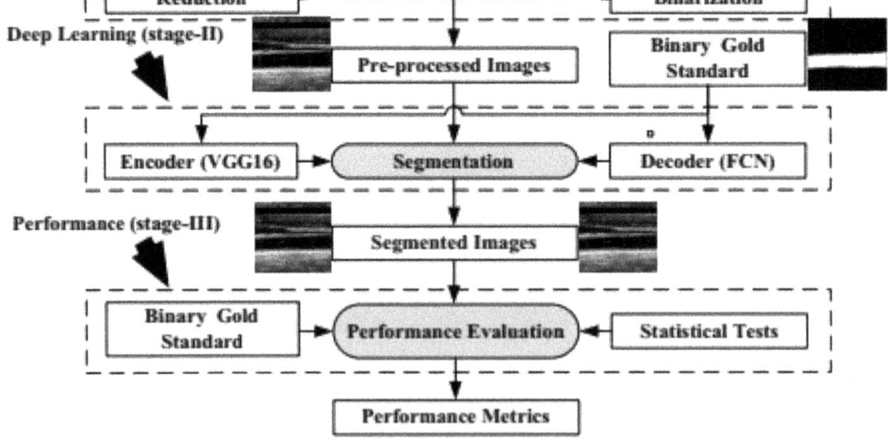

Figure 8.8. Deep learning model for LD measurement from US CCA images (image source: [35]).

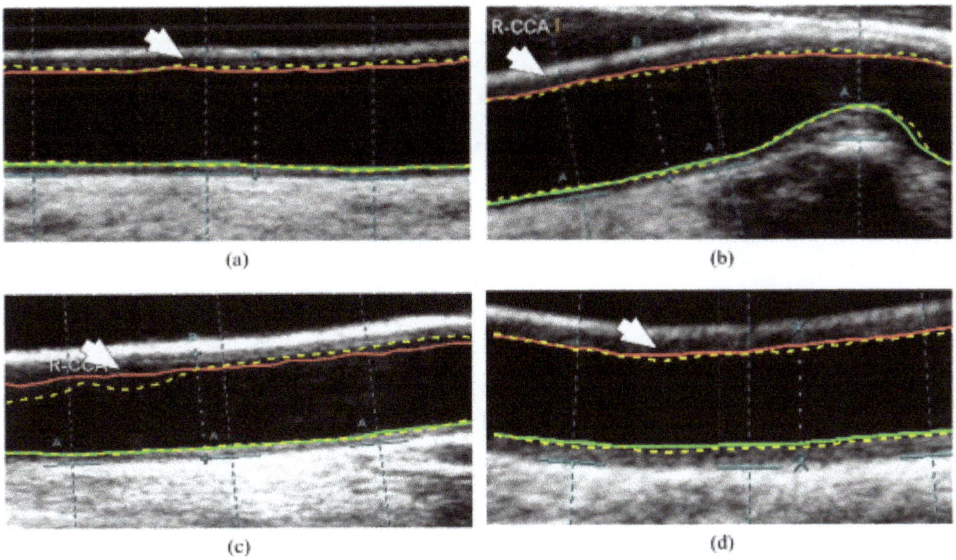

Figure 8.9. DL segmentation is shown in red and green, GT is shown in dashed yellow (image source: [35]).

upsample layers, respectively. The decoder segments the images, which are then fed into the performance evaluation stage for computing the statistics.

Altogether, three DL models were used for three GTs from three radiologists of different experience levels. A total of 407 US CCA images were used for the experiment. The mean LD for DL1 and GT1 was 6.09 ± 0.94 and 6.06 ± 0.91 mm, respectively. For DL2 and GT2, the corresponding mean LD was 6.05 ± 0.91 and 5.91 ± 0.88 mm, respectively. Finally, for DL3 and GT3, the mean LD was 6.09 ± 0.94 and 6.00 ± 0.90 mm, respectively. The mean LD error for DL1, DL2, and DL3 was 0.19 ± 0.27, 0.23 ± 0.23, and 0.21 ± 0.19, respectively. The diagrammatic representation of segmentation is shown in figure 8.9.

8.2.2.2 Carotid intima segmentation measurement using deep learning

Suri and his team [37] used a four-phase process for cIMT measurement from US CCA images. The four phases are: multiresolution, deep learning, boundary extraction and performance measurement. The multiresolution phase function is similar to that discussed earlier in section 8.2.3. It also constitutes cropping, reduction, downsampling and binarization stages. However, the GT images are divided into two groups. The first group is the MA delineated GT images and the second group is the lumen border-delineated GT images. Two DL models of the same configuration as section 8.2.3 are used in the second phase. The first DL model is used for MA border segmentation (DL-MA) and the second model is used for lumen segmentation (DL-L). The MA delineated GT images and lumen border-delineated GT images are fed into DL-MA and DL-L models for segmentation. Finally, the MA segmented and lumen segmented images are fed into the boundary extraction phase. It constitutes of two stages: border extraction and calibration.

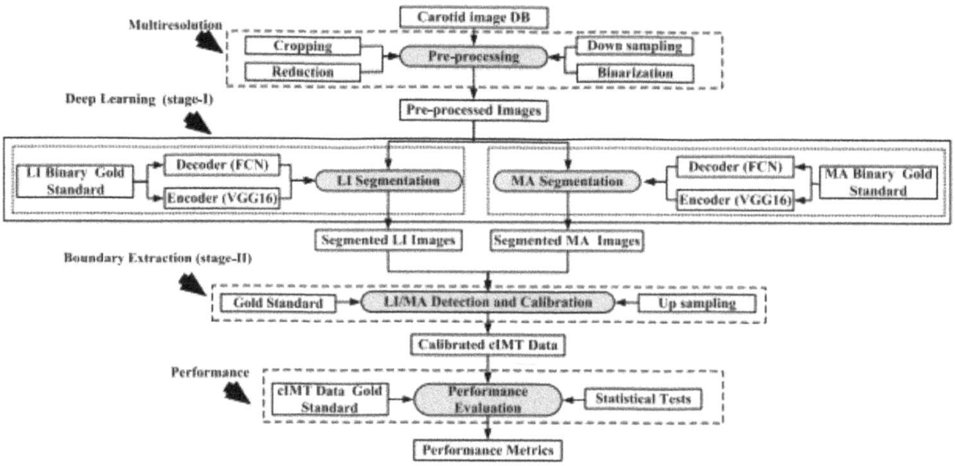

Figure 8.10. DL model for cIMT measurement (image source: [37]).

The LI-far wall and MA-far wall are delineated in the border extraction phase. These LI-far and MA-far walls are then calibrated using a least-squares model [38]. Finally, cIMT is measured between LI-far and MA-far wall using polyline distance method. Two GTs from two different radiologists were used for training. A total of 396 US CCA images were used. The cIMT error for DL1 and DL2 was 0.126 ± 0.134 and 0.126 ± 0.100 mm, respectively. The diagrammatic representation of the model is given in figure 8.10. The results for segmentation is shown in figure 8.11.

8.3 Carotid plaque characterization using machinc lcarning and deep learning approaches

Carotid plaque is made of lipids, cholesterol, calcium and other tissues. In terms of risk, carotid plaque is characterized into two categories: symptomatic and asymptomatic. Symptomatic plaque is the case where the patient shows symptoms of disease. In the case of asymptomatic class, the patient does not show any symptoms, but regular monitoring is required. An example of symptomatic and asymptomatic plaque is given in figure 8.12. Earlier, the grayscale median (GSM) [39], which is the median value of all pixels in the plaque, was used to characterize high-risk and low-risk plaque. It is observed that a GSM value of 32 or less corresponds to high-risk plaque. However, GSM value may vary in images from different OEMs. Diagnosis of symptomatic and asymptomatic plaques using ML/DL techniques can help clinicians categorize patients into high risk and low risk. Based on this diagnosis, better decision making regarding disease and its treatment can be made. There are several computer-based techniques for plaque characterization, which can characterize plaque into symptomatic or asymptomatic, fibrous or non-fibrous etc. These techniques are again divided into two classes: ML and DL. A general discussion is made on these two techniques for plaque characterization followed by examples.

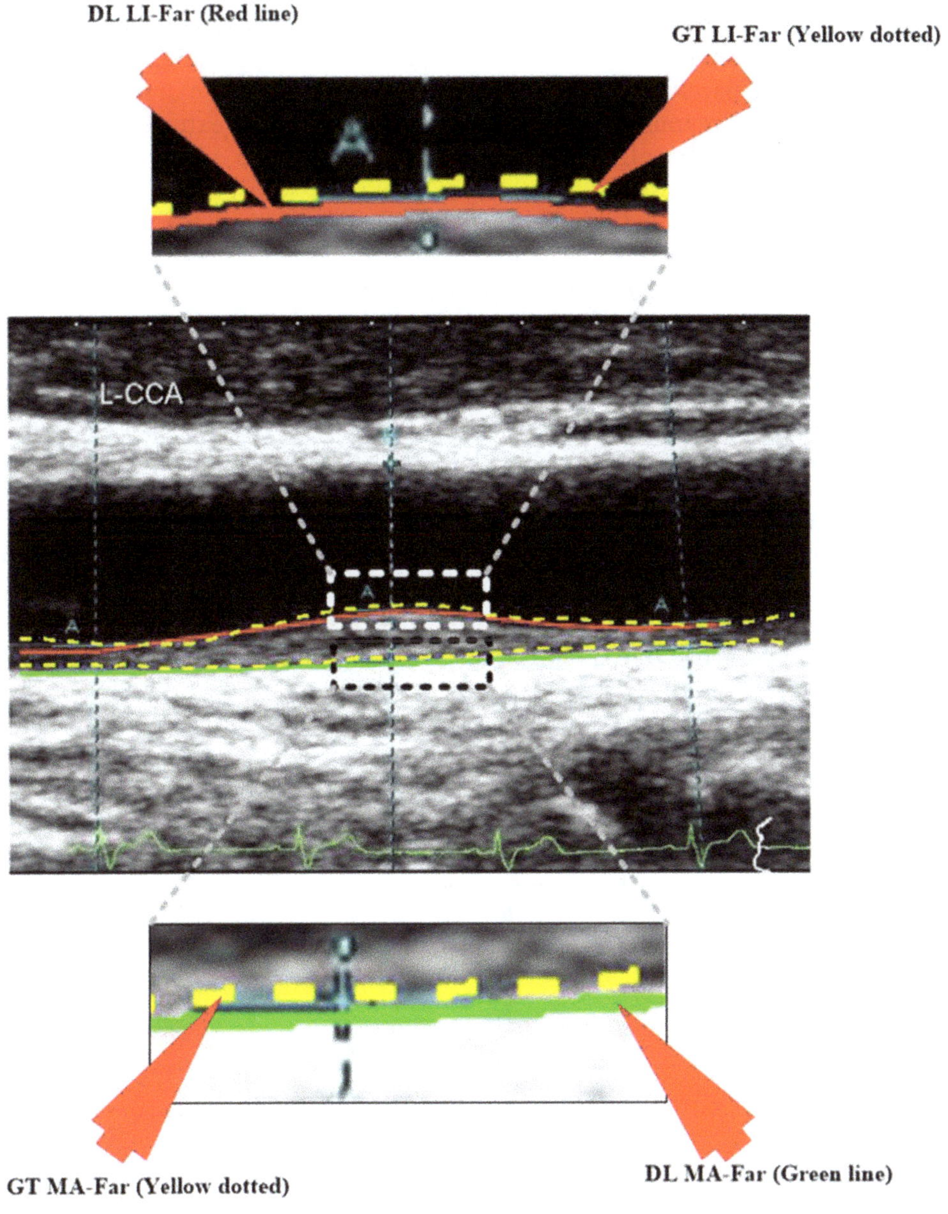

Figure 8.11. DL segmentation is shown in red and green, GT is shown in dashed yellow (image source: [37]).

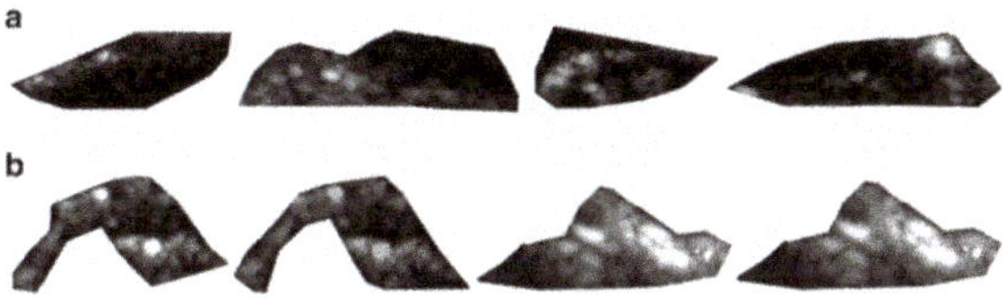

Figure 8.12. Row (a) symptomatic plaque; Row (b) asymptomatic plaque (image source: [39]).

8.3.1 ML-based technique for plaque characterization

The ML-based techniques consist of two stages: feature extraction and training/testing in the ML paradigm. The ROI consisting of plaque is extracted from the carotid image either manually or via automated techniques. Features are extracted from plaque ROI. ML algorithms are applied on these plaque features for characterization. A process model of this approach is shown in figure 8.13. Two techniques using ML methods are given in the next two sections 8.3.1.1 and 8.3.1.2.

8.3.1.1 Plaque classification using support vector machine and AdaBoost

Suri and his team [40] used two ML-based classifiers, AdaBoost [41] and SVM [17], for plaque characterization. 346 carotid plaques images were collected. Plaque ROI was extracted based on echogenic measures. Texture-based features were extracted from these plaque images based on statistical properties of pixels such as standard deviation, entropy, symmetry, run percentage etc. Other measurements based on the grey level co-occurrence matrix and run-length matrix are also collected such as: symmetry, entropy etc. The features extracted are fed into the AdaBoost and SVM classifier. AdaBoost is an ensemble classifier which achieves results by combining accuracy from different classifiers. Some of the classifiers are a least-square (LS) method [42], a maximum likelihood (ML) [43] algorithm, a normal density discriminant function (NDDF) [44], Pocket [45] and Stumps [46]. The SVM classifier used linear, poly and radial basis kernels for computing the separating hyper plane between two classes. SVM recorded better accuracy, sensitivity and specificity results, which are: 82.4%, 82.9% and 82.1%, respectively. AdaBoost recorded

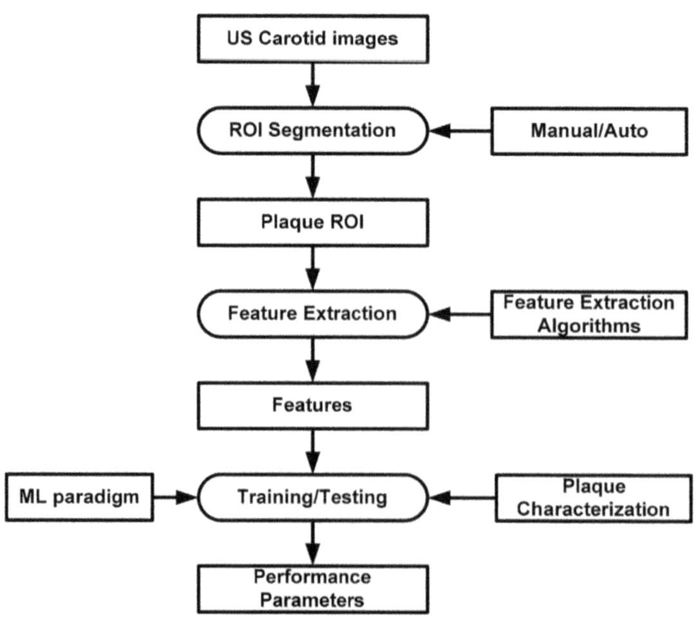

Figure 8.13. Process model showing plaque characterization using ML-based methods.

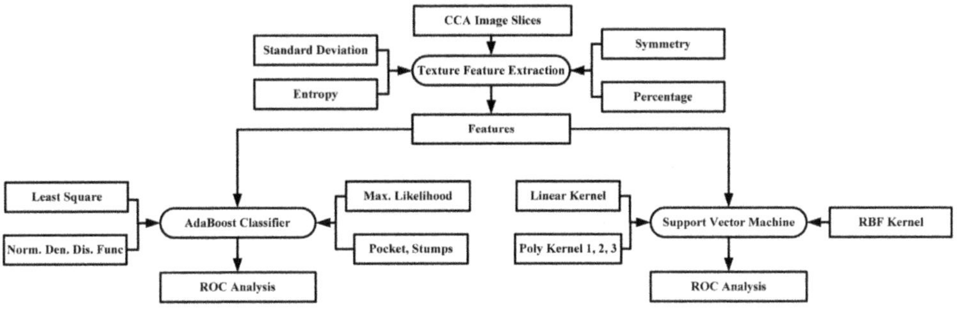

Figure 8.14. Ml model for plaque characterization.

accuracy, sensitivity and specificity of 81.7%, 82.3% and 81.9%, respectively. A model of the methodology is given in figure 8.14.

8.3.1.2 Automated plaque classification using ML-based technique

For the purpose of plaque characterization [47], US CCA images were collected from two datasets: UK and Portugal (146 + 346 = 492). Plaque was extracted using echogenic measures. Grayscale texture features were extracted using GLCM and fuzzy run-length matrix (FRLM) [48]. The classification algorithms which were applied were: SVM, Gaussian mixture model (GMM) [49], radial basis probabilistic neutral network (RBPNN) [50], decision tree [51], K-Nearest Neighbour (KNN) [52], Naïve Bayes Classifier (NBC) [53] and the Fuzzy Classifier model (FCM) [54]. Among all the classifiers, FCM gave the highest accuracy, sensitivity and specificity at 93.1%, 99% and 80%, respectively. The training and testing model is given in figure 8.15.

8.3.2 Plaque characterization using DL model

The DL is an evolving field in the ML domain. The key difference between DL and ML techniques is that DL is capable of automated feature extraction and thus bypassing the hand-made feature extraction model used by ML-based techniques. The US plaque images are fed into the DL model after pre-processing. The DL model trains and tests on labelled plaque images to extract high-level features and perform characterization over them. The process model of the DL plaque characterization system is shown in figure 8.16. An example of usage of a CNN model for plaque characterization is given in section 8.3.3.1.

8.3.2.1 Plaque characterization using CNN

In this work [55], CNNs have been used for plaque classification. For this purpose, 56 images were collected. A patch-based approach was used where these 56 images were translated to 90,000 images for better plaque patch classification. The plaque tissue was divided into three classes: lipid core, fibrous tissue, and calcified tissue. The CNN architecture consisted of four convolution layers with a kernel size of

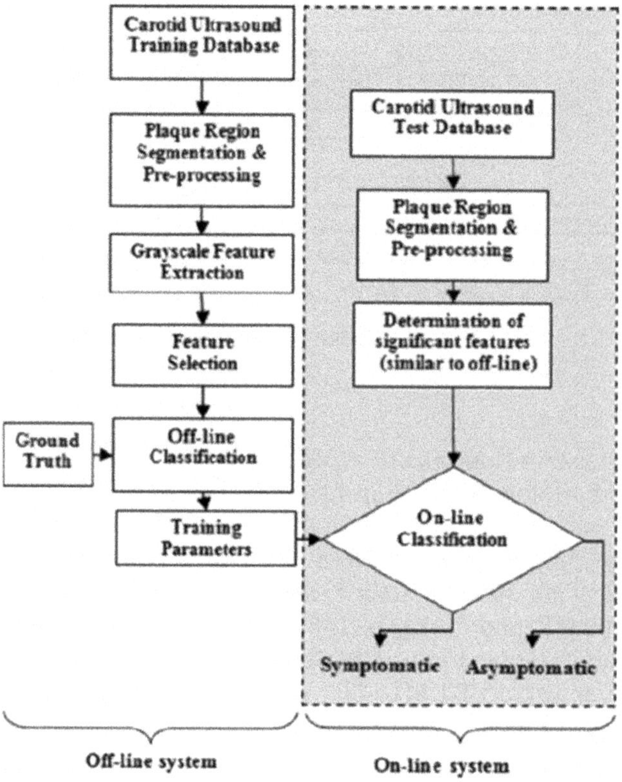

Figure 8.15. Ml training/testing model for plaque characterization (image source: [47]).

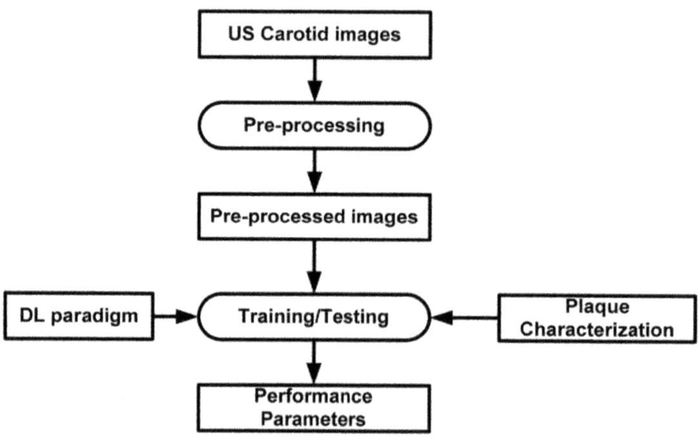

Figure 8.16. DL-based plaque characterization model.

dimension 3×3. It was followed three, fully connected layers, followed by leaky ReLu layer [56]. In addition to CNN, SVM was also used for classification. However, CNN gave a high accuracy of 78.5% when compared with SVM which gave an accuracy of 14.3%.

8.4 Discussion

The main focus of this paper is stroke risk estimation from CCA segmentation and measurement and plaque characterization. Both of methods are based on either ML or DL-based techniques. A thorough investigation is made on each of the methods by ML- and DL-based techniques. The ML-based techniques apply characterization algorithms on features extracted from US CCA images. The DL-based techniques learn directly from the images, skipping the feature extraction part. A comparison on both methods using ML and DL techniques is made in the following subsections.

8.4.1 Benchmarking on CCA segmentation and measurement

The benchmarking of several ML and DL techniques for CCA segmentation and measurement is given in table 8.1.

8.4.2 Benchmarking on CCA plaque characterization

The benchmarking table of ML and DL methods for plaque characterization is given in table 8.2.

A short note on vulnerable plaque

The most common reason for myocardial infarction or sudden death is occurrence of thrombosis in the coronary artery or CCA [57, 58]. Plaque rupture is considered 60%–65% responsible for the thrombosis [59]. There are other causes of coronary or

Table 8.1. Benchmarking table for CCA segmentation and measurement.

S. No.	Method	Features	No. of images	Performance
1	CALEX, fuzzy k-means [26, 27]	Intensity seed points	665	cIMT error: 0.191 ± 0.217 mm
2a	Region-based [29]	Scale-space	404 (JP*) 300 (HK*)	LD (JP*): 6.35 ± 0.95 mm LD (HK*): 6.20 ± 1.35 mm
2b	Boundary-based [29]	Scale-space	404 (JP*) 300 (HK*)	LD (JP*): 5.90 ± 0.97 mm LD (HK*): 5.43 ± 1.24 mm
3	DL for lumen segmentation [35]	Deep features	407	Mean LD error DL1: 0.19 ± 0.27 mm DL2: 0.23 ± 0.23 mm DL3: 0.21 ± 0.19 mm
4	DL for cIMT measurement [37]	Deep features	396	Mean cIMT error DL1: 0.126 ± 0.134 mm DL2: 0.126 ± 0.100 mm

* JP: Japan; HK: Hong Kong.

Table 8.2. Benchmarking table for CCA plaque characterization.

S. No.	Method	Features	No. of images	Performance
1a	SVM [40]	GLCM*, GLRLM*	346	Acc*: 82.4% Sen*: 82.9% Spec*: 82.1%
1b	AdaBoost [40]	GLCM, GLRLM	346	Acc: 81.7% Sen: 82.3% Spec: 81.9%
2	FCM* [47]	GLCM, FRLM*	492	Acc: 93.1% Sen: 99.0% Spec: 80.0%
3	CNN [55]	Deep features	56	Acc: 78.5%

*GLCM: grey level co-occurrence matrix; GLRLM: gray level run-length matrix; Acc: accuracy; Sen: sensitivity, Spec: specificity; FRLM: fuzzy run-length matrix; FCM: fuzzy classifier.

CCA thrombosis, i.e., mental or physical stress, increase in coagulability, etc [60]. The descriptive term to define a vulnerable plaque which is about to rupture (or ruptured plaque) is thin-cap fibroatheroma (TCF). TCF is characterized by a large lipid-rich necrotic core which is separated from the blood stream by a thin layer of fibrous tissue i.e., fibrous cap (FC).

The lipid core is composed of free cholesterol, cholesterol crystals, cholesterol esters derived from lipids, and also lipids from death of foam cells, interplaque haemorrhage, etc [61]. Further, this lipid core also contains prothrombotic oxidized lipids making it highly thrombotic to the circulating blood, making the plaque highly vulnerable. It has been seen that plaque area-wise, the necrotic core is significantly higher (34 ± 17 %) for ruptured plaque than TCF (23 ± 17 %). Also, the mean calcification score of ruptured plaque is significantly higher than TCF. The FC is characterized into three classes. Intact FC is associated with low-risk plaque rupture, thin FC is associated with mild-risk rupture and fissured FC is associated with high-risk rupture. Vulnerable plaques are characterized by thin FC whose size is less than 65 μm.

The presence, number, identification and risk characterization of TCF in coronary and carotid artery can be done by both invasive and non-invasive imaging modalities. For coronary arteries, TCF with large plaque burden can be identified by radio frequency intra vascular ultrasound (RF-IVUS). RF-IVUS technique however, cannot accurately distinguish calcific plaque and non-calcific tissues due to acoustic shadowing. Further, the spectra that differentiates the calcium and necrotic core overlap, adding to inaccuracy. That said, however, RF-IVUS can strongly predict non-culprit lesion-related major adverse cardiovascular events at a median follow-up of 3.4 years, making RF-IVUS validated imaging modality for TCF characterization. The imaging modality used for common carotid artery TCF detection is MRI, which characterizes TF as intact, thin and fissured. The thick FC appears as a juxtaluminal band on low signal, which is identifiable. However, this dark band is absent in thin TCF. The ruptured plaque is identified by absence of the dark band and presence of a bright grey region adjacent to the lumen, which represents the plaque haemorrhage.

In this regard, ML and DL algorithms can be applied for patient risk stratification based on identification and characterization of two features of TCF: necrotic core and FC. The amount of calcification of the necrotic core can be used to characterize the plaque as vulnerable or non-vulnerable. This can be done by using image features such as brightness, contrast, and texture, which can differentiate between a calcified and non-calcified necrotic core. Also, the characterization of the FC is based on the presence or absence of the dark band as well as presence of a grey bright region signifying haemorrhage. ML and DL models can be trained using two classes of images, calcified and non-calcified necrotic cores, to identify vulnerable plaque. Similarly, ML/DL models can be trained using thick, thin and ruptured plaque images to characterize low-risk, mild-risk and high-risk patients.

References

[1] National Center for Health Statistics 2018 https://www.cdc.gov/nchs/nhanes/index.htm accessed 2 October 2018

[2] MEMBERS, WRITING GROUPEmelia J B, Michael J B, Stephanie E C, Mary C, Sandeep R D and Rajat D *et al* 2017 Heart disease and stroke statistics—2017 update: a report from the American Heart Association *Circulation* **135** e146

[3] National Center for Health Statistics 2014 Mortality multiple cause micro-data files, 2014: public-use data file and documentation: NHLBI tabulations http://cdc.gov/nchs/data_access/Vitalstatsonline.htm#Mortality_Multiple accessed 2 October 2018

[4] Heidenreich P A, Justin G T, Olga A K, Javed B, Kathleen D, Michael D E and Finkelstein E A *et al* 2011 Forecasting the future of cardiovascular disease in the United States: a policy statement from the American Heart Association *Circulation* **123** 933–44

[5] Suri J S, Kathuria C and Molinari F 2010 *Atherosclerosis Disease Management* (Berlin: Springer Science & Business Media)

[6] Libby P, Karin E B and Alan R T 2016 Atherosclerosis: successes, surprises, and future challenges *Circ. Res.* **118** 531–4

[7] https://mayfieldclinic.com/pe-carotidstenosis.htm

[8] Sato K, Ogoh S, Hirasawa A, Oue A and Sadamoto T 2011 The distribution of blood flow in the carotid and vertebral arteries during dynamic exercise in humans *J. Physiol.* **589** 2847–56

[9] Dhawan S S *et al* 2010 Shear stress and plaque development *Expert Rev. Cardiovasc. Ther.* **8** 545–56

[10] He X, Peter P C, Kischell E R and Chiang A M 2003 Ultrasound imaging system *U.S. Patent 6,638,226 issued*

[11] Punchard W F B and Robert D P 1988 Magnetic resonance imaging systems *U.S. Patent 4,733,189 issued*

[12] Hsieh J 2009 *Computed Tomography: Principles, Design, Artifacts, and Recent Advances* (Bellingham, WA: SPIE)

[13] Narayanan R, Kurhanewicz J, Shinohara K, David Crawford E, Simoneau A and Suri J S 2009 MRI-ultrasound registration for targeted prostate biopsy *Biomedical Imaging: From Nano to Macro, 2009. ISBI'09. IEEE Int. Symp. on* (Piscataway, NJ: IEEE) pp 991–4
[14] Saba L, Montisci R, Molinari F, Tallapally N, Zeng G, Mallarini G and Suri J S 2012 Comparison between manual and automated analysis for the quantification of carotid wall by using sonography. A validation study with CT *Eur. J. Radiol.* **81** 911–8
[15] El-Baz A and Jasjit S S (ed) 2011 *Lung Imaging and Computer Aided Diagnosis* (Boca Raton, FL: CRC Press)
[16] Haykin S S 2009 *Neural Networks and Learning Machines* 3 (Upper Saddle River, NJ: Pearson)
[17] Cortes C and Vladimir V 1995 Support-vector networks *Mach. Learn.* **20** 273–97
[18] Huang G B, Zhu Q Y and Siew C K 2006 Extreme learning machine: theory and applications *Neurocomputing* **70** 489–501
[19] Yegnanarayana B 2009 *Artificial Neural Networks* (Delhi: PHI Learning Pvt. Ltd)
[20] Hopfield J J 1988 Artificial neural networks *IEEE Circuits Dev. Mag.* **4** 3–10
[21] LeCun Y, Yoshua B and Geoffrey H 2015 Deep learning *Nature* **521** 436
[22] Goodfellow I, Bengio Y, Courville A and Bengio Y 2016 *Deep Learning* 1 (Cambridge, MA: MIT Press)
[23] Virmani R, Allen P B, Kolodgie F D and Farb A 2003 Pathology of the thin-cap fibroatheroma: a type of vulnerable plaque *J. Interv. Cardiol.* **16** 267–72
[24] Torvik A, Svindland A and Lindboe C F 1989 Pathogenesis of carotid thrombosis *Stroke* **20** 1477–83
[25] Roux S, Carteaux J P, Hess P, Falivene L and Clozel J P 1994 Experimental carotid thrombosis in the guinea pig *Thromb. Haemost.* **71** 252–6
[26] Molinari F, Zeng G and Suri J S 2010 A state of the art review on intima–media thickness (IMT) measurement and wall segmentation techniques for carotid ultrasound *Comput. Methods Prog. Biomed.* **100** 201–21
[27] Molinari F, Zeng G and Suri J S 2010 An integrated approach to computer-based automated tracing and its validation for 200 common carotid arterial wall ultrasound images: a new technique *J. Ultrasound Med.* **29** 399–418
[28] Huang Z and Ng M K 1999 A fuzzy k-modes algorithm for clustering categorical data *IEEE Trans. Fuzzy Syst.* **7** 446–52
[29] Araki T, Kumar P K, Harman S S, Nobutaka I, Ajay G, Luca S and Jeny R *et al* 2016 Two automated techniques for carotid lumen diameter measurement: regional versus boundary approaches *J. Med. Syst.* **40** 182
[30] Suri J S, Liu K, Reden L and Laxminarayan S 2002 A review on MR vascular image processing algorithms: acquisition and prefiltering: part I *IEEE Trans. Inform. Technol. Biomed.* **6** 324–37
[31] Suri J S, Haralick R M and Florence H S 2000 Greedy algorithm for error correction in automatically produced boundaries from low contrast ventriculograms *Pattern Anal. Appl.* **3** 39–60
[32] Li C, Xu C, Gui C and Fox M D 2010 Distance regularized level set evolution and its application to image segmentation *IEEE Trans. Image Process.* **19** 3243
[33] LeCun Y, Yoshua B and Geoffrey H 2015 Deep learning *Nature* **521** 436

Applied Problems in the Theory of Electromagnetic Wave Scattering

Online at: https://doi.org/10.1088/978-0-7503-3979-7

Applied Problems in the Theory of Electromagnetic Wave Scattering

Edited by
Oleg I Sukharevsky
Air Force Research Center of Ivan Kozhedub Kharkiv National Air Force University, Kharkiv, Ukraine

IOP Publishing, Bristol, UK

ISBN 978-0-7503-3979-7 (ebook)
ISBN 978-0-7503-3977-3 (print)
ISBN 978-0-7503-3980-3 (myPrint)
ISBN 978-0-7503-3978-0 (mobi)

DOI 10.1088/978-0-7503-3979-7

Version: 20221001

IOP ebooks

British Library Cataloguing-in-Publication Data: A catalogue record for this book is available from the British Library.

Published by IOP Publishing, wholly owned by The Institute of Physics, London

IOP Publishing, No.2 The Distillery, Glassfields, Avon Street, Bristol, BS2 0GR, UK

US Office: IOP Publishing, Inc., 190 North Independence Mall West, Suite 601, Philadelphia, PA 19106, USA

Image credit: Alexey Sukharevskyi.

In cherished memory of our teacher, Professor Il'ya Vladimirovich Sukharevsky

Contents

Preface from editor x

Editor's biography xii

List of contributors xiii

Abbreviations xiv

Introduction xv

1 Using an integral equation method for solving problems of resonant electromagnetic wave scattering 1-1

Sergey Nechitaylo, Valery Orlenko, Oleg Sukharevsky, Vitaly Vasilets and Gennady Zalevsky

1.1 Electromagnetic wave scattering at resonant size impedance screens of finite thickness 1-1

1.1.1 Derivation of integral equation systems 1-2

1.1.2 Method for solving the obtained integral equations 1-4

1.1.3 Electromagnetic wave scattering at a non-perfectly conducting screen of finite thickness 1-6

1.1.4 Computation of fields radiated by a dual-reflector antenna, taking into account the interaction of the two reflectors 1-24

1.2 Magnetic field integral equation for solving problems of resonant electromagnetic wave scattering by perfectly conducting objects 1-41

1.2.1 Deriving the magnetic field integral equation 1-42

1.2.2 Numerical solution to the obtained integral equation 1-45

1.3 The Müller integral equation set for solving problems of resonant electromagnetic wave scattering by dielectric homogeneous objects 1-49

1.3.1 Deriving the Müller integral equation set 1-50

1.3.2 Algorithm for finding a numerical solution to the Müller integral equation set 1-52

1.4 Using an integral equation method for numerical modeling of electromagnetic scattering by metallic, dielectric, and combined resonant objects in applied problems 1-58

1.4.1 Radar scattering characteristics of an unmanned aerial vehicle in the VHF and S frequency bands 1-58

1.4.2 The high-resolution radar range profiles of artillery shells 1-78

1.4.3 The radar detection and identification of metallic and dielectric mines buried in the ground 1-85

1.4.4 Electromagnetic wave scattering by meteorological particles of various shapes 1-97

1.4.5 Electromagnetic scattering and radiation characteristics of an antenna inside a dielectric radome of resonant size 1-103

1.4.6 Electromagnetic wave scattering by nanoparticles in the optical band 1-107

References 1-112

2 Asymptotic methods for solving some applied problems 2-1

Sergey Nechitaylo, Valery Orlenko, Oleg Sukharevsky, Vitaly Vasilets and Gennady Zalevsky

2.1 Electromagnetic wave scattering by a cylindrical object immersed in a dielectric half-space 2-1

2.1.1 Calculation method for a perfectly conducting object 2-1

2.1.2 Calculation features for an impedance object 2-7

2.1.3 The possibility of figuring out the parameters of the scattering cylinder 2-12

2.2 Reconstruction of the 'illuminated' surface part of a convex perfectly conducting scatterer 2-13

2.2.1 Finding the principle curvatures at the point of specular reflection 2-14

2.2.2 The algorithm for reconstructing the 'illuminated' part of the object's surface 2-18

2.2.3 Results of numerical simulation 2-22

2.3 Scattering and radiation characteristics of antenna systems under nose dielectric radomes 2-26

2.3.1 Electromagnetic wave scattering by an antenna system under the nose dielectric radome 2-29

2.3.2 Radiation of an antenna system under a nose dielectric radome 2-40

2.4 The influence of precipitation on the gain of the reflector antennas 2-47

2.4.1 A method for computing the fields radiated by reflector antennas that are partially covered by precipitation layers 2-50

2.4.2 Numerical simulation of the radiation characteristics of reflector antennas that are partially covered with snow 2-55

2.4.3 Numerical simulation of the reflector antenna radiation performance given a non-symmetrical distribution of precipitation deposit over its reflector's surface 2-61

2.4.4 Verification of the proposed method for predicting the radiation performance of reflector antennas whose surface was partially covered by precipitation 2-71

2.5 Near-field calculation for an electrically large reflector antenna 2-74

2.5.1 The method for computing the field radiated by a reflector antenna in the near-field zone 2-75

2.5.2 Near-field computation results 2-79

2.6 Electromagnetic wave scattering by thin wires placed inside dielectric shells 2-83

2.6.1 Scattering of an electromagnetic wave by a thin wire housed inside a thin dielectric shell 2-83

2.6.2 Verification of the method and the results of computation 2-88

2.7 Backscattering of an inflatable dielectric lifting–turning device designed for ground measurement of radar object scattering characteristics 2-98

2.7.1 Method for computing the radar cross section of a thin dielectric shell of large electric sizes 2-100

2.7.2 Computation of the radar cross section for the inflatable lifting–turning device 2-105

2.7.3 Possible measures for reducing parasitic scattering from lifting–turning device 2-109

2.8 A computation method for electromagnetic wave scattering by dielectric toroid formations 2-111

2.8.1 Introduction to electromagnetic scattering by meteorological formations 2-111

2.8.2 A method for computing the electromagnetic waves scattering by dielectric torus 2-113

2.8.3 Results of simulating the radar cross section of the dielectric torus 2-118

2.9 A method for camouflaging ground vehicles to prevent their detection by radar 2-122

2.9.1 Basic relations for the radar cross section of ground objects 2-122

2.9.2 Results of computer simulation 2-127

References 2-133

Preface from editor

Modern computational electrodynamics is shifting steadily toward more emphasis being placed on the word 'computational'. This is a natural process prompted by two causes. The first and most obvious is the development of computers. This has made it possible to solve many electrodynamic problems with the required accuracy, for which just a few decades ago it was only possible to obtain approximate results. The second reason is the development of computational methods that can be algorithmized and used as the basis of special software called an 'electrodynamic solver'. It is clear that the second reason is also related to computational technology, but it is based on electrodynamic methods.

Solvers provide the means for solving the vast majority of various problems that involve the scattering and radiation of electromagnetic waves. However, any solver looks like a 'black box' that receives input data and outputs a result. It would be incorrect to assert that such solvers produce random results, however, it is worth noting that the solver, in its solutions, relies solely on the methods that are embedded within it. In addition, the algorithmization of these methods has become possible given the current developments in computational electrodynamics.

The computational methods discussed in this monograph are original. The methods have been proposed to solve practical problems which, for a number of reasons, modern solvers could not cope with. These reasons have been both technical, for example, a lack of the computing power required by a solver, and methodical, when the methods embedded into the solver cannot solve a problem with the required accuracy or simply yield the wrong results.

This book is mainly concerned with the scattering characteristics of various radar objects or their parts. However, due to the common methodological basis and the proposed theoretical approaches, this book offers some studies regarding the radiation performance of certain types of antenna systems.

The group of authors represents the scientific school of applied electrodynamics working at Ivan Kozhedub Kharkiv National Air Force University, a descendant of the Govorov Military Radiotechnical Academy.

This book can be useful to a wide audience: scientists concerned with the development of electromagnetic wave scattering theory, computational electrodynamics specialists, as well as radiophysics engineers pursuing the development of radar detection and identification algorithms of radar objects. The authors also hope that the proposed methods can one day be algorithmized so that computational electrodynamics will keep shifting toward its 'computational' component.

The authors' contributions to this book are as follows: sections 1.1, 2.4, and 2.5 have been written by Sergey V Nechitaylo and Oleg I Sukharevsky; sections 1.2 and 1.3 have been written by Oleg I Sukharevsky and Gennady S Zalevsky; section 1.4 has been written by Valery M Orlenko, Oleg I Sukharevsky, Vitaly A Vasilets, and Gennady S Zalevsky; sections 2.1 and 2.2 have been written by Oleg I Sukharevsky; sections 2.6 and 2.8 have been written by Oleg I Sukharevsky and Vitaly A Vasilets; section 2.7 has been written by Valery M Orlenko, Oleg I Sukharevsky, Vitaly A Vasilets and Gennady S Zalevsky; and sections 2.3 and 2.9 has been written by Sergey V Nechitaylo, Oleg I Sukharevsky, and Vitaly A Vasilets.

The authors express their gratitude to A Maslovskiy, V Misailov, A Muzychenko, I Ryapolov, I Sukharevsky, and M Surgai, who also took part in the research and whose results are also included in this book.

Professor Oleg I Sukharevsky

Editor's biography

Oleg I Sukharevsky

Professor Oleg I Sukharevsky has worked in the field of applied mathematics and computational electrodynamics since 1972 when he received a diploma degree in computational mathematics from Kharkov Gorky State University and joined the Kharkov Govorov Military Academy of Air Defense as an engineer. Since 1977 Dr Sukharevsky has studied and developed numerical computation methods that were based on solving the two-dimensional integral equations to obtain the scattering characteristics of a nonclosed screen and antenna under a frame radome. In 1983 Dr Sukharevsky earned a candidate of technical science degree (PhD analog) from the All Union Research Institute of Radiotechniques, and by that time he was promoted to the position of a senior researcher. Since then, he has carried on studying scattering phenomena and started the development of scattering theory and techniques for computing the radar cross section of complex objects, special emphasis being placed on objects that had surface fractures (sharp edges) and were covered (partly) by radar absorbing materials. In 1993 he earned a doctor of science degree in radar technology from Kharkov Govorov Military Academy of Air Defense. Since the fall of the Soviet Union in 1991, Dr Sukharevsky has held various academic and faculty positions. He taught applied mathematics and computational electrodynamics at the Govorov Academy, which was reorganized in 1993 into Kharkov Military University, and later, in 2004, into Ivan Kozhedub Kharkiv National Air Force University. He was also employed as principal researcher by the Usikov Institute of Radio Physics and Electronics of the National Academy of Sciences of Ukraine from 1996 to 2001 and from 2004 to 2008. During the turbulent time that ensued after the fall of the Soviet Union, he founded a scientific school that was instrumental in shaping many prominent Ukrainian engineers and scientists. Around that time there was a burst in the development of computer power that introduced a whole new capability for implementing theoretical electrodynamics methods in operational software. The introduction of personal computers facilitated the development of computer models of scattering from a number of aerial, ground, and subsurface objects. Dr Sukharevsky along with alumni of his academic school had acquired wide experience of scattering simulation from real radar objects sounded by not only narrowband but also wideband and ultrawideband probing signals. He has authored and co-authored ten books and more than 200 papers. Oleg Sukharevsky is a Senior Member of the the Institute of Electrical and Electronics Engineers (IEEE) and a Honored Science Worker of Ukraine.

Currently, Dr Sukharevsky is a leading scientific researcher and a professor in the Air Force Research Center at Ivan Kozhedub Kharkiv National Air Force University. His main areas of interest include, but are not limited to, the mathematical theory of diffraction and the scattering characteristics of radar targets.

List of contributors

Sergey V Nechitaylo earned a diploma in radioelectronic devices and systems from Kharkov Military University in 1996, a PhD in radar in 2004, and a doctor of science in radio systems in 2019.

Currently, Dr Nechitaylo is a scientific laboratory head at Ivan Kozhedub Kharkiv National Air Force University and a senior scientist. His scientific interests include antenna theory and the study of the radar observability of various objects.

Valery M Orlenko earned a diploma in radio engineering from Zhitomir Higher Military School of Radio Electronics of Air Defense in 1992 and a PhD in radar in 1998.

Currently, Dr Orlenko is an assistant professor at Ivan Kozhedub Kharkiv National Air Force University and a senior scientist. His scientific interests include the study of the methods for processing wideband radar signals.

Oleg I Sukharevsky earned a diploma in computational mathematics from Kharkiv Gorky State University in 1972, a PhD in 1983, and a doctor of science in radar in 1993.

Currently, Dr Sukharevsky is a leading scientific researcher within the Air Force Research Center at Ivan Kozhedub Kharkiv National Air Force University and a professor. His main areas of interest are the mathematical theory of diffraction and the scattering characteristics of radar targets. He is a senior member of the IEEE.

Vitaly A Vasilets earned a diploma in radio engineering from Zhitomir Higher Military School of Radio Electronics of Air Defense in 1991, a PhD in 1994, and a doctor of science in radar in 2006.

Currently, Dr Vasilets is a leading scientific researcher and a senior scientist in the Air Force Research Center at Ivan Kozhedub Kharkiv National Air Force University. His scientific interests include the study of the scattering characteristics of radar targets.

Gennady S Zalevsky earned a diploma in radio engineering from Zhitomir Higher Military School of Radio Electronics of Air Defense in 1994, a PhD in 2002, and a doctor of science in radio systems in 2016.

Currently, Dr Zalevsky is a professor at Ivan Kozhedub Kharkiv National Air Force University and a senior scientist. His scientific interests include the study of the scattering characteristics for radar targets with resonant sizes.

Abbreviations

ACS	absorption cross section
AE	angel-echo
AHFM	asymptotic high-frequency method
AS	antenna system
CFIE	combined field integral equation
CNR	complex natural resonance
CPCI	conditional probability of correct identification
CPFI	conditional probability of false identification
CW	continuous wave
DP	directivity pattern
ECS	extinction cross section
EFIE	electric field integral equation
EM	electromagnetic
EMF	electromagnetic field
EMW	electromagnetic wave
ES	electronic system
EWP	electromagnetic wave propagation
FC	frequency characteristic
FS	frequency spectrum
GPR	ground penetration radar
HRRP	high-resolution radar range profile
IE	integral equation
LTD	lifting–turning device
MFIE	magnetic field integral equation
MIE	Müller integral equations
MLFMA	multilevel fast multipole algorithm
MLRS	multiple launch rocket system
MoM	method of moments
OSLE	overdetermined set of linear equations
PEC	perfectly electrically conducting
PO	physical optics
RCS	radar cross section
RP	radiation pattern
SCS	spherical coordinate system
SD	scattering diagram
SLE	set of linear equations
SSO	subsurface object
SP	surface-plasmon
TIS	thin inflatable shell
TSCS	total scattering cross section
UAV	unmanned aerial vehicle
USR	unit step response
UWB	ultrawideband
VR	vortex ring
WB	wideband
WR	weather radar

Introduction

This book presents solutions to a number of applied problems of electromagnetic wave (EMW) scattering. The problems considered here are divided in two groups. The first group includes problems of resonant EMW scattering that are to be solved by the integral equation (IE) method. These problems and their solutions are analyzed in the first chapter of the book.

The first chapter begins with a section devoted to scattering by thin screens of finite thickness and finite conductivity. The method we propose here to solve these kinds of problems is based on a modification of the T-matrix method (extended boundary condition method) [1, 2]. For electrically small thin impedance screens, a system of integral representations is obtained for the intensity of total field at any given point in space. In this case, the integrands contain the surface electric current density $\vec{J}$ and the normal component of the electric field e on the screen surface. Having chosen the position of the observation point inside the screen, yet still outside the skin layer[1], we obtain the set of IEs with respect to $\vec{J}$ and e. In this case, the total field intensity is assumed to be equal to zero at the observation point, just as in [1, 2]. The resulting IEs (subsection 1.1.1) are reduced to a set of linear equations (SLE). The intensity of the scattered field is finally calculated using the aforementioned integral representations and the values of $\vec{J}$ and e are thus found. The performance of the method is verified by checking the results of scattered field calculations against experimental data in a given frequency range.

This method is an obvious extension of the electrical field IE (EFIE) method onto the case of impedance scatterers. While the EFIE method is used for perfectly electrically conducting (PEC) scatterers [3, 4], the method discussed here can be used for good conductors too.

The method proposed allows us to estimate the impact of screen thickness and conductivity on the scattering characteristics. Theoretical and experimental studies [3–5] show that the impact of the screen thickness and conductivity, given that screens are electrically small, can affect scattering calculations significantly. The corresponding data for spherical screens are given in this section.

A very important applied problem is accounting for the mutual interference between two screens of resonant sizes. The corresponding study is presented in the section concerned with parabolic and elliptical reflectors operating both in the scattering and transmission modes. In addition, while in the transmission mode, a smaller elliptical reflector paired with a Huygens element illuminating the reflector can, basically, be used as a model of antenna feed for a reflector antenna (parabolic). Computations are carried out for both the near- and far-field (Fraunhofer) antenna zone. We have accounted for the impact of screen thickness and finite conductivity on both the transmitted and scattered fields.

Then, using the method developed for computing fields transmitted by reflector antennas as a basis, we studied the issue of ultrawideband (UWB) signal

[1] The screen thickness is assumed to be much greater than the skin layer value.

transmission by a single reflector antenna. In addition, we used the elementary source as a model for the antenna feed, the frequency response of that source in the far-field zone coinciding with the spectrum density of the required UWB signal, and its directivity pattern (DP) described by that of the Huygens element [6]. We demonstrate that such a UWB antenna transmits pulses of varying shape depending on the direction toward an observation point in the far-field zone.

In contemporary electrodynamics, it is also very important to find solutions to the problems of scattering at a PEC and dielectric objects of resonant size, both standalone and placed in the vicinity of an interface between two lossy dielectric dispersive media.

Sections 1.2 and 1.3 are devoted to these problems. In section 1.2 we develop a computational method for predicting the scattering characteristics of PEC objects. The method is based on solving the magnetic field IE (MFIE). In this section, we derive the MFIE and consider possible issues with its kernel. Then we describe an algorithm for solving the MFIE. One of the typical examples for using MFIE is where wideband (ultrawideband) radar responses of artillery shells are to be obtained that have various electrical sizes and shapes (subsection 1.4.2). The wideband (ultrawideband) radar responses of two types of artillery shells of various electrical sizes and shapes were studied. A computer technique based on MFIE solving was used to obtain the high-resolution radar range profiles (HRRPs) of objects considered in the L and VHF bands, two different bandwidths of the sounding signal being used as an example. Our results show that using sounding signals with a spectrum width of 100–300 MHz in counter-battery radars one can achieve effective non-cooperative radar identification of artillery shells and rockets fired by multiple launch rocket systems (MLRSs). The latter effect can be achieved without using a large database of radar profiles pertaining to various radar aspect angles of shells.

EMW scattering by a dielectric object is a more complicated problem. This problem is addressed in section 1.3. In the case of a homogeneous dielectric object, we use the method based on solving the set of Müller IEs (MIE). In this section we derive the mentioned MIE set. Then the peculiarities of its kernels are discussed, and a numerical algorithm for MIE set solving is described.

One applied problem dealing with dielectric radar objects is that of EMW scattering at hydrometeors (rain drops, snowflakes). Contemporary weather radar (WR) are used for measuring the velocity and heading of weather formations, for detecting weather fronts as well as the boundaries of clouds and precipitation, and for detecting turbulent and convective formations in the atmosphere [7]. In order to be able to perform such tasks, the radar is required to possess information on the scattering properties of precipitation particles [8] (subsection 1.4.4). Given the use of probing signals with wavelengths ranging from a few tenths of centimeters to a few centimeters, one has to deal with the scattering effect of these particles, which are dielectric objects of resonant size ($ka \sim 1$, where k is the wavenumber in the medium the particle is immersed in, and a is the characteristic size of the particle).

As long as one requires *a priori* information regarding the scattering features of buried dielectric objects, such as dielectric landmines, one has to deal with another kind of applied problem (subsection 1.4.3). Yet another problem arises when one needs to predict EM radiation performance and EMW scattering by antenna put under dielectric radomes of resonant sizes. A computational method offering reliable solutions to this problem is considered in subsection 1.4.5. There, we briefly discuss the peculiarities of the algorithm and present some results of our computations. All the above problems and other problems dealing with dielectric object scattering can be solved using an IEs of the Müller type.

All the necessary math expressions have been obtained by applying the integral Lorentz reciprocity theorem [12] to the sought electromagnetic field (EMF) as well as to the fields generated by auxiliary point sources in various parts of space (the latter having been filled with media of varying electromagnetic (EM) properties), and by eventually combining these particular results together [10, 11, 13–15]. The relative simplicity and clear physical sense of the process leading to the IE set as an objective and the algorithmic efficiency of the computation method give us grounds to choose in favor of the reciprocity theorem instead of another famous method that is based on fundamental solution matrices and tensor integral representations (generalized Stratton–Chu formulas) [10, 11, 13].

Based on the analysis of simulation results regarding scattering from landmines, we have developed a method for mine detection and identification. The method is based on finding the landmine's own complex natural resonances (CNRs) [14]. These CNRs depend on the shape and sizes of subsurface objects (SSO).

Returning to buried PEC and in particular dielectric objects (landmines) we consider (subsection 1.4.3) the practical implementation peculiarities of the detection and identification algorithms, and elaborate on the database structure for storing the objects to be detected and identified. The database must include not only the identification signatures of the designated type objects themselves, but also take into account the varying depth of the object placement as well as the varying electrical and physical features of the surrounding soil. The latter requirement makes the practical implementation of our method complicated, yet its application allows improving the conditional probability of correct identification (CPCI) of landmines and reducing the false alarm rate, which is due to the presence of foreign objects in the soil.

Yet another applied problem, a solution to which can be obtained using an MIE set, is the problem of visible light scattering (wavelength ranging from 300 nm through 900 nm) at metallic (silver, gold) nanoparticles [15]. At present, the effects of resonant scattering at such particles of noble metals are being studied quite commonly (subsection 1.4.6).

Finding the scattering resonances can bring additional advantages when manufacturing optical antennas, sensors, and visible light absorbers. It should be noted that silver and gold possess very peculiar scattering properties given the visible light wavelengths. The real part of the relative permittivity of silver can be negative and, by absolute value, can be as high as several dozens, whereas its imaginary part is a positive less than unit value across almost the entire visible light range.

It should also be noted that the sources available or known to us contain very little information when the 3D scattering at nanoparticles is concerned [15]. In this book, we make an attempt to fill this gap. We are going to consider the application of a method developed previously by us that applies the MIE set [15] to a plasmonic metal scatterer. We are also going to obtain the expressions relating the total scattering cross section (TSCS), extinction cross section (ECS), and absorption cross section (ACS) (since the metals do cause losses) by means of the optical theorem in the 3D case. We present examples of computations featuring TSCS and ACS values for the case of silver nanoparticles that are shaped as ellipsoids of revolution given various radiation conditions and various polarizations of the probing signal.

When one deals with compound resonance-sized aerial objects containing both PEC and dielectric design elements, it is necessary to develop special algorithms. These would be based on solving the Fredholm boundary IEs of the second kind, which puts them in the context of the method of analytical regularization (subsection 1.4.1). Given PEC elements the MFIE is employed and given dielectric scatterers the MIE set is employed. The algorithms proposed in this monograph present important novelties when the prediction of the radar scattering characteristics of the scatterers of various curvature radii, including scatterers that are electrically thin, is concerned. The description is accompanied by an analysis of the convergence of the developed algorithms. For the validation, a comparison is presented of the results for the model object obtained by different methods (in preliminary section 1.3). The results of modeling the radar scattering characteristics of an unmanned aerial vehicle (UAV) and its individual metallic and dielectric elements in the VHF and S frequency bands are demonstrated and discussed. The scattering by two types of UAV propellers at the VHF and S band frequencies is studied separately and discussed in this subsection.

In the second part of the book, chapter 2, there are example solutions to some applied 2D and 3D scattering problems, that can be solved by means of asymptotical (short wave) methods.

The first such problem (section 2.1) is that of 2D EMW scattering at both perfectly conducting and impedance (non-perfectly conducting) cylindrical objects that are immersed in a dielectric half-space (the parameters of which coincide with those of soil in particular). One can encounter such a task while searching for various types of underground pipelines using ground penetration radar (GPR). The study of the scattering features of such objects can improve the performance of the GPR designed for their detection as well as help in designing methods for figuring out certain properties of cylindrical underground objects. This part of the book includes relatively simple asymptotical expressions (for the case of electrically large cylindrical SSOs) useful for obtaining coarse engineering estimations, these expressions relating the field scattered by objects with the task layout parameters such as the object depth in the soil, the object's curvature radius at the specular reflection point, and the object orientation in space.

We also point out that it is possible to apply the results obtained in this book (given the two orthogonal polarizations of the illumination wave) to finding the aforementioned parameters of the object. It should be noted that the formulas we obtain can be interpreted easily using simple physical concepts, and the formulas for the impedance objects become those for PEC objects given that the scatterer's impedance W approaches zero.

When considering certain applied radar and EM problems, it becomes necessary to figure out the geometrical shape of the scatterer from its radar echo.

To find solutions to such tasks, a whole variety of different approaches has been proposed so far. Among them, for instance, is an approach that relates the object's impulse response to the second derivative of the function describing the profile of the object's surface [16, 17]. The works [18, 19] also developed an approach that was based on using polarization data for the same purpose.

The work by C I Bennet [20] described the method of sequential advancement in time and an iterative approach to restoring the shape of a PEC convex body of revolution, the dimensions of which fit into the resonant scattering region. Such a restoration was based on solving a non-stationary IE with respect to the surface current density. However, the applicability of such methods is limited to simple geometrical shapes only (bodies of revolution) and by an object's size, which needs to be resonant with respect to the wavelength. This book (section 2.2) also features a new method for restoring the 'illuminated' part of a convex perfectly conducting object, the geometrical dimensions and curvature radii of which are much greater than the illumination signal wavelength. The method [12, 14] is based on comparing, at every moment of discrete time, the computed values of the high-frequency approximation of a unit step response to that obtained in an experiment so that a discrepancy between the two can be analyzed. Data obtained by means of full polarization reception allows us to determine the principal curvature radii of the object's surface at the point of 'specular reflection' as well as the orientation of the principal directions. In our method, we obtain such information by integrating the Fock equation [21] that includes the surface current density, and by finding two basic terms of the scattered field ray asymptotic for the general case of a bistatic radar layout [12]. The aforementioned curvature radii and principal directions provide the basis to substitute a paraboloid for the surface itself, the apex of the paraboloid being tangent to the surface at the 'specular' reflection point. Such a paraboloid is assumed to be an initial approximation of the surface in the vicinity of this particular point. Then we take a few steps in discrete time, at every step the surface we restore being approximated by a ruled surface, which is later elaborated by introducing a correction. The correction is found by comparing the actual unit step response of the object to the unit step response computed using the object's surface restored by that moment in time. The actual unit step response can be found using knowledge of actual probing and echo signals. For instance, one can solve the convolution IE to obtain the unit step response of the scattering object $H(t)$:

$$\Omega(t) = \int_{-\infty}^{t} \Omega'_0(t)H(t-\tau)\ \mathrm{d}\tau.$$

Here $\Omega'_0(t)$ is the time derivative of the probing signal and $\Omega(t)$ is the echo signal. This procedure runs until the boundary of the 'illuminated' object part is reached.

Sections 2.3–2.8 are dedicated to applied problems of antenna theory.

In particular, in section 2.3 we detail the method for computing the fields that are scattered and transmitted by an antenna placed under a dielectric fairing (radome).

A great variety of aerodynamic objects (aircraft, guided missiles with homing heads) that are of interest in radar are equipped with nose fairings that house antenna systems (ASs) underneath. The presence of an AS under a fairing can alter the radar cross section (RCS) figure of the entire radar object significantly. Thus it is important to develop methods for computing and evaluating the RCS figures of AS underneath such fairings, so that one can account for those RCS figures while designing the radar detection algorithms of such aerodynamic objects. Geometrical optics methods that are commonly used for computing the scattering from such systems do not usually account for EM interaction between the antenna and the radome [22].

At the same time, such EM interactions between the antenna reflector and the radome can alter the RCS figure of the entire AS. In addition, such a change can be resonant in nature, which can affect the RCS figure of the radar object itself [22]. Thus it is necessary to have a method at hand, at least a coarse one, that would be capable of providing us with an estimate of the 3D model of an AS, and which would take into account at least single wave bounces between the antenna and its radome.

In a number of recent works [23–30] the authors describe the models for computing the DPs of antennas placed under dielectric radomes. However, no methods are specified for computing the RCS of antennas under particular kinds of radomes, namely, nose cones (or nose fairings) (task #1). Along with the problem of estimating the RCS of the 'antenna–nose cone' system, there is also another applied problem of computing the DP of the 'antenna–nose cone' system (task #2). From the standpoint of the antenna maker, the ease of nose cone manufacturing is also a very important problem. In this section, we consider three different shapes of nose cones: ogive, compound cone (consisting of several truncated cones put together), and pyramidal (with a regular octagon placed at the base of the pyramid). Given that each of these types of nose cones is used as a cover for an AS, we compute both the RCS and DP of the AS beneath. As a first approximation for the scattered field computation, we assume that the antenna itself (a phased antenna array, for instance) is just an octagonal plate, we also assume that the field scattered from such a plate (as is commonly done in engineering evaluation) is multiplied by a factor of ≈ 0.9. Such a representation gives a decent approximation of the field scattered by the antenna (given that the wavelength of the signal that illuminates the antenna was far greater than the operating wavelength of the antenna). Thus while simulating the signal transmitted by the AS, we assume that the antenna has an octagonal aperture (with a certain amplitude and phase distribution) cut from a perfectly absorbing screen.

In our computations we employ the method of mirror images [12], yet it is generalized to the case of ambient space containing non-homogeneities of various kinds, namely, dielectrics, conductors, and magnetics. The methods we use to solve both tasks #1 and #2 employ basically the same 'key' tasks. In addition, the rest of the 'ideology' behind all the computations employed for the 'antenna–radome' system is the same regardless of the type of radome. Some computational specifics are presented (if necessary) as the material presentation unfolds. It should be noted that the tip of a nose cone contains a spherical metallic cap that is integrated smoothly into the nose cone's surface. While computing the passage of an EMW through the nose cone, the impact of this cap is accounted for by means of a ray tracing technique (in the approximation of geometrical optics), the latter means that if an incident wave hits the cap, then it does not pass through the nose cone. The field scattered by the nose cone itself also takes into account the impact of the spherical cap. In this case, we use a physical optics (PO) approach.

Wave propagation through the nose cone (including waves bouncing from its inner surface) is simulated using a geometrical optics method. Here, we used the locality principle while computing the field of a wave that has passed through the nose cone's wall. In other words, we compute the field of a wave, which has passed through the wall at particular point, as if it has passed through the plane and parallel layers of material of which the nose cone is made. In our computations this layer is always tangent to the actual nose cone's surface at the point of wave incidence being considered at every particular moment. This approach makes it possible to compute waves transmitted and scattered by ASs placed underneath multi-layered nose cones. In this case, we just need to use the solution of another 'model' problem of EMW passing through and reflecting off multiple layer plates. As an example, in this section we consider the field transmitted by an AS underneath a two-layer conical radome. We provide the analysis of the transmission losses caused by the radome given that the antenna rotates under the radome. We also present the results for various values of radome layer thickness.

Section 2.4 is dedicated to the analysis of the impact that precipitation (rain, snow), which may build up on the surface of weather radar reflector antennas, can have on their parameters.

Multi-function networks of weather radar stations are used widely in many countries to provide a warning against dangerous atmospheric events [31], such as torrential rain, hailstorms, and blizzards. In this case, antennas (which are reflector antennas in many cases) are exposed to precipitation accretion building up on their surfaces. Since a weather radar's purpose is to measure the absolute reflectivity of precipitation particles, one needs to take into account the impact of precipitation on antenna performance in order to correctly calibrate those parameters of the radar that affect its power measurements.

According to measurements presented in [32], a layer of precipitation (snow, for instance) distorts the information regarding the specific RCS and spatial distribution of precipitation significantly, which leads to incorrect estimates regarding the type and quantity of precipitation. Some experimental study results can be found in [33], which reveal the significant impact of rain and snow formations on the surface of a reflector antenna on its transmission performance. In this section, we propose a method for predicting the transmission performance of reflector antennas, the surfaces of which are partially covered by a precipitation layer (water or snow). We also present the results of the experimental verification of the veracity of this method obtained by measuring the decrease in antenna gain due to the presence of precipitation formation on the antenna reflector.

In section 2.5 we present a method for computing the field transmitted by a reflector antenna in the near field zone. The method is based on integral representations obtained by means of the reciprocity theorem and approximate (asymptotic) computation of fields appearing in these representations. In addition, we account for the scattering components provided by the edges of the reflector. For this purpose we use the available solution to the model problem of wave scattering at the wedge (in the vicinity of the wedge). We use a reflector antenna (a nonsymmetrical cut from the paraboloid of revolution) that had a feed shaped like a pyramidal horn to demonstrate how the DP is formed as the observation point is gradually brought away from the antenna aperture.

Section 2.6 addresses the problem of EMW scattering at thin wires placed inside a dielectric shell. Such applied problems arise, for instance, when one needs to predict the lowest possible scattering from the wing of a UAV that has thin wires laid in some manner inside it. The method developed here makes it possible to predict the RCS of a dielectric wing that has conductors inside it given various layouts of these conductors. In addition, one can also figure out ways for reducing the RCS of such a structure. The method is based on combining different integral representations obtained from the reciprocity theorem, the scattered field, and some representation that follows from the Pocklington's IE [34] for the case of bent wires.

Using methods similar to those we employed for estimating the scattering component of nose cones in the total field scattered by the antenna–radome system, in section 2.7 we develop a method for predicting the backscattering intensity of an inflatable dielectric lifting–turning device (LTD) [35]. This device was designed for elevating real radar objects while their scattering parameters are being measured at testing ranges. In the conditions of a testing range, it is very important to reduce the signal scattered by the LTD that supports the object under test as much as possible. While measuring the RCS of reasonably heavy objects, one can use a thin inflatable shell (TIS) or air balloon to elevate and hold the object at specified height. The basic advantages of the TIS include low weight and simplicity of operation, as well as its mobility. The TIS we consider here is made up of toroidal balloons, every balloon being an air inflated electrically large thin dielectric shell. Even a provisional estimate of the LTD's RCS has practical importance, because it allows one to assess its effect on the RCS of the object under test and figure out the measures for decreasing such an effect.

In section 2.8 an asymptotic method is proposed for obtaining the scattering characteristics of atmospheric entities [36]. This method correlates with the method used in section 2.7. One of the possible physical mechanisms of the so-called 'angel-echo' (AE) is considered. Such AEs are due to discrete reflections of radar signals from objects that cannot be observed visually. One of the reasons these AEs exist is the vortex flows that bring some volume of air having certain radiophysical parameters into an area of space filled with air with different radiophysical parameters. Toroidal vortices (vortex rings) have a fairly long lifetime and they can travel long distances before they disappear. The computation of diagrams for radio waves scattered by vortex rings using contemporary universal software, which is designed to solve electrodynamics problems and which is based on the method of moments (MoM), requires significant computational resources and time. A fast asymptotic method for predicting the scattering of EMW by dielectric toroidal formations of large radius is developed. We simulate the corresponding wave scattering and compared the results with those obtained using the Altair Feko software package. A good correspondence between the two is shown. For some interesting practical cases, we compute the monostatic and bistatic scattering areas of vortex rings. The results of the calculations coincide quite well with the results of known experimental works.

In section 2.9 we consider a method for optimizing the partial coverage ('camouflage') of a ground object. The method is based on the approach developed previously by the authors [12] for predicting the scattering characteristics of ground objects (possibly coated with a radar absorber). The method uses the partial application of a radar absorber in order to control the mean RCS of an object given some range of aspect angles, which brings the scatterer's RCS as close as possible to the mean RCS of a ground patch that was 'shadowed' by that scatterer. In this way the proposed method allows one to make the ground object almost invisible ('low-contrast') for radars. The numerical calculations for a model of a tank are carried out given different sectors of aspect angles and ground surfaces of various degrees of roughness.

References

[1] Kyurkchan A G and Smirnova N I 2009 Solution of wave diffraction problems by the null-field method *Acoust. Phys.* **55** 691–7

[2] Wriedt T 2007 Review of null-field method with discrete sources *J. Quant. Spectrosc. Radiat. Transfer* **106** 535–45

[3] Zalevsky G S, Nechitaylo S V, Sukharevsky O I and Sukharevsky O I 2010 Em wave scattering by perfectly conducting disk of finite thickness *2010 International Conference on Mathematical Methods in Electromagnetic Theory (Kyiv, 6–8 September)* pp 1–4

[4] Sukharevsky I O, Nechitaylo S V, Ivanchenko D D and Melezhik P N 2012 Electromagnetic wave scattering by a curved screen of finite thickness *Telecommun. Radio Eng.* **71** 413–22

[5] Ivanchenko D D and Sukharevsky I O 2010 Backscattering measurements for metallic unclosed spherical screens *Telecommun. Radio Eng.* **69** 423–8

[6] Goldstein L D and Zernov N V 1971 *Electromagnetic Fields and Waves* (Moscow: Sovetskoe Radio) p 664

[7] Sukharevsky O I, Nechitaylo S V, Voitovich O A and Khlopov G I 2015 Radiation characteristics of single-dish antennas partially covered by a layer of water *Radioelectron. Commun. Syst.* **58** 61–8

[8] Sukharevsky O I, Nechitaylo S V, Voitovich O A and Khlopov G I 2015 Influence of the snow cover on radiation characteristics of reflector antennas *J. Commun. Technol. Electron.* **60** 594–602

[9] Harrington R F 1989 Boundary integral formulations for homogeneous material bodies *J. Electromagn. Waves Appl.* **3** 1–15

[10] Sirenko Yu K, Sukharevsky I V, Sukharevsky O I and Yashina N P 2000 *Fundamental and Applied Problems of Electromagnetic Wave Scattering* (Kharkiv: Krok)

[11] Sukharevsky O I, Zalevsky G S and Vasilets V O 2016 Modeling of ultrawideband (UWB) impulse scattering by aerial and subsurface resonant objects based on integral equation solving *Advanced Ultrawideband Radar Systems* ed J D Taylor (Boca Raton/London: CRC Press/Taylor and Francis) ch 5 pp 195–235
[12] Sukharevsky O I, Vasylets V O, Nechitaylo S V and Orlenko V M 2014 *Electromagnetic Wave Scattering by Aerial and Ground Radar Objects* ed O I Sukharevsky (Boca Raton/London: CRC Press/Taylor and Francis) p 334
[13] Sukharevsky O I and Zalevsky G S 1997 EM wave scattering by resonance-size buried objects *Telecommun. Radio Eng.* **51** 80–7
[14] Sukharevsky O I, Zalevsky G S and Muzychenko A V 2012 Method of radar detection and identification of metal and dielectric objects with resonant sizes located in dielectric medium *Radioelectron. Commun. Syst.* **55** 393–404
[15] Sukharevsky O I and Zalevsky G S 2018 3-D electromagnetic scattering by ellipsoidal silver nanoparticles in optical band *2018 IEEE 17th Int. Conference on Mathematical Methods in Electromagnetic Theory (Kyiv, 3–5 July)* pp 152–5
[16] Bojarski N N 1967 *Three-dimensional Electromagnetic Short Pulse Inverse Scattering* (Syracuse: Syracuse University Research Corporation)
[17] Lewis R M 1969 Physical optics inverse diffraction *IEEE Trans. Antennas Propag.* **17** 308–14
[18] Chaudhuri S K and Boerner W M 1976 A monostatic inverse scattering model based on polarization utilization *Appl. Phys.* **11** 337–50
[19] Chaudhuri S K and Boerner W M 1977 Polarization utilization in profile inversion of a perfectly conducting prolate spheroid *IEEE Trans. Antennas Propag.* **25** 505–11
[20] Bennet C L 1981 Time domain inverse scattering *IEEE Trans. Antennas Propag.* **AP-29** 213–9
[21] Fock V A 1965 *Electromagnetic Diffraction and Propagation Problems* (Oxford: Pergamon)
[22] Sukharevsky O I, Vasilets V A and Nechitaylo S V 2017 Scattering and radiation characteristics of antenna systems under nose dielectric radomes *Prog. Electromagn. Res.* B **76** 141–57
[23] Lansink Rotgerink J, van der Ven H, Voigt T, Jehamy E, Schick M and Schippers H 2016 Modelling of effects of nose radomes on radar antenna performance *10th European Conference on Antennas and Propagation (EuCAP)* p 1–5
[24] Shangji L, Huiyuan Z, Chang D and Lizhong S 2016 Electrical properties analysis of conical radome *2016 IEEE Int. Conference on Ubiquitous Wireless Broadband (ICUWB)* p 1–4
[25] Mrdakovic B Lj and Kolundzija B M 2016 A method for full wave analysis of electrically large transparent radomes *2016 IEEE Int. Symposium on Antennas and Propagation (APSURSI)* p 1331–2
[26] Sheret T, Parini C and Allen B 2016 Efficient design of a radome for minimized transmission loss *IET Microw. Antennas Propag.* **10** 1662–6
[27] Rim J-W, Koh I-S and Choi K S 2016 IPO analysis of performance of arbitrary shaped radome *10th European Conference on Antennas and Propagation (EuCAP)* p 1–5
[28] Nair R U, Suprava M and Jha R M 2015 Graded dielectric inhomogeneous streamlined radome for airborne applications *Electron. Lett.* **51** 862–3
[29] Kim J H, Park Y B and Song S-C 2016 Electromagnetic analysis of a tangent-ogive dielectric radome with a metallic cap *2015 Int. Workshop on Antenna Technology (iWAT)* p 367–8
[30] Dwivedi S and Mishra V 2013 Numerical analysis for directivity improvement of radome structure *2013 IEEE Applied Electromagnetics Conf. (AEMC)* p 1–2

[31] Bazlova T A *et al* 2002 *Meteorological Automated Radar Networks* (Saint-Petersburg: Gidrometeoizdat) p 332

[32] Belov Ye N, Voytovich O A, Makulina T A, Rudnev G A, Khlopov G I and Khomenko S I 2010 Software and hardware complex for investigation of meteorological radar echoTelecommun. Radio Eng. **69** 1517–27

[33] Shimba M, Sato T and Koike H 1988 The antenna pattern degradation by snow accretion on the reflector surface *IEEE Antennas Propagat. Soc. Int. Symp.* vol 2 (Syracuse, NY: Syracuse University) p 871–4

[34] Sukharevsky O I, Nechitaylo S V and Vasilets V O 2018 Scattering characteristics of thin-wire conductors placed inside dielectric shell *Telecommun. Radio Eng.* **77** 671–87

[35] Sukharevsky O *et al* 2020 Backscattering calculation of inflatable dielectric lifting-turning device for ground measurement of radar object scattering characteristics *Proc. 2020 IEEE Ukrainian Microwave Week (UkrMW-2020), 10th Int. Kharkiv Symp. on Physics and Engineering of Microwaves, Millimeter and Submillimeter Waves (MSMW)* vol 3 *(Kharkiv, 21–25 Sept)* pp 964–9

[36] Sukharevsky O I, Misailov V L and Vasilets V O 2020 Calculation method for electromagnetic wave scattering from dielectric toroidal atmospheric entities *Radioelectron. Commun. Syst.* **63** 596–605

IOP Publishing

Applied Problems in the Theory of Electromagnetic Wave Scattering

Oleg I Sukharevsky

Chapter 1

Using an integral equation method for solving problems of resonant electromagnetic wave scattering

Sergey Nechitaylo, Valery Orlenko, Oleg Sukharevsky, Vitaly Vasilets and Gennady Zalevsky

1.1 Electromagnetic wave scattering at resonant size impedance screens of finite thickness

A number of works [1–3] dedicated to solving the problem of electromagnetic wave (EMW) scattering at unclosed thin screens of resonant sizes employ the method of E-field integral equation (IE). The solution to the IE obtained in every particular case involves a transfer of boundary conditions from the screen's surface itself onto an auxiliary surface, which is placed inside the screen, using a technique employed, for example, in [4, 5]. The kernel of the IE obtained as a result of such a transfer does not have singularities since the integration (source) points and the observation (field) point belong to different surfaces.

In the works mentioned above, the problem of EMW scattering at a screen of finite thickness was solved under an assumption of a perfectly conducting screen surface, which is obviously an idealization. The elements of real antennas are made of materials that possess high yet finite conductivity (for instance, the conductivity of aluminum is 3.54×10^7 Sm m^{-1}, and that of duralumin D16 (an alloy of Al–Cu–Mg) is 1.74×10^7 Sm m^{-1}), which may affect the antenna's directivity properties and scattering characteristics. Therefore, this factor needs to be taken into consideration while designing the elements of antenna systems (ASs). This factor requires even more attention given that an antenna includes reflectors (particular kinds of screens) that, compared to its operating wavelength, are of resonant sizes.

Thus this subsection is concerned with the development of a method for computing the fields scattered by non-perfectly conducting screens of finite thickness given the resonant wavelength band. To obtain the method, we derive the necessary

doi:10.1088/978-0-7503-3979-7ch1

IE sets for the E-field given the impedance scatterers and then we find a way to solve these equations efficiently on a computer.

1.1.1 Derivation of integral equation systems

Let us consider a scatterer V bounded by a closed surface S, the scattered being put in the infinite uniform isotropic medium with permittivity ε_0, μ_0 (figure 1.1). We further assume that the scatterer is a not perfect yet good conductor.

We can represent the total EM field (EMF) at point M outside the scatterer (figure 1.1) in the following form

$$\vec{E}(M) = \vec{E}^0(M) + \vec{E}^s(M),$$
$$\vec{H}(M) = \vec{H}^0(M) + \vec{H}^s(M),$$

where $\vec{E}^s(M)$, $\vec{H}^s(M)$ is the field scattered by the non-perfectly conducting object V, and $\vec{E}^0(M)$, $\vec{H}^0(M)$ is the incident field.

Our objective is to find the field $\vec{E}^s(M)$, $\vec{H}^s(M)$.

To solve the problem at hand, we use a well-known integral representation of the EMF at point M (figure 1.1) that involves the electric field strength vector $\vec{E}$ [6, 7]

$$\vec{E}(M) = \vec{E}^0(M) - \oint_S \left\{ j\omega\mu_0 g(r)\vec{J} + \left(\left(\vec{n}_Q \times \vec{E} \right) \times \vec{\nabla} g(r) \right) + \left(\vec{n}_Q \cdot \vec{E} \right) \vec{\nabla} g(r) \right\} \mathrm{d}s_Q, \tag{1.1}$$

where $g(r) = \dfrac{1}{4\pi} \dfrac{e^{jk_0 r}}{r}$, r is the distance between the field point M and the source points Q that belong to the scatterer's surface S (figure 1.1),

$\vec{J}(Q) = \vec{n}_Q \times \vec{H}$ is the equivalent surface current density,

$\vec{n}_Q$ is the internal normal to the surface S at point Q, and

k_0 is the wavenumber in free space.

All EMFs considered in the sections 1.1–1.4 have time dependences expressed by the multiplier $\exp(-j\omega t)$ (where $j^2 = -1$, $\omega = 2\pi f$ is cyclic frequency, f is linear frequency, and t is running time).

Now, we require the boundary conditions to hold at the impedance surface S. The relation between the complex amplitudes of the tangential components of the EMF

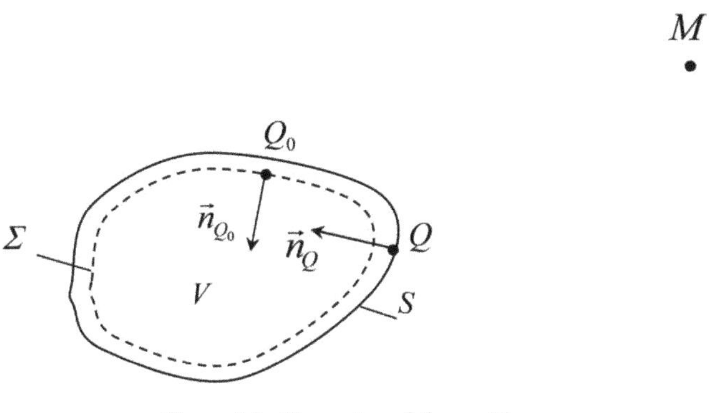

Figure 1.1. Geometry of the problem.

at an interface between two half-spaces, one of which is a good conductor, was formulated by Leontovich [8] in the form of an approximate boundary condition known as the Leontovich–Shchukin's boundary condition.

That particular boundary condition is usually presented in the following form:

$$\vec{n} \times \vec{E} = W\left(\vec{n} \times \left(\vec{n} \times \vec{H}\right)\right) \text{ at } S, \tag{1.2}$$

where $W = Z_0\sqrt{\dfrac{\mu}{\varepsilon}} = \sqrt{\dfrac{\mu_0\mu}{\varepsilon_0\varepsilon}}$ is the surface impedance of the scatterer's surface S,

ε, μ are the relative permittivity and permeability, respectively, of the material of which the scatterer is made, and

$\vec{n}$ is the unit vector of the internal normal to surface S.

We present the surface impedance as follows [9]:

$$W = (1 - j)\sqrt{\frac{\omega\mu_0\mu}{2\sigma}}, \tag{1.3}$$

where σ is the conductivity of the material of which the scatterer V is made.

Taking into account (1.2) we can rewrite the integral representations (1.1) in the following form:

$$\vec{E}(M) = \vec{E}^0(M) - \oint\limits_S \left\{j\omega\mu_0 g(r)\vec{J} + W\left(\left(\vec{n}_Q \times \vec{J}\right) \times \vec{\nabla}g(r)\right) + \left(\vec{n}_Q \cdot \vec{E}\right)\vec{\nabla}g(r)\right\} \mathrm{d}s_Q. \tag{1.4}$$

In our particular problem, the scatterer V was made of a metal that had finite conductivity. It is known [10] that the EMF incident on a metal excites a process that resembles a plane uniform wave, its front being parallel to the interface of two media and the amplitude of its oscillating EMF being exponentially dumped as the wave propagates into the metal and away from the surface. However, the latter considerations hold true only given that the surface curvature radii are much greater than the depth of the field penetration into the metal.

The depth of field penetration into metal is described by the so-called skin effect. Hence, we can limit the field integration within metal depending on the skin depth Δ_0. At high frequencies of incident field, such a skin depth Δ_0 becomes very small; therefore we may assume that there would be no EMF inside the metal beyond a certain depth.

So, we set a surface $\sum$ inside the object V that would be following the object's surface S at a depth Δ inside the object, this depth being significantly greater than the skin depth Δ_0. Now, we move the field point M to point Q_0 that belongs to surface $\sum$, and we demand that the total field $\vec{E}\left(Q_0\right)$ equals zero at surface $\sum$.

In this case, expression (1.4) transforms into a vector IE of the first kind with respect to the density of the equivalent surface current $\vec{J}(Q) = \vec{n}_Q \times \vec{H}$ and the normal component of the electric field $e(Q) = \vec{n}_Q \cdot \vec{E}(Q)$ at surface S:

$$\vec{E}^0\left(Q_0\right) = \oint\limits_S \left\{j\omega\mu_0 g(r)\vec{J} + W\left(\left(\vec{n}_Q \times \vec{J}\right) \times \vec{\nabla}g(r)\right) + e(Q)\vec{\nabla}g(r)\right\} \mathrm{d}s_Q, \quad Q_0 \in \Sigma. \tag{1.5}$$

We notice that now the observation (field) point Q_0 and the integration (source) point Q belong to different surfaces, the distance r between them never becoming equal to zero and the integral in the right-hand part of (1.5) becoming the proper one. The latter condition simplifies the algorithm for finding numerical solution of IE (1.5).

We multiply the left- and right-hand sides of equation (1.5) by $\vec{n}_{Q_0}$ from the left vector to obtain the following expression:

$$\oint_S \left\{j\omega\mu_0 g(r)\left(\vec{n}_{Q_0} \times \vec{J}\right) + \left(\vec{n}_{Q_0} \times \vec{\nabla} g(r)\right)e(Q)\right\} \mathrm{d}s_Q$$
$$+ \oint_S W\left\{\left(\vec{n}_{Q_0} \times \vec{J}\right)\left(\vec{\nabla} g(r) \cdot \vec{n}_Q\right) - \left(\vec{n}_{Q_0} \times \vec{n}_Q\right)\left(\vec{\nabla} g(r) \cdot \vec{J}\right)\right\} \mathrm{d}s_Q \tag{1.6}$$
$$= \vec{n}_{Q_0} \times \vec{E}^0\left(Q_0\right).$$

Next, we multiply the left- and right-hand sides of equation (1.5) by $\vec{n}_{Q_0}$ scalar to obtain the following expression:

$$\oint_S \left\{j\omega\mu_0 g(r)\left(\vec{n}_{Q_0} \cdot \vec{J}\right) + \left(\vec{n}_{Q_0} \cdot \vec{\nabla} g(r)\right) e(Q)\right\} \mathrm{d}s_Q$$
$$+ \oint_S W\vec{\nabla} g(r)\left(\left(\vec{n}_{Q_0} \cdot \vec{J}\right)\vec{n}_Q - \left(\vec{n}_{Q_0} \cdot \vec{n}_Q\right)\vec{J}\right) \mathrm{d}s_Q = \vec{n}_{Q_0} \cdot \vec{E}^0\left(Q_0\right). \tag{1.7}$$

Expression (1.6) is a vector one. To turn it into a set of scalar equations, we generate three scalar products of equation (1.6) with local basis vectors ($\vec{v}_{1,Q_0}$, $\vec{v}_{2,Q_0}$, $\vec{v}_{3,Q_0}$) using one local vector at a time at point Q_0. The basis vector corresponds to the coordinate system, in which the calculations are carried out.

As a result, we obtain the three scalar equations presented below:

$$\oint_S \left\{j\omega\mu_0 g(r)\left(\vec{f}_p \cdot \vec{J}\right) + \left(\vec{f}_p \cdot \vec{\nabla} g(r)\right) e(Q)\right\} \mathrm{d}s_Q$$
$$+ \oint_S W\left\{\left(\vec{f}_p \cdot \vec{J}\right)\left(\vec{\nabla} g(r) \cdot \vec{n}_Q\right) - \left(\vec{f}_p \cdot \vec{n}_Q\right)\left(\vec{\nabla} g(r) \cdot \vec{J}\right)\right\} \mathrm{d}s_Q \tag{1.8}$$
$$= \vec{f}_p \cdot \vec{E}^0\left(Q_0\right), \quad p = \overline{1,3}$$

where $\vec{f}_p = \vec{v}_{p,\,Q_0} \times \vec{n}_{Q_0}$.

Expressions (1.7) and (1.8) represent the set of scalar IEs containing the sought surface current density $\vec{J}$ at the impedance scatterer. One can now compute the field scattered by the screen in question by substituting the current $\vec{J}$ obtained as a solution to the latter set into expression (1.4).

1.1.2 Method for solving the obtained integral equations

We represent the surface current density $\vec{J}$ at point Q belonging to the surface S as a sum of mutually orthogonal components

$$\vec{J} = J_1\vec{\tau}_1 + J_2\vec{\tau}_2, \tag{1.9}$$

where $\vec{\tau}_1$, $\vec{\tau}_2$ are mutually orthogonal unit vectors that are both tangential to the surface S.

We denote the quantity of observation (field) points, for which we are going to solve the set of scalar IEs (1.7) and (1.8), as N_0. Then, any particular field point will be denoted as Q_{n_0} $(n_0 = \overline{1, N_0})$.

Now, we break down surface S into N segments $\left(S = \sum_{n=1}^{N} S_n\right)$, the surface current density (1.9) being assumed constant within every particular segment.

Having introduced such a piecewise constant approximation of equivalent surface current density $\vec{J}$ and of the normal component of electric field $e(Q)$, and having accounted for expression (1.9), we obtain the set of linear equations (SLE) from the IE set (1.7) and (1.8). This SLE includes the components of the equivalent surface current density for every surface segment where this density does not vary. This later SLE can be presented in matrix form as follows:

$$\mathbf{A}\vec{J}_e = \vec{B}, \tag{1.10}$$

where

$$\vec{J}_e = \begin{pmatrix} \vec{J}_1 \\ \vec{J}_2 \\ \vec{\xi} \end{pmatrix}, \tag{1.11}$$

$$\vec{J}_1 = \begin{pmatrix} J_{1,1} \\ \vdots \\ J_{1,n} \\ \vdots \\ J_{1,N} \end{pmatrix}, \quad \vec{J}_2 = \begin{pmatrix} J_{2,1} \\ \vdots \\ J_{2,n} \\ \vdots \\ J_{2,N} \end{pmatrix}; \quad \vec{\xi} = \begin{pmatrix} e_1(Q_1) \\ \vdots \\ e_n(Q_n) \\ \vdots \\ e_N(Q_N) \end{pmatrix} = \begin{pmatrix} \vec{n}_{Q_1} \cdot \vec{E}(Q_1) \\ \vdots \\ \vec{n}_{Q_n} \cdot \vec{E}(Q_n) \\ \vdots \\ \vec{n}_{Q_N} \cdot \vec{E}(Q_N) \end{pmatrix}$$

$$\begin{cases} \sum\limits_{n=1}^{N} A_{n,n_0}^{11} J_{1,n} + \sum\limits_{n=1}^{N} A_{n,n_0}^{12} J_{2,n} + \sum\limits_{n=1}^{N} A_{n,n_0}^{13} e_n = B_{1,n_0} \\ \sum\limits_{n=1}^{N} A_{n,n_0}^{21} J_{1,n} + \sum\limits_{n=1}^{N} A_{n,n_0}^{22} J_{2,n} + \sum\limits_{n=1}^{N} A_{n,n_0}^{23} e_n = B_{2,n_0} \\ \sum\limits_{n=1}^{N} A_{n,n_0}^{31} J_{1,n} + \sum\limits_{n=1}^{N} A_{n,n_0}^{32} J_{2,n} + \sum\limits_{n=1}^{N} A_{n,n_0}^{33} e_n = B_{3,n_0} \\ \sum\limits_{n=1}^{N} A_{n,n_0}^{41} J_{1,n} + \sum\limits_{n=1}^{N} A_{n,n_0}^{42} J_{2,n} + \sum\limits_{n=1}^{N} A_{n,n_0}^{43} e_n = B_{4,n_0} \end{cases}, \quad n_0 = \overline{1, N_0}, \tag{1.12}$$

$$\begin{aligned} A_{n,n_0}^{pq} = \int_{S_n} \Big\{ & j\omega\mu_0 g(r)\alpha_{n_0,n}^{pq} + W\Big[\alpha_{n_0,n}^{pq}\big(\vec{\nabla}g(r)\cdot\vec{n}_{Q_n}\big) \\ & - \beta_{n_0,n}^{pq}\big(\vec{\nabla}g(r)\cdot\vec{\tau}_{q,n}\big)\Big]\Big\}\, ds_Q, \end{aligned} \tag{1.13}$$

$$p = \overline{1,4}, \quad q = 1,2,$$

$$A_{n,\,n_0}^{p3} = \begin{cases} \int_{Sn} \vec{f}_{p,\,n_0} \cdot \vec{\nabla} g(r) \mathrm{d}s_Q, & \text{if } p \neq 4 \\ \int_{Sn} \vec{n}_{Q_{n_0}} \cdot \vec{\nabla} g(r) \mathrm{d}s_Q, & \text{if } p = 4 \end{cases}, \tag{1.14}$$

$$B_{p,\,n_0} = \begin{cases} \vec{f}_{p,\,n_0} \cdot \vec{E}^0\left(Q_{n_0}\right), & \text{if } p \neq 4 \\ \vec{n}_{Q_{n_0}} \cdot \vec{E}^0\left(Q_{n_0}\right), & \text{if } p = 4 \end{cases}, \tag{1.15}$$

$$\alpha_{n_0,\,n}^{pq} = \begin{cases} \vec{f}_{p,\,n_0} \cdot \vec{\tau}_{q,\,n}, & \text{if } p \neq 4 \\ \vec{n}_{Q_{n_0}} \cdot \vec{\tau}_{q,\,n}, & \text{if } p = 4 \end{cases}, \tag{1.16}$$

$$\beta_{n_0,\,n}^{p} = \begin{cases} \vec{f}_{p,\,n_0} \cdot \vec{n}_{Q_n}, & \text{if } p \neq 4 \\ \vec{n}_{Q_{n_0}} \cdot \vec{n}_{Q_n}, & \text{if } p = 4 \end{cases}, \text{ and} \tag{1.17}$$

$$\vec{f}_{p,\,n_0} = \vec{v}_{p,\,Q_{n_0}} \times \vec{n}_{Q n_0}. \tag{1.18}$$

Matrix **A** in (1.10) has a dimensionality of $(4N_0 \times 3N)$. Given the number of field points N_0 that is equal to or greater than the number N of surface segments, within which the surface current density is held constant, one obtains the overdetermined set of linear equations (OSLE).

Solutions to the latter OSLE are components of the equivalent surface current density that are excited by the primary incident field $\vec{E}^0$ on the surface of the impedance scatterer.

These components of equivalent surface current density obtained as solutions to the OSLE are then used to compute the field scattered by the screen in question using expression (1.4).

1.1.3 Electromagnetic wave scattering at a non-perfectly conducting screen of finite thickness

The problem of wave scattering at spherical perfectly conducting screens of finite thickness has been solved in a number of works [1–3, 11]. Additionally, the authors in [3, 11] clearly demonstrate that in the resonant scattering waveband the radar cross section (RCS) of such screens depends on their thickness. In addition, the screen's RCS dependence on its thickness becomes stronger with an increase in the screen's curvature (two screens with different curvatures can be compared given that their apertures are equal). This particular dependence is caused by the surface currents flowing from the illuminated parts of the curved screen's surface into the shadow zone. In some cases, such shadow currents become so significant that they can be of the same order of magnitude as the currents flowing over the illuminated parts of the screen's surface. The sum of fields excited by these two kinds of currents is different from the field excited by the surface currents flowing only over the illuminated surface part of infinitely thin screens [12].

Unlike the approach presented in [1–3, 11], expressions (1.7) and (1.8), in addition to their capability of dealing with a finite screen thickness, also allow us to account for the finite conductivity of the material of which the screen is made.

Let us consider a screen generated by the intersection of a spherical surface that has non-zero thickness h and inner radius a, and a circular cylinder of radius c that has axis Oz as its axis of symmetry (figure 1.2). Let us present the surface of the spherical screen as a union of its three parts $S = S_1 \cup S_2 \cup S_3$ (figure 1.2).

The inner surface S_1 (front face) of the screen (figure 1.2) can be formalized as follows:

$$\left.\begin{aligned} x &= a \sin\theta \cos\varphi \\ y &= a \sin\theta \sin\varphi \\ z &= -a\cos\theta \end{aligned}\right\}, \quad \begin{pmatrix} 0 \leqslant \varphi \leqslant 2\pi \\ 0 \leqslant \theta \leqslant \vartheta_1 \end{pmatrix}, \tag{1.19}$$

where

$$\vartheta_1 = \arcsin(c/a). \tag{1.20}$$

The outer surface S_2 (rear face) can also be formalized using the same expressions (1.19) and (1.20) except that a needs to be replaced by $a + h$. In addition, the angle ϑ_1 in expression (1.19) transforms into the angle $\vartheta'_1 = \arcsin(c/(a + h))$.

The side surface S_3 (edge) can be formalized as follows:

$$\left.\begin{aligned} x &= c\cos\varphi \\ y &= c\sin\varphi \\ -(a+h)\cos\vartheta'_1 &\leqslant z \leqslant -a\cos\vartheta_1 \end{aligned}\right\}, \quad (0 \leqslant \varphi \leqslant 2\pi). \tag{1.21}$$

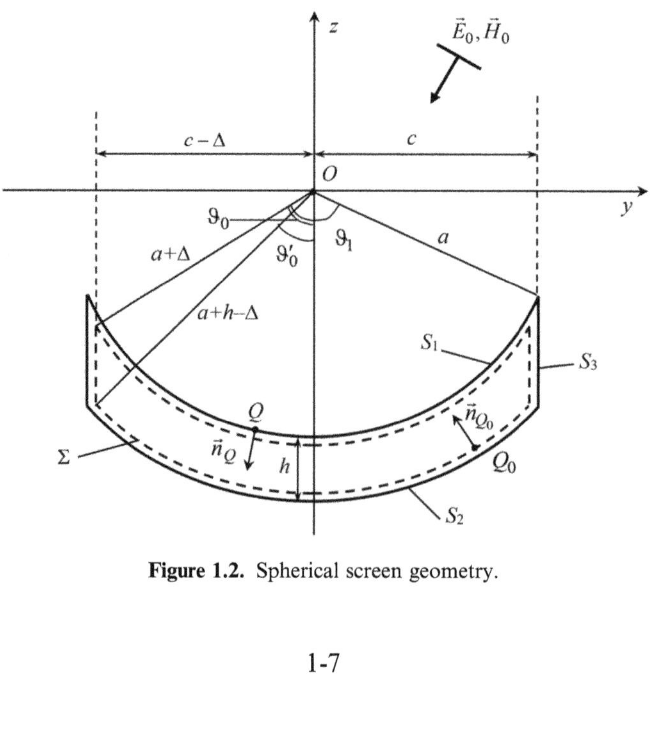

Figure 1.2. Spherical screen geometry.

Auxiliary surface Σ is set in such a manner that it will follow the surface S yet placed inside the object at the depth Δ, the total EMF being equal to zero at this auxiliary surface. The author of [13] stressed the fact that although the boundary conditions can be moved (due to good analytic properties) from the object's surface S onto the auxiliary surface Σ inside the object, the approximate numerical solution to equations (1.5), although devoid of singularity, still suffers from certain specific instability. If one increases the distance Δ between the two surfaces (Σ and S) one makes the boundary conditions less stringent at the object's surface. Therefore, in our computations we set the value Δ to be small enough to keep the boundary conditions stringent and we control the solution stability by comparing the results of our computations obtained given varying values of Δ. To be precise, at the beginning we carried out three rounds of computations given three different values of Δ, namely Δ was changed from $8\Delta_0$ to $10\Delta_0$ and finally to $12\Delta_0$ (where Δ_0 is the skin depth). Our computations showed us that the RCS of the screen with parameters, $k_0c = 3$ and $\vartheta_1 = 45°$ varied by less than 0.4% given that Δ was changed from $\Delta = 12\Delta_0$ to $\Delta = 10\Delta_0$. When Δ was changed from $\Delta = 12\Delta_0$ to $\Delta = 8\Delta_0$, the variation in the screen's RCS varied by less than 1.1%. Thus, in further computations we set the parameter Δ to be equal to $10\Delta_0$, which corresponds to the incident field attenuation by factor of e^{10}.

Now, the equivalent surface current density $\vec{J}$ excited on the entire screen's surface S can be computed as follows:

$$\vec{J}^{1(2)} = J_\theta^{1(2)}(\theta, \varphi)\vec{\tau}_\theta + J_\varphi^{1(2)}(\theta, \varphi)\vec{\tau}_\varphi \quad \text{at} \quad S_{1(2)}, \tag{1.22}$$

$$\vec{J}^{3} = J_z^{3}(z, \varphi)\vec{\tau}_z + J_\varphi^{3}(z, \varphi)\vec{\tau}_\varphi \quad \text{at} \quad S_3, \tag{1.23}$$

where $\vec{\tau}_\theta = (-\cos\theta\cos\varphi, -\cos\theta\sin\varphi, -\sin\theta)$, $\vec{\tau}_\varphi = (-\sin\varphi, \ \cos\varphi, 0)$, and $\vec{\tau}_z = (0,0,1)$.

Similar to the technique used in subsection 1.1.2, we break down each part of the surface S_m (two faces and the edge) into N_m segments ($S_m = \sum_{n=1}^{N_m} S_{mn}$, ($m = 1,2,3$)). Then, the components of the equivalent surface current density $J_{\theta n}^{S_m}$, $J_{\varphi n}^{S_m}$, $J_{zn}^{S_m}$ will be assumed to be constant within each surface segment.

However, the segments with a constant level of surface current density across the surfaces S_1 and S_2 are not set at random. We need these segments to represent real phenomena. Thus we draw a system of rings at the two faces, S_1 and S_2 (figure 1.3), generated by the intersection of the screen's surface with co-axial circular cones, the apices of which coincide with the center of the circular screen and the axes of which coincide with the axis Oz. The half angles of the cones are equal to $\theta_i = i\Delta\theta$ ($i = 1,\ldots, N$; $\Delta\theta = \vartheta_1/N$).

Each cone's intersection with the screen's surface produces a circumference, and two neighboring circumferences form a circular ring $S_{1(2)}$. Next, we break down this ring into equal segments by angle φ. The number of angular segments increases proportionally to the length of the ring's outer circumference (figure 1.3), and the total number of segments can be controlled by the number of angular segments in

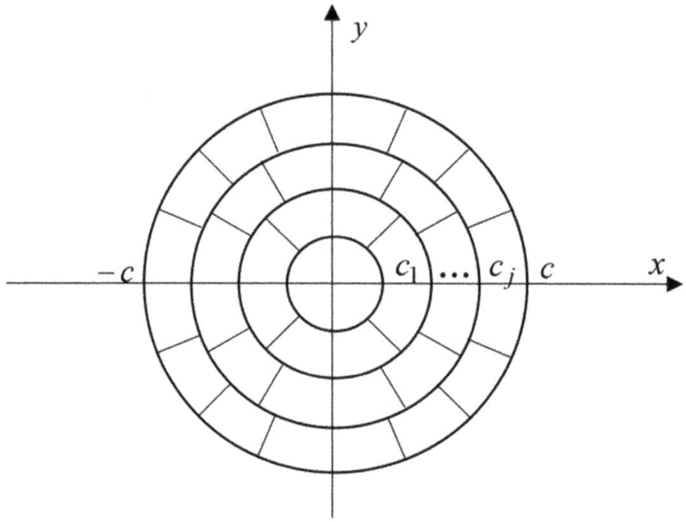

Figure 1.3. The method for breaking down the screen faces S_1 and S_2 into segments with constant surface current density.

the innermost ring. In our computations we adopted the method in which the innermost ring was broken down by angle φ into four equal angular segments.

The side surface S_3 is also broken down into 'ribbons' generated by intersecting the surface S_3 with planes parallel to the plane xOy, the planes being spaced by distance $\Delta b = h/M$ from one another. As a next step, we break down the 'ribbons' thus obtained, every ribbon being Δb wide, into equal segments by angle ϕ. For consistency, we set the number of segments in each ribbon equal to the number of angular segments in the outermost ring of the screen's front face S_1.

Now, we employ the expressions (1.10)–(1.18) obtained in subsection 1.1.2.

Taking into account expressions (1.22) and (1.23), we can rewrite the SLE (1.10) in this particular case as

$$\mathbf{A}_1\vec{J}_e^{\,1} + \mathbf{A}_2\vec{J}_e^{\,2} + \mathbf{A}_3\vec{J}_e^{\,3} = \vec{B}, \tag{1.24}$$

where

$$\vec{J}_e^{\,m} = \left(J_{\theta 1}^m \ldots J_{\theta N_m}^m \; J_{\varphi 1}^m \ldots J_{\varphi N_m}^m \; e_1^m \ldots e_{N_m}^m\right)^T, \quad m = \overline{1,2},$$

$$\vec{J}_e^{\,3} = \left(J_{\varphi 1}^3 \ldots J_{\varphi N_3}^3 J_{z1}^3 \ldots J_{zN_3}^3 e_1^3 \ldots e_{N_3}^3\right)^T,$$

$$\mathbf{A}_m = \begin{pmatrix} \mathbf{A}_{m1} \\ \vdots \\ \mathbf{A}_{mn_0} \\ \vdots \\ \mathbf{A}_{mN_0} \end{pmatrix}, \quad m = \overline{1,3},$$

$$\mathbf{A}_{mn_0} = \begin{pmatrix} A^{11}_{m1,\,n_0} \cdots A^{11}_{mN,\,n_0} \, A^{12}_{m1,\,n_0} \cdots A^{12}_{mN,\,n_0} \, A^{13}_{m1,\,n_0} \cdots A^{13}_{mN,\,n_0} \\ A^{21}_{m1,\,n_0} \cdots A^{21}_{mN,\,n_0} \, A^{22}_{m1,\,n_0} \cdots A^{22}_{mN,\,n_0} \, A^{23}_{m1,\,n_0} \cdots A^{23}_{mN,\,n_0} \\ A^{31}_{m1,\,n_0} \cdots A^{31}_{mN,\,n_0} \, A^{22}_{m1,\,n_0} \cdots A^{32}_{mN,\,n_0} \, A^{33}_{m1,\,n_0} \cdots A^{33}_{mN,\,n_0} \\ A^{41}_{m1,\,n_0} \cdots A^{41}_{mN,\,n_0} \, A^{22}_{m1,\,n_0} \cdots A^{42}_{mN,\,n_0} \, A^{43}_{m1,\,n_0} \cdots A^{43}_{mN,\,n_0} \end{pmatrix},$$

$$A_{mn_0}\vec{J}^{\,m}_{e} = \begin{cases} \sum_{n=1}^{N_m} A^{11}_{mn,\,n_0} J^{m}_{\theta\,n} + \sum_{n=1}^{N_m} A^{12}_{mn,\,n_0} J^{m}_{\varphi\,n} + \sum_{n=1}^{N_m} A^{13}_{mn,\,n_0} e^{m}_{n} \\ \sum_{n=1}^{N_m} A^{21}_{mn,\,n_0} J^{m}_{\theta\,n} + \sum_{n=1}^{N_m} A^{22}_{mn,\,n_0} J^{m}_{\varphi\,n} + \sum_{n=1}^{N_m} A^{23}_{mn,\,n_0} e^{m}_{n} \\ \sum_{n=1}^{N_m} A^{31}_{mn,\,n_0} J^{m}_{\theta\,n} + \sum_{n=1}^{N_m} A^{32}_{mn,\,n_0} J^{m}_{\varphi\,n} + \sum_{n=1}^{N_m} A^{33}_{mn,\,n_0} e^{m}_{n} \\ \sum_{n=1}^{N_m} A^{41}_{mn,\,n_0} J^{m}_{\theta\,n} + \sum_{n=1}^{N_m} A^{42}_{mn,\,n_0} J^{m}_{\varphi\,n} + \sum_{n=1}^{N_m} A^{43}_{mn,\,n_0} e^{m}_{n} \end{cases}, \quad m = \overline{1,\,2}, \tag{1.25}$$

$$A_{3n_0}\vec{J}^{\,3}_{e} = \begin{cases} \sum_{n=1}^{N_3} A^{11}_{3n,\,n_0} J^{3}_{\varphi n} + \sum_{n=1}^{N_3} A^{12}_{3n,\,n_0} J^{3}_{z\,n} + \sum_{n=1}^{N_3} A^{13}_{3n,\,n_0} e^{3}_{n} \\ \sum_{n=1}^{N_3} A^{21}_{3n,\,n_0} J^{3}_{\varphi n} + \sum_{n=1}^{N_3} A^{22}_{3n,\,n_0} J^{3}_{z\,n} + \sum_{n=1}^{N_3} A^{23}_{3n,\,n_0} e^{3}_{n} \\ \sum_{n=1}^{N_3} A^{31}_{3n,\,n_0} J^{3}_{\varphi n} + \sum_{n=1}^{N_3} A^{32}_{3n,\,n_0} J^{3}_{z\,n} + \sum_{n=1}^{N_3} A^{33}_{3n,\,n_0} e^{3}_{n} \\ \sum_{n=1}^{N_3} A^{41}_{3n,\,n_0} J^{3}_{\varphi n} + \sum_{n=1}^{N_3} A^{42}_{3n,\,n_0} J^{3}_{z\,n} + \sum_{n=1}^{N_3} A^{43}_{3n,\,n_0} e^{3}_{n} \end{cases}. \tag{1.26}$$

Considering the screen's geometry, a convenient set of unit vectors $\vec{v}_{p,\,Q_{n0}}$ $(p = \overline{1,\,3})$ that enter into expression (1.8) comprises the following unit vectors:

$$\vec{v}_{1,\,Q_{n0}} = \vec{v}_{\theta_0} = (-\cos\theta_0 \cos\varphi_0,\ -\cos\theta_0 \sin\varphi_0,\ -\sin\theta_0), \tag{1.27}$$

$$\vec{v}_{2,\,Q_{n0}} = \vec{v}_{\varphi_0} = (-\sin\varphi_0,\ \cos\varphi_0,\ 0), \tag{1.28}$$

$$\vec{v}_{3,\,Q_0} = \vec{v}_{z_0} = (0,\ 0,\ 1), \tag{1.29}$$

where θ_0, φ_0 are angular coordinates of the observation (field) point Q_{n_0}.

Then, elements of the matrix $\mathbf{A}_{mn_0}$ can be written as follows

$$A^{pq}_{mn,\,n_0} = \int_{S_{mn}} \left[j\omega\mu_0 g(r) \alpha^{pq}_{mn_0,\,n} + W g'(r) \left(\alpha^{pq}_{mn_0,\,n} r^0_n - \beta^{pq}_{n_0,\,n} \gamma_{mq,\,n} \right) \right] \mathrm{d}s_Q, \tag{1.30}$$

$$p = \overline{1, 4}, \quad q = 1, 2,$$

$$A^{p3}_{mn,\, n_0} = \begin{cases} \displaystyle\int_{Smn} \left(\vec{f}_{p,\, n_0} \cdot \vec{\nabla} g(r)\right) \mathrm{d}s_Q, & \text{if } p \neq 4 \\ \displaystyle\int_{Smn} \left(\vec{n}_{Q_{n0}} \cdot \vec{\nabla} g(r)\right) \mathrm{d}s_Q, & \text{if } p = 4 \end{cases}, \tag{1.31}$$

where

$$\alpha^{pq}_{mn,\, n_0} = \begin{cases} \vec{f}_{p,\, n_0} \cdot \vec{\tau}_{mq,\, n}, & \text{if } p \neq 4 \\ \vec{n}_{Q_{n0}} \cdot \vec{\tau}_{mq,\, n}, & \text{if } p = 4 \end{cases}, \tag{1.32}$$

$$\beta^{p}_{n_0,\, n} = \begin{cases} \vec{f}_{p,\, n_0} \cdot \vec{n}_{Q_n}, & \text{if } p \neq 4 \\ \vec{n}_{Q_{n0}} \cdot \vec{n}_{Q_n}, & \text{if } p = 4 \end{cases}, \tag{1.33}$$

$$\vec{f}_{1,\, n_0} = \vec{v}_{\theta_{n0}} \times \vec{n}_{Qn_0}, \quad \vec{f}_{2,\, n_0} = \vec{v}_{\varphi_{n0}} \times \vec{n}_{Qn_0}, \quad \vec{f}_{3,\, n_0} = \vec{v}_{z_{n0}} \times \vec{n}_{Qn_0}, \tag{1.34}$$

$$\vec{\tau}_{mq,\, n} = \begin{cases} \vec{v}_{\theta,\, n}, & \text{if } q = 1, \quad m = \overline{1, 2} \\ \vec{v}_{\varphi,\, n}, & \text{if } q = 2, \quad m = \overline{1, 2} \\ \vec{v}_{\varphi,\, n}, & \text{if } q = 1, \quad m = 3 \\ \vec{v}_{z,\, n}, & \text{if } q = 2, \quad m = 3 \end{cases}, \tag{1.35}$$

$$\vec{n}_{Q_{n0}} = \begin{cases} \vec{v}_{\rho_0}, & \text{if } Q_{n_0} \in \Sigma_1 \\ -\vec{v}_{\rho_0}, & \text{if } Q_{n_0} \in \Sigma_2 \\ (-\cos\varphi_0, -\sin\varphi_0, 0), & \text{if } Q_{n_0} \in \Sigma_3 \end{cases}, \tag{1.36}$$

$$\vec{r}^{\,0}_{n,\, n_0} = \frac{\vec{Q}_n - \vec{Q}_{n_0}}{r}, \quad r = \left| \vec{Q}_n - \vec{Q}_{n_0} \right|, \tag{1.37}$$

$$r^0_n = \vec{r}^{\,0}_{n,\, n_0} \cdot \vec{n}_{Q_n}, \tag{1.38}$$

$$\gamma_{mq,\, n} = \vec{r}^{\,0}_{n,\, n_0} \cdot \vec{\tau}_{mq,\, n}, \tag{1.39}$$

and

$$q = 1, 2; \quad n = \overline{1, N}; \quad n_0 = 1, N_0; \quad m = \overline{1, 3}.$$

Let us note that because of our particular choice of auxiliary surface Σ the values of r never become equal to zero, which provides for the non-singular character of integrands in (1.30).

The coefficients $A^{pq}_{mn,\, n_0}$ in the equation set (1.24) obtained above are double integrals of smooth functions. They are computed by the five-point composite Gauss quadrature [14]. In addition, the computation error for these integrals is less than 0.2%.

The entire SLE (1.24) is solved by the collocation method. The number of collocation points Q_{n0} is set to be greater than the number of unknowns, which define the approximate solution to the equivalent surface current density. In this case, we obtain the OSLE, which can be solved by means of a least squares method. In our computations, we set the number of collocation points to be 1.5 times greater than the number of unknowns.

The total number of segments with a constant level of surface current density in our computations depends on the screen's surface area.

The solution to the OSLE gives us the approximation of equivalent surface current density, which is used subsequently for computing the field scattered by the screen in question using expression (1.4).

Consistency check of the method proposed above and study of the impact of the screen edge's shape on the screen's RCS

The scattering impact of the screen edge's shape is studied here by comparing the fields scattered by spherical antenna reflectors, the edges of which are shaped in two different ways. The two kinds of reflector edge shapes are shown in figure 1.4. Actually, the antenna reflector with the edge of first kind has already been considered in this section above.

A spherical screen with an edge of the second kind is generated by an intersection of a spherical surface that has non-zero thickness h and inner radius a, and the cone, the apex of which coincides with the center of the spherical screen. The half angle of the cone in figure 1.4 is denoted as ϑ_1. As was the case in this section above, the entire

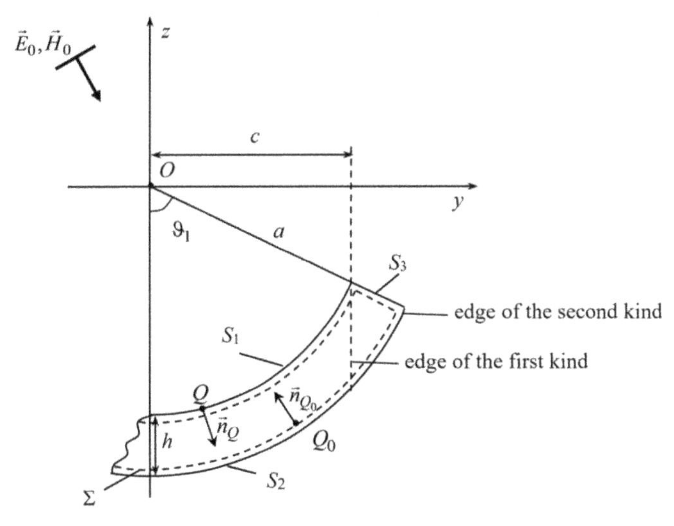

Figure 1.4. Shapes of the antenna reflector's edge adopted in computations.

surface of this spherical screen is represented as a union of three parts: $S = S_1 \cup S_2 \cup S_3$ (figure 1.4).

For further computations we adopted a spherical coordinate system (SCS).

The 'illuminated' inner surface (front face) S_1 in the SCS can be formalized using expressions (1.19) and (1.20).

The 'shadowed' outer surface (rear face) S_2 can be formalized using the same expressions except that parameter a needs to be replaced by parameter $a + h$.

The side surface (edge) S_3 can be formalized as follows:

$$\left.\begin{array}{l} x = \rho \sin \vartheta_1 \cos \varphi \\ y = \rho \sin \vartheta_1 \sin \varphi \\ z = -\rho \cos \vartheta_1 \end{array}\right\}, \quad (a \leqslant \rho \leqslant a + h). \tag{1.40}$$

Auxiliary surface Σ is also set as a union $\Sigma = \Sigma_1 \cup \Sigma_2 \cup \Sigma_3$. In addition, auxiliary surface parts Σ_1 and Σ_2 are the segments of spherical surfaces with radii of $a + \Delta$ and $a + h - \Delta$, respectively. These two segments can be described by equation (1.19) if the following substitutions are introduced: $a \to a + \Delta$ (for Σ_1), $a \to a + h - \Delta$ (for Σ_2), and $\vartheta_1 \to \vartheta_0 \left(\vartheta_0 = \vartheta_1 - \arcsin\left(\dfrac{\Delta}{a + \Delta} \right) \right)$.

The part Σ_3 of the auxiliary surface can be described by expressions (1.40) if the following substitution is introduced, $\vartheta_1 \to \vartheta_0$, and if limits of the coordinate ρ are set as $a + \Delta \leqslant \rho \leqslant a + h - \Delta$.

Now, the equivalent surface current density $\vec{J}$ excited on entire screen's surface S can be computed as follows: at

$$\vec{J}^{1(2)} = J_\theta^{1(2)}(\theta, \varphi)\vec{v}_\theta + J_\varphi^{1(2)}(\theta, \varphi)\vec{v}_\varphi \quad \text{at} \quad S_{1(2)}, \tag{1.41}$$

$$\vec{J}^3 = J_\rho^3(\rho, \varphi)\,\vec{v}_\rho + J_\varphi^3(\rho, \varphi)\,\vec{v}_\varphi \quad \text{at} \quad S_3, \tag{1.42}$$

where $\vec{v}_\theta$, $\vec{v}_\varphi$ are the unit vectors of SCS at the source point Q, which belongs to surface S:

$\vec{v}_\rho = (\sin\theta \cos\varphi,\ \sin\theta \sin\varphi,\ -\cos\theta)$,
$\vec{v}_\theta = (-\cos\theta \cos\varphi,\ -\cos\theta \sin\varphi,\ -\sin\theta)$,
$\vec{v}_\varphi = (-\sin\varphi,\ \cos\varphi,\ 0)$.

Considering the screen's geometry, a set of local basis vectors that would be convenient for obtaining the SLE in (1.8) comprises the unit vectors of SCS:

$$\vec{v}_{1,\,Q_0} = \vec{v}_{\theta_0} = (-\cos\vartheta_0 \cos\varphi_0,\ -\cos\vartheta_0 \sin\varphi_0,\ -\sin\vartheta_0), \tag{1.43}$$

$$\vec{v}_{2,\,Q_0} = \vec{v}_{\varphi_0} = (-\sin\varphi_0,\ \cos\varphi_0,\ 0), \tag{1.44}$$

$$\vec{v}_{3,\,Q_0} = \vec{v}_{\rho_0} = (\sin\vartheta_0 \cos\varphi_0,\ \sin\vartheta_0 \sin\varphi_0,\ -\cos\vartheta_0) \tag{1.45}$$

where ϑ_0, φ_0 are angular coordinates of the observation (field) point Q_0.

Subsequently, having introduced the piecewise constant approximation of equivalent surface current density we obtain the OSLE, which is described by expressions (1.24)–(1.26) (subsection 1.1.3).

Elements of the OSLE matrix are also described by expressions (1.30)–(1.33) and (1.35)–(1.39), in which, considering equations (1.43)–(1.45), the vectors $\vec{f}_{p,\,n_0}$ ($p = \overline{1,3}$) need to be replaced by

$$\vec{f}_{1,\,n_0} = \vec{v}_{\theta_0} \times \vec{n}_{Qn_0},\quad \vec{f}_{2,\,n_0} = \vec{v}_{\varphi_0} \times \vec{n}_{Qn_0},\quad \vec{f}_{3,\,n_0} = \vec{v}_{\rho_0} \times \vec{n}_{Qn_0}. \tag{1.46}$$

The rest of the procedure for finding the equivalent surface current density and the field scattered by the screen is similar to that in subsection 1.1.3.

The scattering impact of the screen edge's shape and the computation method consistency check are studied here by comparing the computed RCS values σ [15] for the screens having edges of two kinds with the corresponding results of the experimental RCS measurement presented in [16].

To that end, we substitute into the right-hand part of OSLE (1.24) the plane EMW

$$\vec{E}^0(\vec{X}) = \vec{p}^{\,0} e^{jk_0\left(\vec{R}^0 \cdot \vec{X}\right)}, \tag{1.47}$$

where $\vec{p}^{\,0}$ is the unit vector of the incident wave polarization, and $\vec{R}^0$ is the unit vector of the wave incident direction.

Additionally, we checked our computation results against those obtained by Vinogradov [12] for the infinitely thin perfectly conducting screen.

The work [16] contains the results of experimental RCS measurements for an aluminum spherical screen given its illumination in the axial direction. The results were obtained at Karazin Kharkiv National University. The study was carried out for the frequency band of illumination signal from 26.2 GHz through 38.4 GHz, which corresponds to the values of k_0c from 6.89 through 10.12. The specimen under study had the following geometric parameters: the radius of the spherical screen was $a = 18.08$ mm; the aperture radius of the spherical screen was $c = 12.55$ mm; and the screen thickness was $h = 0.15$ mm. The conductivity of aluminum was assumed in the computations to be equal to $3.54 \cdot 10^7$ Sm m^{-1}.

Figure 1.5 shows the normalized RCS $\frac{\sigma}{\pi c^2}$ versus parameter k_0c for the spherical screen mentioned above. The plot in figure 1.5 features several curves. Curve #1 presents experimental RCS measurements, curves #2 and #3 present the results of computation using our method presented above given a screen with edges of the first and second kind, respectively, and curve #4 presents the RCS computed in [12] for the case of a perfectly conducting infinitely thin screen.

It is obvious from figure 1.5 that the measured RCS (curve #1) is basically confined between the curves #2 and #3, which correspond to our computation method. Some deviations between the experimental and our computational results given small values of $\frac{\sigma}{\pi c^2}$ (these small values of normalized RCS correspond to the interval $7.25 < k_0c < 8.25$) can be explained by certain screen manufacturing

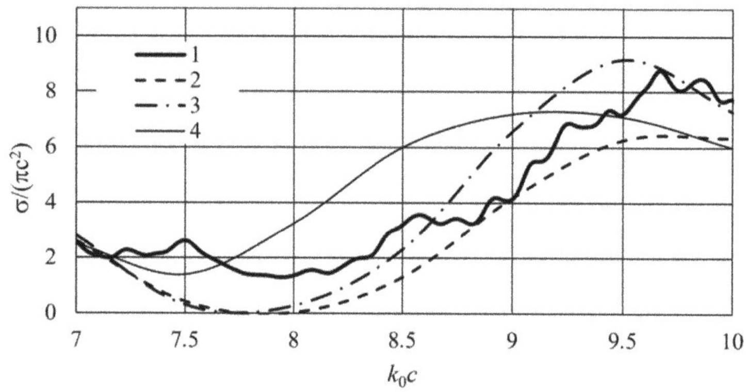

Figure 1.5. Normalized RCS of a spherical screen (1—experimental data [16]; 2—our computation results for an aluminum screen with an edge of the first kind; 3—our computation results for an aluminum screen with an edge of the second kind; 4—computation results for a perfectly conducting infinitely thin screen [12]).

inaccuracy, by errors caused by weak radar echoes, frequency measurement errors, and the effect of the fixtures used to hold the screen on the field scattered by it [16].

It can be seen from figure 1.5 that the RCS computed for the case of a perfectly conducting infinitely thin screen [12] (curve #4) differs significantly from both the experimental RCS and from the RCS computed by our method, given a non-perfectly conducting screen of finite thickness. This particular fact makes it important to account for a real screen's thickness and the finite conductivity of the material of which the screen is made. The latter is particularly true when the RCS computation is concerned with resonant EMW scattering.

The difference between curves #2 and #3 reveals the necessity of accounting for the shape of the screen's edge, in particular in the resonant waveband, since the RCSs of two identical screens that differ only by the shape of their edges can vary significantly. For instance, given $k_0c = 9.5$, the two spherical screens, which differ only in the shapes of their edges, have RCS values differing by 32%. This fact needs to be taken into account when designing reflector ASs operating in the resonant waveband.

Study of the effect of the material an electrically thin screen is made of on the screen's RCS in the resonant waveband

The authors of [3] emphasize that the scattering characteristics computed for electrically thin perfectly conducting screens differ significantly from those computed for infinitely thin perfectly conducting screens where the resonant waveband is concerned. We have already demonstrated that the method for computing the RCS of screens presented above in sections 1.1 and 1.2 is suitable for analyzing the effect of the screen's thickness and material on its RCS [17].

Here, we study the effect of the material the screen is made of using as examples several spherical screens that have different aperture radii and different thicknesses. It turns out that normalized RCS $\frac{\sigma}{\pi c^2}$, and not regular RCS σ, is more convenient as long as a comparison of the scattering features of different screens was concerned [15].

First, we fixed the screen's aperture radius to $c = 0.19$ m and we considered two distinctly different screen shapes, which, from now on, we refer to as a shallow screen and deep screen. These two shapes differed by the angle ϑ_1 (figure 1.2), for the shallow one this angle was $\vartheta_1 = 30°$ and for the deep one this angle was $\vartheta_1 = 45°$. In addition, both the shallow and deep screens had the edge of the first kind (figure 1.4).

Now, for these two screens, we computed the normalized RCS versus parameter k_0c (related to the frequency of incident wave) given that both screens were illuminated in the axial direction. In addition, the parameter k_0c varied in the range of $1 \leqslant k_0c \leqslant 9.5$. We repeated our computations several times for different screen materials: a perfect conductor, aluminum with a conductivity of 3.54×10^7 Sm m^{-1}, and duralumin 2024 (D16) with a conductivity of 1.74×10^7 Sm m^{-1}.

Figures 1.6 and 1.7 show the results of our RCS computations for both the shallow and deep screens given that both screens are 1 mm thick. Curve #1 corresponds to the screen made of aluminum, curve #2 corresponds to the screen

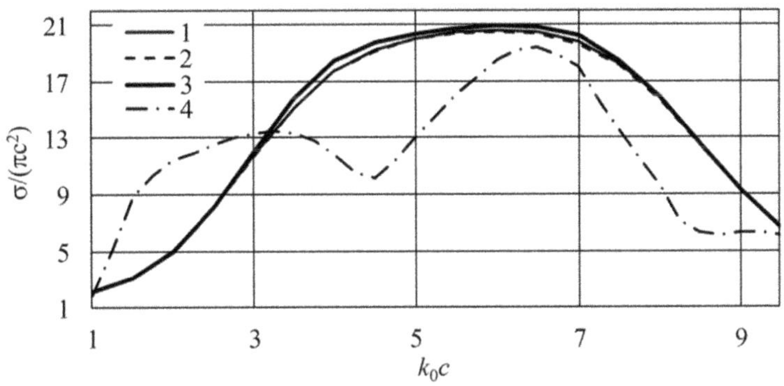

Figure 1.6. The RCS of a shallow spherical screen ($\vartheta_1 = 30°$), which was 1 mm thick (1—aluminum screen; 2—duralumin D16 screen; 3—perfectly conducting screen; 4—infinitely thin (zero thickness) perfectly conducting screen [12]).

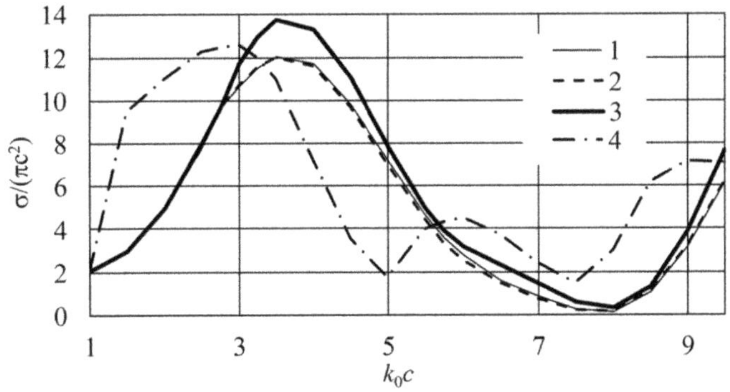

Figure 1.7. The RCS of a deep spherical screen ($\vartheta_1 = 45°$), which is 1 mm thick (1—aluminum screen; 2—duralumin D16 screen; 3—perfectly conducting screen; 4—infinitely thin (zero thickness) perfectly conducting screen [12]).

made of duralumin 2024, and curve #3 corresponds to the screen made of a perfect conductor (the thickness of this particular screen Δ was assumed to be the same as that for the aluminum screen). In addition, curve #4 in both figures 1.6 and 1.7 presents the computed RCS data [12] for a perfectly conducting infinitely thin screen.

Analysis of curves presented in figures 1.6 and 1.7 shows that as far as the resonant scattering waveband is concerned ($1 < k_0c < 9.5$), a non-zero thickness of the screen (given both a perfectly conducting, curves #1 in figures 1.6 and 1.7, and a non-perfectly conducting, curves #2 and #3 in figures 1.6 and 1.7, material of the screen) changes the screen's RCS quite significantly compared to that of the infinitely thin (zero thickness) screen (curve #4 in figures 1.6 and 1.7). The latter phenomenon is caused by the significant effect of surface currents spilling over to the shadowed parts of the screen's surface. This effect is particularly pronounced for the curved (spherical in particular) screen. The behavior of surface currents given thin (but not infinitely thin) spherical screens made of perfectly conducting material of various sizes was studied by I Sukharevsky in [3].

In the latter work, I Sukharevsky demonstrates that where resonant wave scattering from an electrically thin screen is concerned (the frequency related parameter $k_0c = 3$, in particular), the level of 'shadow' currents is of the same order of magnitude as the level of currents excited on the 'illuminated' part of the screen's surface, which results in scattering characteristics of the screen that are significantly different from those given the infinitely thin screen. In addition, the author of [3] also notes that in the case of planar screens, this difference between the fields scattered by electrically thin and infinitely thin screens is less pronounced. The author of [1] compared the results of the backscattering characteristic computation for the case of perfectly conducting disks of various sizes that had finite thickness versus infinitely thin disks. The results presented in [1] show that if the frequency related parameter k_0a fell within the limits from 2 through 6, namely $2 \leqslant k_0a \leqslant 6$, where a is the disk radius, then the maximum variation between the RCSs of the disks with finite thickness and those which were infinitely thin did not exceed 10%. A greater variation between the RCSs of these two kinds of disks was observed given smaller values of that frequency related parameter k_0a when the disk's thickness becomes comparable to its radius. The study of the surface current behavior at the surface of a perfectly conducting screen with illumination along the screen's axis carried out by I Sukharevsky in [3] demonstrates that the effect of surface currents spilling over to the shadowed part of the surface becomes less pronounced if the screen is flat. That same effect is amplified if the screen is curved. Thus, in case of the disk (which is basically a flat screen), the shadow components of surface current density are significantly less than those excited at the illuminated parts of the surface. This insignificance of shadow currents brings the frequency dependence of RCS, given a disk of finite thickness much closer to that of the infinitely thin disk. In addition, these two RCS dependences on the frequency related parameter do not possess peaks of resonant nature. Similar effects were also observed in the case of wave scattering from non-perfectly conducting screens of finite thickness.

At the same time, the same result suggests that one needs to take into consideration the material of which the screen is made, in particular where the screen's scattering in the resonant waveband is concerned. In that latter case, the deviation between the RCSs of the perfectly conducting (curve #3) and non-perfectly conducting (curves #1 and #2) spherical screens becomes more pronounced as the screen's curvature increases. For instance, given that RCS belonged to the region of its higher magnitudes ($3 < k_0 c < 5$), then the shallow ($\vartheta_1 = 30°$) screen's RCS deviation between the cases of perfectly and non-perfectly conducting materials attained the values of 2.5% through 4.5% (figure 1.6). At the same time, the deep ($\vartheta_1 = 45°$) screen's RCS deviation went as high as 12% to 17% (figure 1.7).

The effect of surface currents spilling over to the shadowed surface part, which is observed in the case of curved screens, is also affected by the conductivity of the material of which the screen was made. The higher the conductivity of the material, the closer the RCS of a well conducting screen to that of a perfectly conducting screen. Thus the two curves corresponding to the screens made of aluminum and duralumin (curves #1 and #2 in figures 1.6 and 1.7) are evidently pretty close to each other since the values of these two materials' conductivity were not that far apart (the conductivity of aluminum was assumed to be equal to 3.54×10^7 Sm m^{-1}, and that of duralumin D16 (alloy of Al–Cu–Mg) was assumed to be equal to 1.74×10^7 Sm m^{-1}). The surface impedance of these two materials differed only by a factor of 1.42.

The frequency response of impedance screens of finite thickness in the resonance band

In a number of problems of antenna design, it is necessary to calculate the scattering of the ultrawideband (UWB) signal by screens of resonant sizes. Such a situation arises when the whole spectrum of the UWB signal emitted by the reflector antenna, or a part of it, occupies the frequency range in which the characteristic dimensions of the reflector fall into the resonant wave scattering domain.

Also, the key model problem that provides for the prediction of radiation performance of reflector antennas in the waveband corresponding to the resonant scattering of EMWs from the antenna reflector is the problem of wave scattering at a screen of finite thickness and conductivity. That particular problem has been solved in subsections 1.1.1 and 1.1.2 of this chapter.

Here we present an example of studying the radiation performance of a single-reflector AS given that it radiates an UWB signal. The antenna has a spherical reflector (figure 1.8) [18], which is made of aluminum (conductivity equal to 3.54×10^7 Sm m^{-1}). The geometrical parameters of the reflector are as follows: $c = 0.19$ m; $h = 1$ mm.

Here we consider two different antennas that differ by the 'depth' of the reflector given that the aperture radii of their reflectors are the same:

- antenna with shallow reflector, $\vartheta_1 = 30°$ (the ratio of $\frac{a}{c} = 2$, figure 1.8);
- antenna with deep reflector, $\vartheta_1 = 45°$ (the ratio of $\frac{a}{c} = 1.414$, figure 1.8).

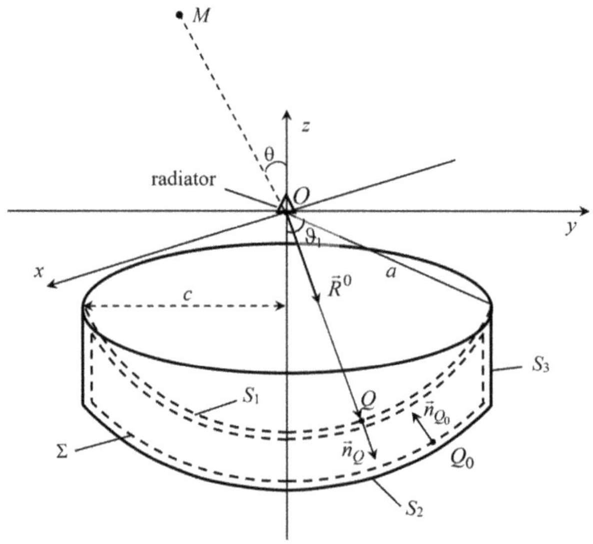

Figure 1.8. AS with a spherical reflector.

In our example, we employed an elementary Huygens radiator as an antenna feed and it was placed at the reflector's focal point. At the reflector's surface, the feed generated an EMF with the following electric intensity [19]:

$$E_\theta^0(Q) = \frac{g(r)}{\lambda_0}(1 + \cos\vartheta)\cos\varphi, \tag{1.48}$$

$$E_\varphi^0(Q) = \frac{g(r)}{\lambda_0}(1 + \cos\vartheta)\sin\varphi, \tag{1.49}$$

where r is the distance from the antenna feed to point Q on the reflector's surface.

Using the method developed in subsections 1.1.1 and 1.1.2, we computed the cross-sections of antenna directivity patterns (DPs) for the antenna described above, the DP cross-sections belonged to the yOz plane given various radiation frequencies from the range of 200 MHz through 1 GHz, which corresponded to the frequency related parameter k_0c ranging from $1 < k_0c < 4$. According to [20], such a signal is the UWB one, its bandwidth index is $\eta = 2\frac{f_{\max} - f_{\min}}{f_{\max} + f_{\min}} = 1.33$.

Figure 1.9 shows cross-sections of the DPs obtained for the shallow antenna (first variant). These cross-sections have been normalized with respect to their maxima, which were attained given $k_0c = 4$. Angle θ is the angle between the Oz-axis and the direction toward the observation (field) point M (figure 1.8).

Analysis of the plots presented in figure 1.9 shows that antenna radiation gain strongly depends on the frequency related parameter through entire its range of $1 < k_0c < 4$. This circumstance prevents the use of the DP concept in the case where one needs to radiate wideband (WB) signals in the form of short pulses.

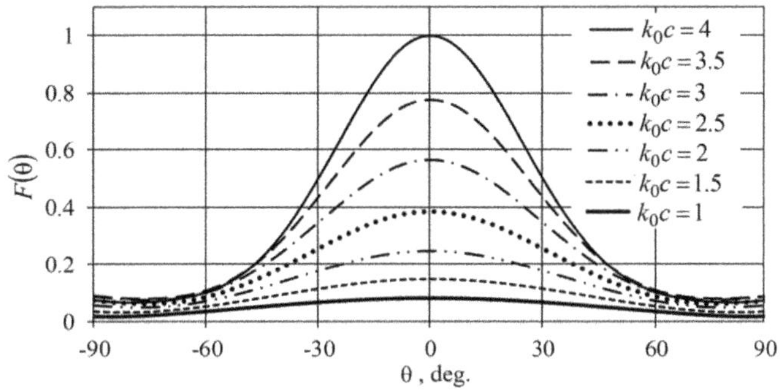

Figure 1.9. Normalized cross-sections of the antenna DPs given signals at various frequencies radiated by the feed.

However, the radiating (receiving) properties of such antennas can be represented comprehensively by the so-called space–frequency dependence of electric $\vec{E}(\omega, \vec{r})$ (or magnetic $\vec{H}(\omega, \vec{r})$) field intensity that is radiated (received) by the antenna (here, $\vec{r}$ is the radius-vector of the reception (or transmission) point, ω is the circular frequency). In contrast to antennas radiating narrowband signals close to monochromatic continuous wave (CW) signals, for which circular frequency ω can be considered as a fixed parameter, the UWB antennas have to have their directivity characterized by two parameters: angle and frequency. Hence the dependence $\vec{E}(\omega, \vec{r})$, which defines the intensity of the radiated field at every circular frequency ω and at every point of space $\vec{r}$. This particular dependence allows one to predict the change in the shape of the pulse radiated by the antenna compared to that radiated by the antenna feed. In that regard the space–time dependence of the field intensity $\vec{E}(t, \vec{r})$ turns out to be more convenient to handle the pulsed signals radiated by the UWB antenna. The latter space–time dependence of electric field intensity $\vec{\mathcal{E}}(t, \vec{r})$ is related to the original space–frequency dependence $\vec{\mathcal{E}}(\omega, \vec{r})$ via an inverse Fourier transform:

$$\vec{\mathcal{E}}(t, \vec{r}) = \int \vec{E}(\omega, \vec{r})e^{-j\omega t}\, d\omega. \tag{1.50}$$

In our computations it turns out that the analog of formula (1.50) that included the wavenumber k_0 instead of circular frequency ω was more convenient:

$$\vec{\mathcal{E}}(c_c t, \theta) = \int_{k_1}^{k_2} \vec{E}(k_0, \theta)e^{-jk_0 c_c t}\, dk_0, \tag{1.51}$$

where c_c is the light velocity, and k_1 and k_2 are the wave numbers corresponding to the signal spectrum boundaries.

Thus, using the above method, we computed the fields radiated by the reflector antenna in question given a set of fixed values of frequency related parameter k_0c from the range of $1 < k_0c < 4$, the individual values of k_0c being spaced apart by an increment of 0.03125, and given a set of fixed radiation directions θ similar to those presented in figure 1.9. The fields thus obtained represented the frequency responses of the antenna given certain radiation directions θ.

Next, using the inverse Fourier transform (1.51) of the antenna frequency responses mentioned above, we obtained electric field amplitudes in the time domain given various radiation directions θ. Figures 1.10 and 1.11 present the time domain dependences of normalized field amplitude $| A/A_{0\,\max} |$ versus the frequency related parameter $c_c t$ given a shallow reflector antenna (the first variant).

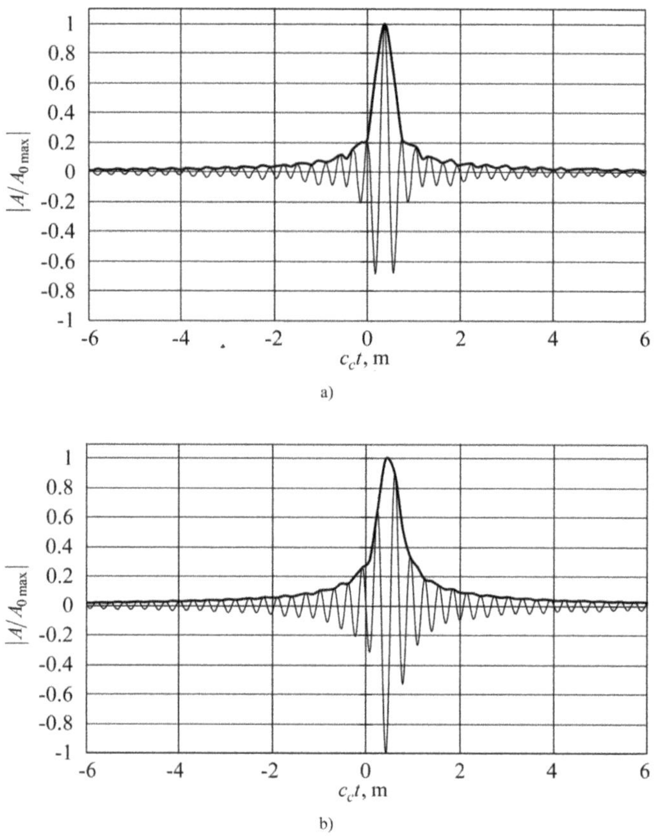

Figure 1.10. Time domain signals radiated by the antenna feed (a) and by the antenna itself (b) in the axial direction given an antenna with a shallow reflector ($\vartheta_1 = 30°$).

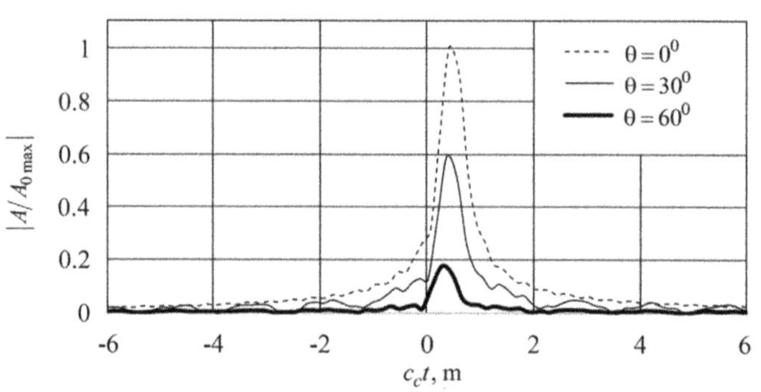

Figure 1.11. Envelopes of the signals radiated by the shallow antenna ($\vartheta_1 = 30°$) in three different directions: $\theta = 0°, 30°, 60°$.

Figure 1.10 shows the signal that irradiated the antenna reflector (the maximum of antenna feed radiation was pointed toward the center of the reflector) (figure 1.10(a)), and the signal that was actually radiated by the antenna in the axial direction $\theta = 0°$ (figure 1.10(b)).

The bold curve in figure 1.10 shows the signal envelope. Having compared the plots (a) and (b), one can see that once the pulse radiated by the antenna feed bounces off the reflector it stretches out. For instance, the width of the pulse radiated by the antenna as measured by its −3 dB amplitude level exceeds that of the pulse that irradiates the center of the antenna reflector by 23% (given that the antenna radiation direction was $\theta = 0°$).

Figure 1.11 shows the results of computing the envelopes of pulses radiated by the 'shallow' antenna in the directions of $\theta = 0°, 30°, 60°$. All the envelopes were normalized with respect to the maximum radiation amplitude attained in the axial direction.

The plots in figure 1.11 characterize the directivity properties of the shallow antenna. It can be seen that the maximum amplitude of signal radiated in the direction of $\theta = 30°$ was reduced by factor of 0.6 with respect to that radiated in the axial direction. Given the radiation direction of $\theta = 60°$, the maximum signal amplitude was reduced by factor of 0.18.

Similar computations were carried out for the second antenna that was characterized by the same aperture radius as the first one but it had a more 'deep' reflector. Computation results for the deep antenna are presented in figure 1.12.

Analysis of the above plots shows that when the signal radiated by the antenna feed bounced off the deeper reflector of the antenna it stretched out a bit more than in the case of the shallow antenna reflector. For instance, the width of the pulse radiated by the antenna as measured by its −3 dB level (the curve corresponding to $\theta = 0°$ in figure 1.12) exceeds that of the pulse that irradiates the center of the antenna reflector (figure 1.10(a)) by 26.5%.

In addition, the directivity of the 'deep' antenna was somewhat worse than that of the 'shallow' antenna. It can be seen that the maximum amplitude of the signal radiated in the direction of $\theta = 30°$ was reduced by a factor of 0.634 (which is 6%

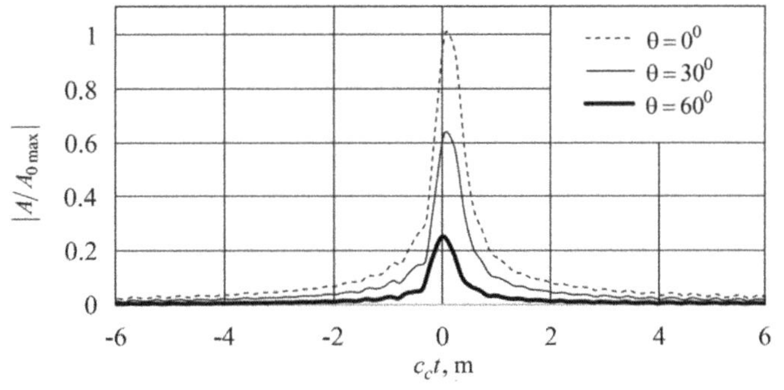

Figure 1.12. Envelopes of the signals radiated by the deep antenna ($\vartheta_1 = 45°$) in three different directions: $\theta = 0°, 30°, 60°$.

more than in the case of the shallow antenna) with respect to that radiated in the axial direction. Given the radiation direction of $\theta = 60°$, the maximum signal amplitude was reduced by a factor of 0.253 (which is 29% more than in the case of the shallow antenna).

Now, we consider the signals radiated by the 'deep' antenna. However, we narrow the signal spectrum down a bit to the frequency range from 500 MHz through 1 GHz, which corresponds to the frequency related parameter k_0c ranging from 2 through 4, $2 < k_0c < 4$ [18]. The bandwidth index of the signal with this particular spectrum width is $\eta = 0.67$.

The computation results are presented in figures 1.13 and 1.14. Figure 1.13 shows the signal that irradiated the antenna reflector, the maximum of antenna feed radiation being pointed toward the center of reflector (figure 1.13(a)), and the signal

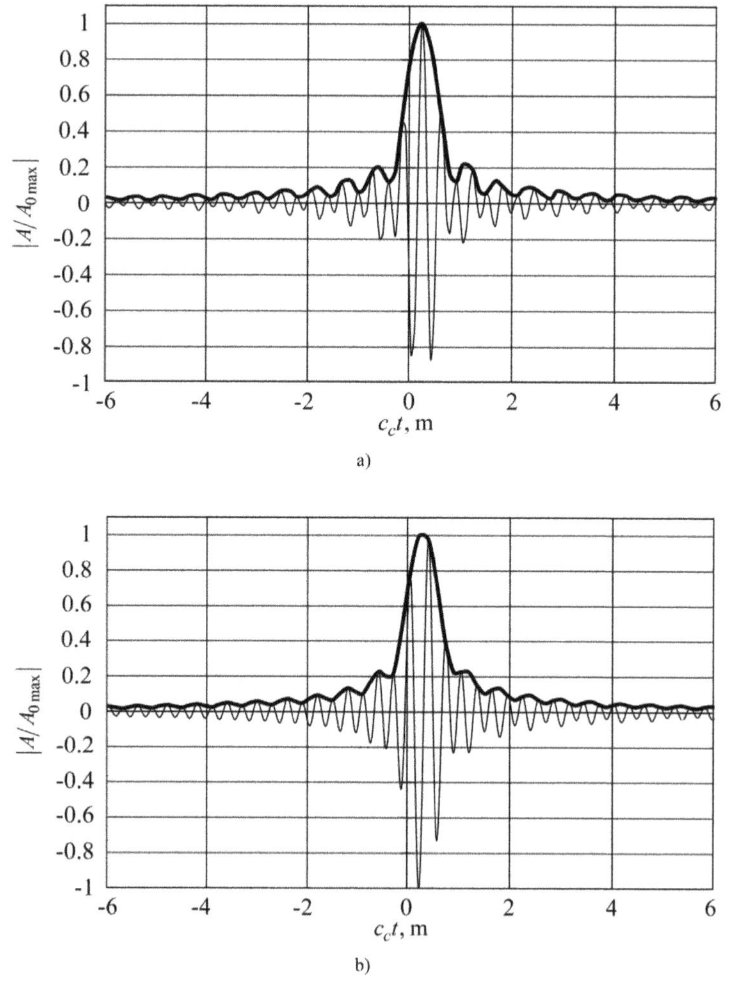

Figure 1.13. Time domain signals radiated by the antenna feed (a) and by the antenna itself (b) in the axial direction given an antenna with a 'deep' reflector, $\vartheta_1 = 45°$; the signal spectrum ranged from 0.5 GHz to 1 GHz.

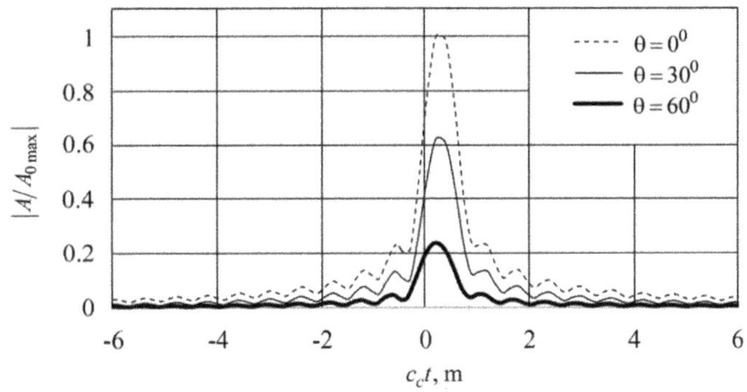

Figure 1.14. Envelopes of the signals radiated by the 'deep' antenna ($\vartheta_1 = 45°$) in three different directions: $\theta = 0°$, 30°, 60°; the signal spectrum ranged from 0.5 GHz to 1 GHz.

that was actually radiated by the antenna in the axial direction $\theta = 0°$ (figure 1.13(b)).

Figure 1.14 shows the results of computing the envelopes of pulses radiated by the 'deep' antenna in the directions $\theta = 0°$, 30°, 60°. All the envelopes were normalized with respect to the maximum radiation amplitude attained in the axial direction.

Analysis of the above plots shows that given a signal spectrum confined to the frequency range from 0.5 GHz through 1 GHz, the temporal stretch of the pulse radiated by the antenna compared to the pulse radiated by antenna feed is less pronounced than in the case of a wider signal spectrum. For instance, the width of pulse radiated by the antenna as measured by its −3 dB level exceeds that of the pulse that irradiates the center of the antenna reflector by 6.5% (given that the antenna radiation direction was $\theta = 0°$).

At the same time, antenna directivity remained at the same level as was the case given a wider spectrum of signal radiated by the antenna, which ranged from 0.2 GHz through 1 GHz.

1.1.4 Computation of fields radiated by a dual-reflector antenna, taking into account the interaction of the two reflectors

A number of reflector ASs consist of several reflectors (screens). Such ASs, for example, include the dual-reflector antennas of Cassegrain or Gregorian designs [21]. As long as ASs with reflectors of large electrical sizes are involved, there is a whole variety of analysis methods that have been developed to date. However, in some cases the characteristic dimensions of the reflectors are comparable to the signal wavelength. Such cases arise in the design of portable ASs (communications or airborne ASs, etc).

The elements of the AS can significantly influence each other, particularly in the resonant (with respect to the characteristic dimensions of the antenna reflector) waveband. As soon as the EMW from the antenna feed hits the secondary reflector, it induces electric currents on its surface. These currents appear not only on the side of the secondary reflector that faces the feed, but, due to diffraction, these currents spill over to the other side of secondary reflector, the one shadowed from the feed. In addition, the surface of secondary reflector also has currents induced by the field radiated by the primary reflector dish. In turn, the currents flowing on the surface of the primary reflector dish are induced by the EMF radiated by the secondary reflector and by the leaking through the primary field of the feed. Thus the EMF radiated by the dual-reflector AS is the result of interference between the primary field radiated from the feed and the fields radiated by surface currents on the secondary and primary reflectors. Rigorous accounting for the interaction of reflectors and the feed is a complicated problem, particularly where the resonant waveband is concerned. In that latter case (as shown in subsection 1.1.3), one has to take into account the finite thickness and conductivity of the reflectors. Therefore, in this subsection we develop a calculation method that will take into account the mutual influence of antenna elements when assessing its radiation properties in the resonant waveband.

A method for computing the radiation characteristics of dual-reflector ASs if the reflectors fall into the resonant waveband, based on an iterative procedure

We start our analysis from the IE set (1.10) presented in subsection 1.1.2, which is written with respect to the equivalent surface current density $\vec{J}(Q) = \vec{n} \times \vec{H}(Q)$ and with respect to the normal component of electric field intensity $e(Q) = \vec{n}_Q \cdot \vec{E}(Q)$ at the surface S of a non-perfectly conducting scatterer.

As a next step, we consider the AS consisting of two reflectors (figure 1.15): a secondary dish S_1 and a primary dish S_2. Now, the scatterer S in equation (1.10) can be presented as a superposition of two surfaces: $S = S_1 + S_2$. We proceed by placing the field point Q_0 on an auxiliary surface Σ_1, which is inside the surface of the secondary dish S_1, and then we move it to the auxiliary surface Σ_2, being inside the surface of primary dish S_2. As before, the auxiliary surface $\Sigma_{1(2)}$ is set to be following the surface $S_{1(2)}$ underneath the interface of $S_{1(2)}$ with free space at the distance Δ. As a result, we obtain the following set of equations:

$$\begin{cases} \mathbf{A}_{11}\vec{J}_e^{S_1} + \mathbf{A}_{12}\vec{J}_e^{S_2} = \vec{B}^{S_1} \\ \mathbf{A}_{21}\vec{J}_e^{S_1} + \mathbf{A}_{22}\vec{J}_e^{S_2} = \vec{B}^{S_2}, \end{cases} \tag{1.52}$$

where $\mathbf{A}_{\xi\eta}$ ($\xi = 1,2$; $\eta = 1,2$) are the rectangular matrices that, taken together, constitute the matrix $\mathbf{A}$ of SLE in (1.10). The order of matrices is as follows:

$$\mathbf{A} = \begin{pmatrix} \mathbf{A}_{11} & \mathbf{A}_{12} \\ \mathbf{A}_{21} & \mathbf{A}_{22} \end{pmatrix}. \tag{1.53}$$

In this case, the vectors $\vec{J}_e^{S_1}$ and $\vec{J}_e^{S_2}$ denote the isolated parts of the equivalent surface current density $\vec{J}_e^{S_1}, \vec{J}_e^{S_2}$, and normal components of the electric field intensity $e(Q) = \vec{n}_Q \cdot \vec{E}(Q)$ at the surface of dishes S_1 and S_2, respectively. $\vec{B}^{S_1}$ and $\vec{B}^{S_2}$ are the right-hand parts of the SLE (1.10) for dishes S_1 and S_2, respectively.

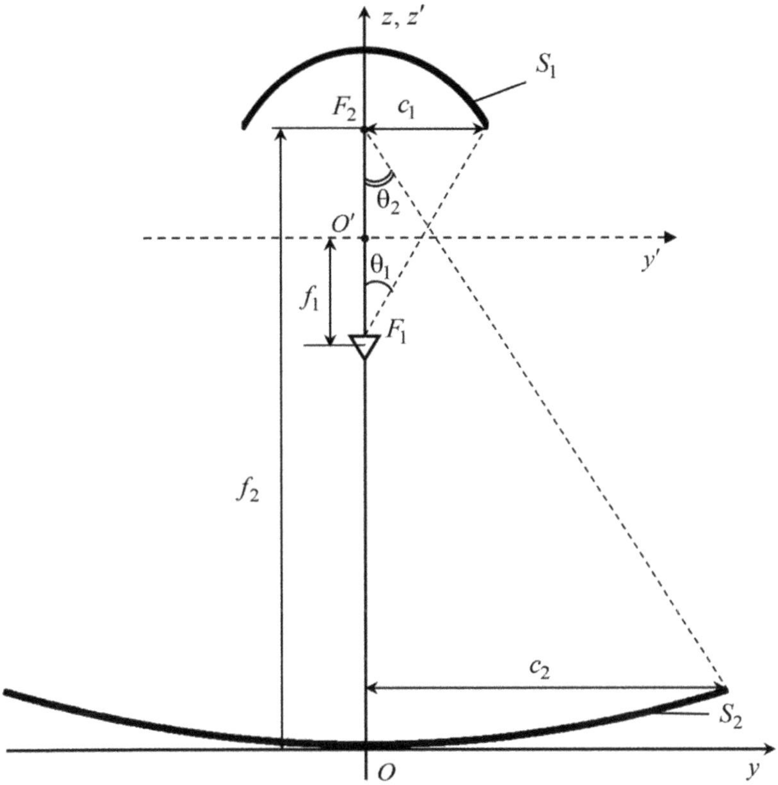

Figure 1.15. Geometry of a Gregorian antenna.

We are going to find a solution for SLE (1.52) using an iterative process, which is similar to that presented in [22, 23], in the following manner:

$$\mathbf{A}_{11}\vec{J}^{S_1}_{e(\xi)} = \vec{B}^{S_1} - \mathbf{A}_{12}\vec{J}^{S_2}_{e(\xi-1)}, \tag{1.54}$$

$$\mathbf{A}_{22}\vec{J}^{S_2}_{e(\xi)} = \vec{B}^{S_2} - \mathbf{A}_{21}\vec{J}^{S_1}_{e(\xi)}, \tag{1.55}$$

where $\xi = \overline{1, \xi_{\text{st}}}$; ξ_{st} is the number of iterations necessary to obtain stable values of current densities at the surfaces of two dishes.

We start our iterations from the assumption of $\vec{J}^{S_2}_{e(0)} = 0$, which means that the set (1.54) is solved under the condition that currents at the surface of the secondary dish are induced by primary incident field radiated by the feed. This primary field enters the $\vec{B}^{S_1}$ and it does not depend on the field scattered back from the primary dish.

Currents at the surface of primary dish are induced by both a primary incident field (created by the feed, which is represented by component $\vec{B}^{S_2}$ in the right-hand part of equation (1.55)), and the field scattered back to it from the secondary dish (component $\mathbf{A}_{21}\vec{J}^{S_1}_{e(1)}$ in equation (1.55)).

At further iterations, we also take into account the influence of the field scattered back from the primary dish onto the currents induced on the surface of secondary dish S_1 (component $\mathbf{A}_{12}\vec{J}^{S_2}_{e(\xi-1)}$ in equation (1.54)).

A study of the effect of the mutual interference between reflectors of a dual-reflector AS on its radiation performance in the resonant waveband

We study the effect of the mutual interference between the primary and secondary dishes included in a dual-reflector AS using a Gregorian antenna design as an example [24].

A Gregorian antenna employs the two screens: the primary (bigger) one, which is shaped as a paraboloid of revolution; the secondary (smaller) one, which is shaped as ellipsoid of revolution (figure 1.15). The secondary reflector, which is the ellipsoidal one, has two focal points. We refer to these foci here by their distance from the secondary reflector: the far focus (the one further away from the secondary reflector) and the near focus (the one closer to the secondary reflector).

In classical design of a Gregorian antenna, the near focus of the secondary reflector coincides with the focus of the primary reflector (point F_2 in figure 1.15), and the far focus of the secondary reflector is the point where the antenna feed is placed (point F_1 in figure 1.15). The purpose of the secondary reflector is to direct and re-reflect a spherical EMW radiated by the feed toward the primary parabolic reflector. In addition, the virtual source of the incident (primary) spherical wave is the focal point F_2. Then, the primary dish transforms that incident spherical wave after it is reflected back by the secondary dish into a plane EMW.

The above considerations explain the principles of operation and the radiation characteristics of an antenna given that its reflectors have large geometric sizes compared to the operating wavelength. In this case, the problem of scattering of the EMF radiated by the antenna feed at its reflectors can be solved quite effectively by approximate methods, for example, by methods of geometric or physical optics. However, when the antenna reflectors become smaller or when the operating wavelength becomes longer and the reflector sizes fall into the resonant waveband, one needs rigorous methods to solve the above problem. Therefore, below we are going to employ iterations to compute the equivalent currents induced on the surface of antenna reflectors using the computation method for the screens of finite thickness and conductivity described in subsection 1.1.2 (expressions (1.10)–(1.18)).

Considering the problem geometry (figure 1.15), it is convenient to use a cylindrical coordinate system for further computations. Now, the local basis vectors $\vec{v}_{p,\,Q_{n0}}$ $(p = \overline{1,3})$ (in expression (1.8) of subsection 1.1.1), which are used to express matrices $\mathbf{A}_{\xi\eta}$ $(\xi = 1,2;\ \eta = 1,2)$, take the following form:

$$\vec{v}_{1,\,Q_{n0}} = \vec{v}_{\theta_0} = (-\cos\theta_0\cos\varphi_0,\ -\cos\theta_0\sin\varphi_0,\ -\sin\theta_0), \tag{1.56}$$

$$\vec{v}_{2,\,Q_{n0}} = \vec{v}_{\varphi_0} = (-\sin\varphi_0,\ \cos\varphi_0,\ 0), \tag{1.57}$$

$$\vec{v}_{3,\,Q_{n0}} = \vec{v}_{z_0} = (1,0,0), \tag{1.58}$$

where θ_0, φ_0, z_0 are coordinates of observation (field) point Q_{n0} in a cylindrical coordinate system.

Computation of the currents induced on the surface of the secondary reflector
Let us consider the secondary reflector, which is generated by a cross-section of the surface of the ellipsoid of revolution (its semi-major and semi-minor axes being b and a, respectively) and of the right circular cylinder of radius $c1$ (figure 1.16). Here, we use Cartesian coordinate system $Oxyz$ tied to the primary reflector (figures 1.15 and 1.16). The center of the ellipsoid of revolution that depicts the face of the secondary reflector has been shifted from the coordinate system origin by a distance equal to $f_2 - f_1$ along the Oz axis.

Now, we represent the whole surface of the elliptical screen as a union of three distinct parts $S_1 = S_{11} \cup S_{12} \cup S_{13}$ (figure 1.16).

The concave face (the inner surface) S_{11} irradiated by the antenna feed of the whole screen S_1 (figure 1.16) can be described as follows:

$$\begin{cases} x = \rho \cos \varphi \\ y = \rho \sin \varphi \\ z = \dfrac{b}{a}\beta + \Delta f \end{cases}, \tag{1.59}$$

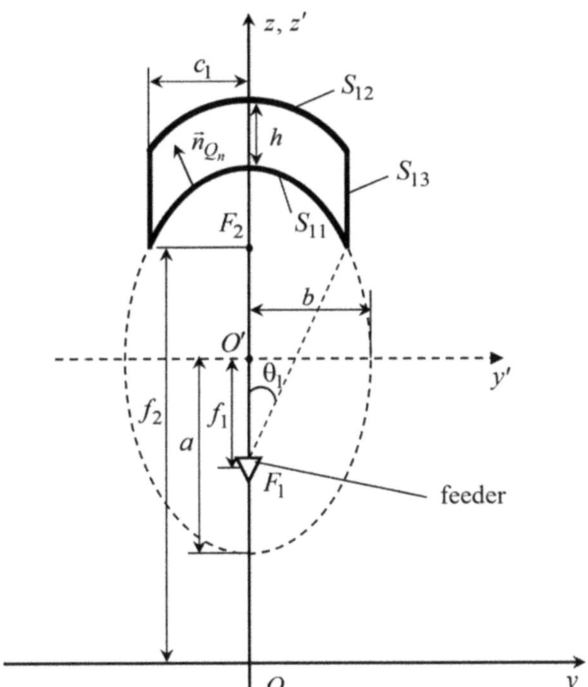

Figure 1.16. Geometry of the secondary reflector.

where $0 \leqslant \varphi \leqslant 2\pi$,

$$0 \leqslant \rho \leqslant c_1,$$

$$\Delta f = f_2 - f_1,$$

$$\beta = \sqrt{a^2 - \rho^2}.$$

The convex back side (the outer surface) S_{12} of the whole screen S_1 (figure 1.16) can be described in a similar way:

$$\begin{cases} x = \rho \cos \varphi + hn^1_{Q_n} \\ y = \rho \sin \varphi + hn^2_{Q_n} \\ z = \dfrac{b}{a}\beta + \Delta f + hn^3_{Q_n} \end{cases}, \tag{1.60}$$

where $0 \leqslant \rho \leqslant c_1 - \dfrac{hc_1 b}{\sqrt{c_1^2 f_1^2 + a^4}}$,

$$0 \leqslant \varphi \leqslant 2\pi,$$

$\vec{n}_{Q_n} = \left(n^1_{Q_n}, n^2_{Q_n}, n^3_{Q_n}\right)$ is the unit vector of internal normal to the surface S_{11} (the one directed to the screen's inside), and h is the thickness of the screen.

The flat side surface S_{13} (figure 1.16) is described as follows:

$$\begin{cases} x = c_1 \cos \varphi \\ y = c_1 \sin \varphi \\ z_1 \leqslant z \leqslant z_2 \end{cases}, \tag{1.61}$$

where z_1, z_2 are the values of the coordinate z at the edges of surfaces S_{11} and S_{12} (given that $\rho = c_1$),

$$0 \leqslant \varphi \leqslant 2\pi.$$

Considering expression (1.9), we shall seek the surface current density at S_{1i} ($i = \overline{1,3}$) in the following form:

$$\vec{J}^{S_{1i}} = J_{i1}\vec{\tau}_{i1} + J_{i2}\vec{\tau}_{i2}, \quad i = \overline{1,3}, \tag{1.62}$$

where $\vec{\tau}_{i2} = \vec{v}_\varphi = (-\sin \varphi,\ \cos \varphi,\ 0)$, $\vec{\tau}_{i1} = \vec{n}_Q \times \vec{v}_\varphi$,

$\vec{n}_Q$ is the unit vector of internal normal to the surface S_{1i} at point Q.

Instead of using an approximate model of antenna feed, we simulate the feed in a rigorous way. To that end, we take two electric dipoles, each having the vector-moment $\vec{p} = (0,1,0)$. These dipoles were placed at the focal axis of the ellipsoid and they were spaced by a distance of $\lambda/4$ from each other. The dipoles were energized to have a phase difference of $\pi/2$. Thus, the total field radiated by these two dipoles takes the form

$$\vec{E}^0(Q_{n_0}) = \vec{E}_1^0(Q_{n_0}) - j\vec{E}_2^0(Q_{n_0}) \tag{1.63}$$

$$\vec{E}_i^0(Q_{n_0}) = \frac{1}{4\pi\varepsilon_0}\left(\left(k_0^2 + \frac{jk_0 r - 1}{r^2}\right)\vec{p}^{\perp} + \frac{2(1 - jk_0 r)}{r^2}\vec{p}^{\parallel}\right), \quad (i = 1,\ 2), \tag{1.64}$$

where $\vec{p}^{\parallel} = \vec{r}^0(\vec{p} \cdot \vec{r}^0)$,

$$\vec{p}^{\perp} = \vec{p} - \vec{p},$$

$$r = \left| \vec{Q}_{n_0} - \vec{Q}_i^* \right|,$$

$$\vec{r}^0 = \frac{\vec{Q}_{n_0} - \vec{Q}_i^*}{r},$$

and $\vec{Q}_i^*$ is the radius-vector of the point where the ith dipole is.

When the two dipoles are spaced apart by the distance of $\left| \vec{Q}_1^* - \vec{Q}_2^* \right| = \lambda/4$, we obtain the radiation, which is directed toward the secondary reflector whereas there is zero radiation toward the center of the primary reflector.

Now, the density $\vec{J}_{(\xi)}^{S_1}$ of currents induced on the surface S_1 at the ξth iteration can be computed using the method described in subsection 1.1.2 by means of expressions (1.10)–(1.18). In addition, at the first iteration we used the field radiated by the two dipoles (1.63) as a primary field incident on the reflector. At further iterations, this incident field is replaced by a superposition of the field radiated by the dipoles and of the field scattered back from the primary reflector.

Later, components of the surface current densities thus obtained will be used to compute the field scattered by the screen by means of expression (1.4).

Below, we present our computation results with regard to the field scattered by elliptical reflectors of various configurations at the first iteration. Various configurations differed by the point of feed placement as shown in figure 1.17, as well as by the size and geometry of the reflector.

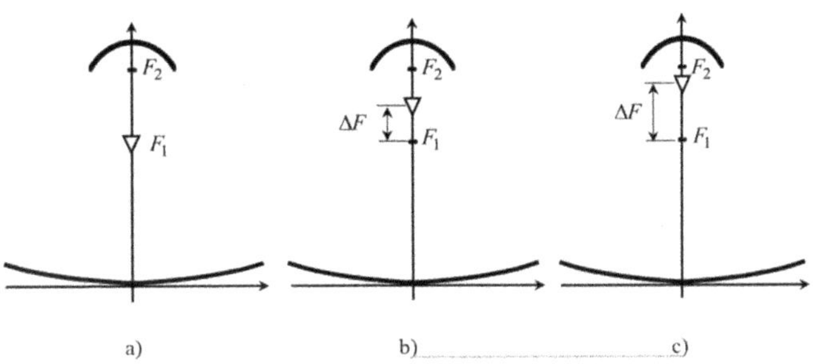

Figure 1.17. Geometry of different options for antenna feed placement, the aperture radius of the secondary dish being $k_0 c_1 = 4$.

We considered the following options of the feed placement:

- the antenna feed was placed at the far focus (point F_1) of the elliptical secondary reflector, which means that its shift with respect to that point is $\Delta F = 0$ (figure 1.18(a));
- the antenna feed was shifted from the point F_1 toward the secondary dish by the distance $\Delta F = f_1$, which means that now the feed is placed right in the middle between the two foci (figure 1.18(b));
- the antenna feed was shifted from the point toward the secondary dish by the distance $\Delta F = 1.5f_1$, which puts it closer to the near focus F_2 (figure 1.18(c)).

We carried out computations given an antenna with the following features: $k_0c_1 = 4$; $k_0f_1 = 3.45$ ($f_1 = 0.069$ m); $a = 0.098$ m; $b = 0.12$ m. In addition, in the case of an antenna that had its feed placed exactly at the far focus of an elliptical secondary dish (figure 1.17(a)), the angular span of the secondary dish as seen from the feed's location was $\theta_1 = 30°$. Given other antenna configurations where the feed is shifted toward the secondary dish, the angular span of the secondary dish increased.

Figure 1.18 shows normalized scattering diagrams (SD) of the elliptical screen in question (not yet accounting for the field radiated by the antenna feed).

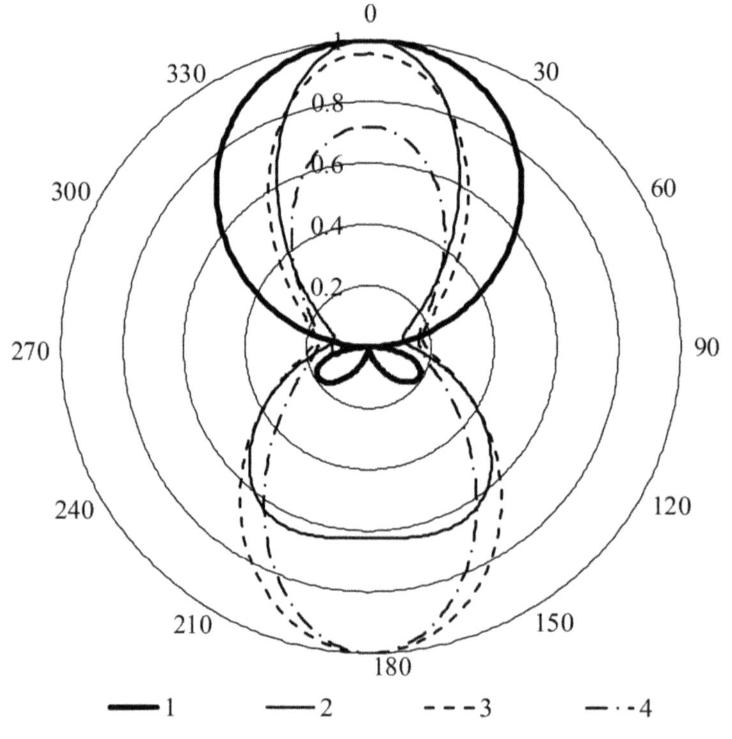

Figure 1.18. SDs of elliptical screen (the secondary dish of the antenna) with an aperture radius of $k_0c_1 = 4$. Curve #1 is the radiation DP of the antenna feed; curve #2 is the SD of the screen given that the feed is placed at its far focus F_1 (figure 1.7(a)); curve #3 is the SD of the screen given that the feed is put right in the middle between its two foci F_1 and F_2 (figure 1.17(b)); curve #4 is the SD of the screen given that the feed is shifted closer to its near focus F_2 by the distance of $\Delta F = 1.5f_1$ (figure 1.17(c)).

Diagrams are presented for the yOz plane given various options of the feed placement (figure 1.17). Our particular SDs show that the dependence of the absolute value of the scattered field projection onto the direction

$$\vec{p}' = \frac{\vec{p} - \vec{R}^0(\vec{p} \cdot \vec{R}^0)}{\left|\vec{p} - \vec{R}^0(\vec{p} \cdot \vec{R}^0)\right|}$$

($\vec{R}^0$ is the unit vector of the observation (field) point).

Curve #1 in figure 1.18 shows the radiation DP of the antenna feed, curve #2 shows the SD of the screen given that the feed was placed at its far focus F_1 ($\Delta F = 0$ figure 1.17(a)), curve #3 shows the SD of the screen given that the feed was shifted from the far focus F_1 of the secondary dish closer to the latter by a distance $\Delta F = f_1$ (figure 1.17(b)), and curve #4 shows the SD of the screen given that the feed was shifted from the far focus F_1 of the secondary dish by distance $\Delta F = 1.5f_1$, which brings it closer to its near focus (figure 1.17(c)).

Analysis of the SDs of figure 1.18 shows that unlike the optical waveband, the effect of exposing the screen to EM radiation in the resonant waveband might be counter-intuitive. This means that instead of blocking it off and the EMW reflecting it back, the screen might pass it on as well (curve #2). In this case, the maximum of the reflector's SD appears at its 'shadow' side. As the antenna feed gets closer to the secondary dish, its screening performance regarding the incident wave becomes better (curves #3 and #4 in figure 1.18). The latter is indicated by the increased lobes of the SD at the 'illuminated' side of the reflector (and toward the feed and the primary dish). Such conditions are common for the antennas that feature smaller angular spans of secondary dishes, as seen from the feed's location, or the smaller values of angle θ_1 (figure 1.16).

Such an effect of forward wave scattering by secondary screens needs to be taken into account, in particular if the antenna is to be operating in the resonant waveband, the latter depending on the sizes of the antenna reflectors with respect to the operating wavelength.

Figure 1.19 presents SDs similar to those presented in figure 1.18 above, except that the antenna featured a secondary screen with different characteristics, namely $k_0c_1 = 3$, $k_0f_1 = 2.6$ ($f_1 = 0.052$ m), $a = 0.0735$ m, and $b = 0.09$ m. The corresponding options of antenna feed placement are the same as those in figure 1.17.

Given that the antenna feed is located at the far focus of the secondary dish (figure 1.17(a)), the angular span of the secondary dish characterized by angle θ_1 (figure 1.16) attains its minimum of $\theta_1 = 30°$. For other options when the feed becomes shifted closer to the secondary dish, the angle span of the dish θ_1 increases.

Analysis of the SD in figure 1.19 shows that even though the physical size of the secondary reflector's aperture was smaller than that in figure 1.18, the effect of forward scattering was less pronounced, which means that secondary reflector's capability of scattering back the wave radiated by the antenna feed toward the primary reflector improved compared to the case shown in figure 1.18.

The latter can be explained by the fact that the feed gets closer to the dish in terms of relative distance expressed in the number of wavelengths. For instance, the first

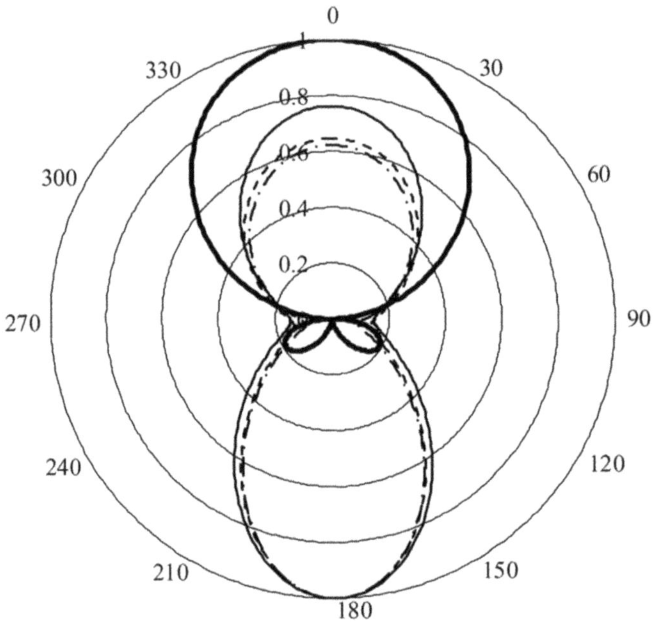

Figure 1.19. SDs for the case of an elliptical screen (the secondary dish of the antenna) with the aperture radius of $k_0c_1 = 3$, curve #1 is the radiation DP of the antenna feed; curve #2 is the screen's SD given that the feed is placed at its far focus F_1 (figure 1.17(a)); curve #3 is the screen's SD given that the feed is put right in the middle between its two foci F_1 and F_2 (figure 1.17(b)), and curve #4 is the screen's SD given that the feed is shifted closer to its near focus F_2 by the distance of $\Delta F = 1.5f_1$ (figure 1.17(c)).

option of the feed placement (figure 1.17(a)) had the feed at a relative distance $\approx 1.5\lambda$ from the secondary dish given the case of $k_0c_1 = 4$, whereas that same distance decreased to $\approx 1.1\lambda$ given the case of $k_0c_1 = 3$. Under other options of the feed placement (figures 1.17(b) and (c)), this relative distance from the feed to the secondary dish decreased even further to become less than the incident field wavelength given the case of $k_0c_1 = 3$.

Computation of the surface currents at the primary reflector

Let us consider the primary reflector of an antenna, which is generated by a cross-section of the surface of a paraboloid of revolution (its focal distance being f_2) and of the right circular cylinder of radius c_2 (figure 1.15).

We represent the whole surface of the parabolic screen as a union of three distinct parts: $S_2 = S_{21} \cup S_{22} \cup S_{23}$.

The concave face (the inner surface) S_{21} of the whole surface S_2, which is exposed to the field reflected by the secondary screen, can be expressed as follows:

$$\begin{cases} x = \rho \cos \varphi \\ y = \rho \sin \varphi \\ z = \dfrac{\rho^2}{4f_2} \end{cases}, \tag{1.65}$$

where $0 \leqslant \varphi \leqslant 2\pi$,

$$0 \leqslant \rho \leqslant c_2.$$

The convex back side (the outer surface) S_{22} of the whole screen S_2 can be described in a similar way:

$$\begin{cases} x = \rho \cos \varphi + hn^1_{Q_n} \\ y = \rho \sin \varphi + hn^2_{Q_n} \\ z = \dfrac{\rho^2}{4f_2} + hn^3_{Q_n} \end{cases}, \tag{1.66}$$

where $0 \leqslant \rho \leqslant c_2\left(1 - \dfrac{h}{\sqrt{c_2{}^2 + 4f_2{}^2}}\right)$,

$$0 \leqslant \varphi \leqslant 2\pi,$$

$\vec{n}_{Q_n} = \left(n^1_{Q_n}, n^2_{Q_n}, n^3_{Q_n}\right)$ is the unit vector of the internal normal to the surface S_{21} (the one directed to the screen's inside), and h is the thickness of the screen.

The flat side surface S_{23} is described as follows:

$$\begin{cases} x = c_2 \cos \varphi \\ y = c_2 \sin \varphi \\ z_1 \leqslant z \leqslant z_2 \end{cases}, \tag{1.67}$$

where z_1 and z_2 are the values of coordinate z at the edges of surfaces S_{21} and S_{22} (given that $\rho = c_2$),

$$0 \leqslant \varphi \leqslant 2\pi.$$

As before, we shall seek the surface current density at S_{2i} ($i = \overline{1, 3}$) in the following form:

$$\vec{J}^{S_{2i}} = J_{i1}\vec{\tau}_{i1} + J_{i2}\vec{\tau}_{i2}, \quad i = \overline{1, 3}, \tag{1.68}$$

where $\vec{\tau}_{i2} = \vec{e}_\varphi = (-\sin \varphi,\ \cos \varphi,\ 0)$,

$$\vec{\tau}_{i1} = \vec{n}_Q \times \vec{e}_\varphi,$$

and $\vec{n}_Q$ is the unit vector of internal normal to the surface S_{2i} at point Q.

Density $\vec{J}^{S_2}_{(\xi)}$ of currents induced on the surface S_2 at the ξth iteration are also computed using the method described in subsection 1.1.2 by means of expressions (1.9)–(1.18). In this case, the primary field that is incident on the primary reflector's surface, which enters the right-hand part of the SLE, is a combination of two incident fields: one radiated by the feed (simulated by two dipoles), $\vec{E}^0\left(Q_{n_0}\right)$, and the one scattered back by a secondary reflector, $-A_{21}\vec{J}^{S_1}_{e(\xi)}$.

Optimization of the arrangement of components in a dual-reflector Gregorian antenna for its operation in the resonant waveband

The computation results presented above in this subsection concerning the fields scattered by the secondary dish in the resonant waveband suggest that one could use the same technique to optimize the arrangement of the basic components in dual-reflector antennas (by finding, for example, the best location of the feed with respect to the secondary reflector) in order to improve the performance of these antennas in terms of their DPs.

Figures 1.20–1.29 show the computed field DPs given various design option of dual-reflector antennas. These design options differ by size of the secondary reflector aperture (expressed as k_0c_1) and by the location of the feed (expressed as ΔF).

All the antenna designs considered below feature the same primary reflector that had the following geometric parameters: $k_0c_2 = 12$; $\theta_2 = 32.917°$ (figure 1.15).

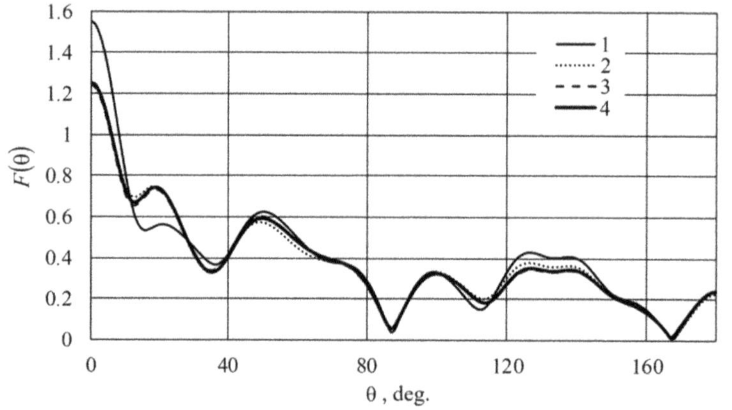

Figure 1.20. Field DPs of a dual-reflector antenna for various numbers of iterations given the following antenna features: $\theta_1 = 30°$, $k_0c_1 = 4$, $\Delta F = 0$. The number of the curve represents the number of iterations.

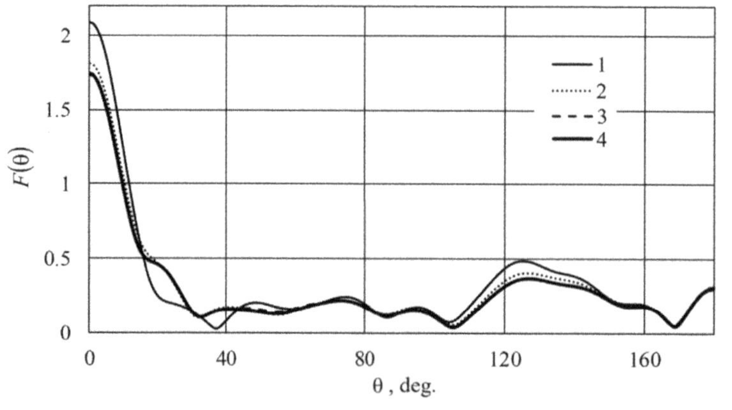

Figure 1.21. Field DPs of a dual-reflector antenna for various numbers of iterations given the following antenna features: $\theta_1 = 30°$, $k_0c_1 = 4$, $\Delta F = f_1$. The number of the curve represents the number of iterations.

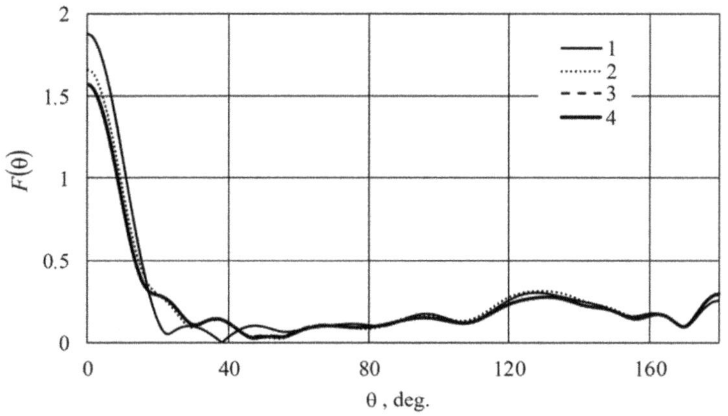

Figure 1.22. Field DPs of a dual-reflector antenna for various numbers of iterations given the following antenna features: $\theta_1 = 30°$, $k_0c_1 = 4$, $\Delta F = 1.5f_1$. The number of the curve represents the number of iterations.

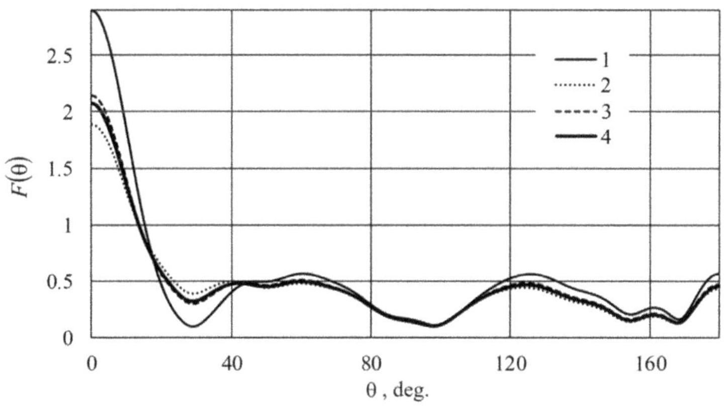

Figure 1.23. Field DPs of a dual-reflector antenna for various numbers of iterations given the following antenna features: $\theta_1 = 30°$, $k_0c_1 = 3$, $\Delta F = 0$. The number of the curve represents the number of iterations.

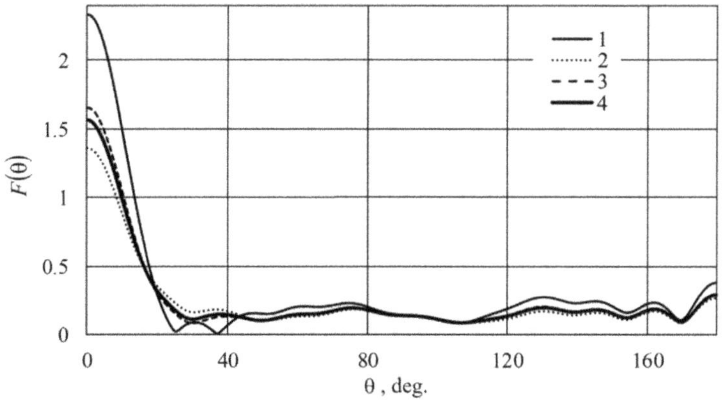

Figure 1.24. Field DPs of a dual-reflector antenna for various numbers of iterations given the following antenna features: $\theta_1 = 30°$, $k_0c_1 = 3$, $\Delta F = f_1$. The number of the curve represents the number of iterations.

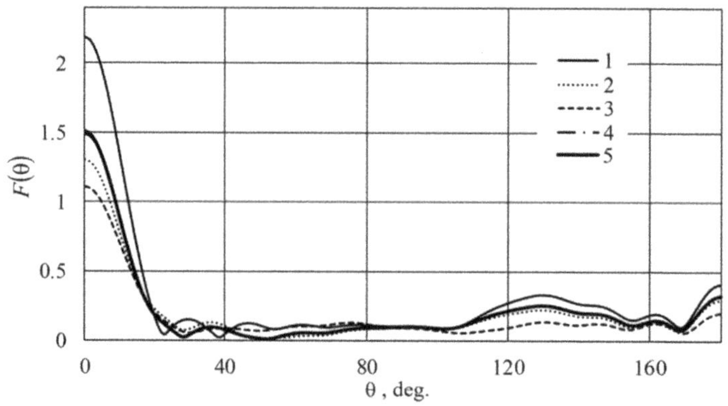

Figure 1.25. Field DPs of a dual-reflector antenna for various numbers of iterations given the following antenna features: $\theta_1 = 30°$, $k_0c_1 = 3$, $\Delta F = 1.5f_1$. The number of the curve represents the number of iterations.

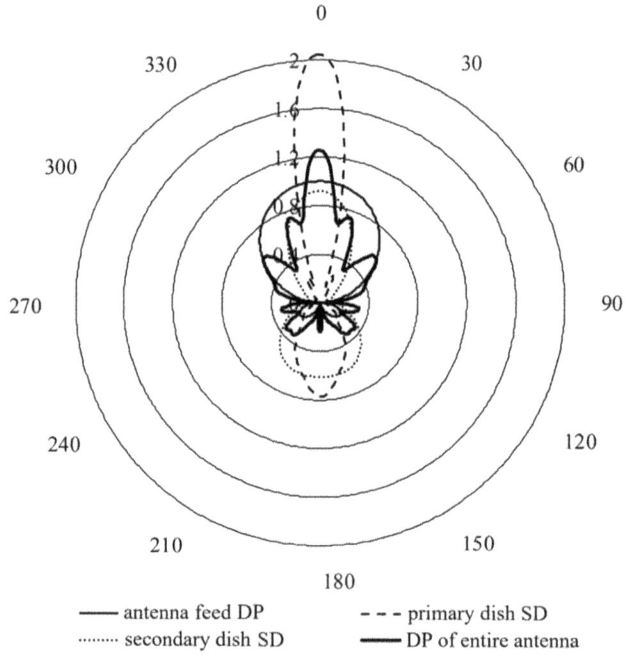

Figure 1.26. Step-by-step formation of the DP of a dual-reflector antenna given that $\theta_1 = 30°$, $k_0c_1 = 4$, and $\Delta F = 0$ computed as a result of iteration #4.

The focal distance that determined the location of the secondary reflector with respect to the primary one was equal to $f_2 = 0.4062$ m ($k_0f_2 = 20.31$).

These antenna designs differed only by the sizes of the secondary reflectors and by the shifts of the antenna feed with respect to them, as was the case in this subsection where we computed the SDs of various secondary dishes. We assumed that the Gregorian antenna featured the same options of secondary dishes as those in this

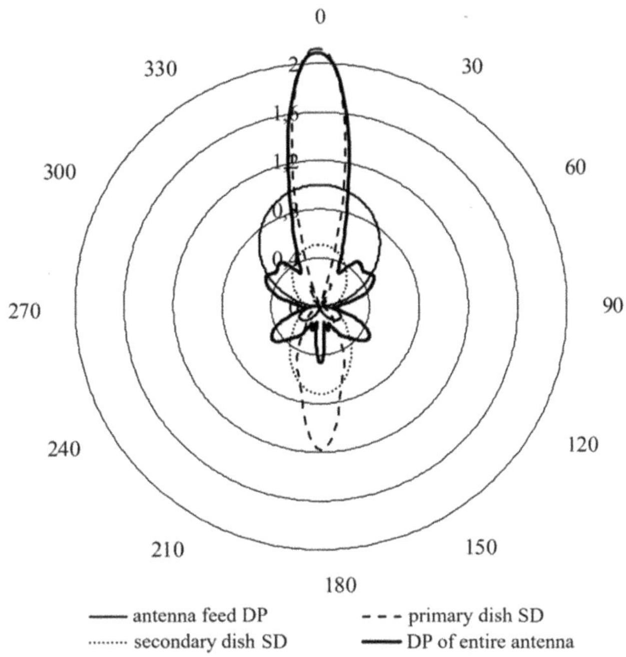

Figure 1.27. Step-by-step formation of the DP of a dual-reflector antenna given that $\theta_1 = 30°$, $k_0c_1 = 3$, and $\Delta F = 0$ computed as a result of iteration #4.

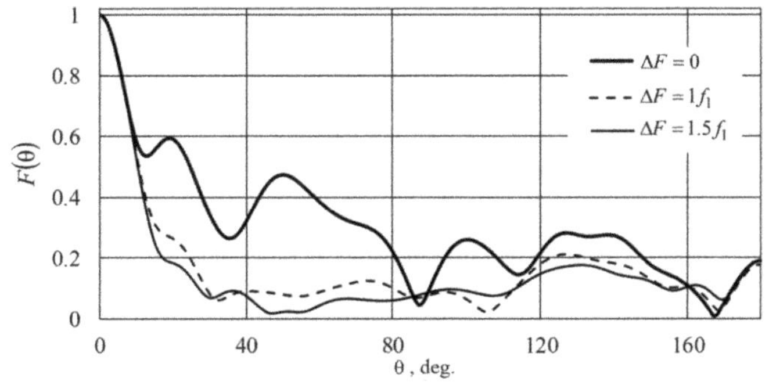

Figure 1.28. DPs of a dual-reflector Gregorian antenna featuring a secondary reflector with the aperture of $k_0c_1 = 4$ ($\theta_1 = 30°$) given various shifts ΔF of its feed from the far focus of an elliptical secondary reflector.

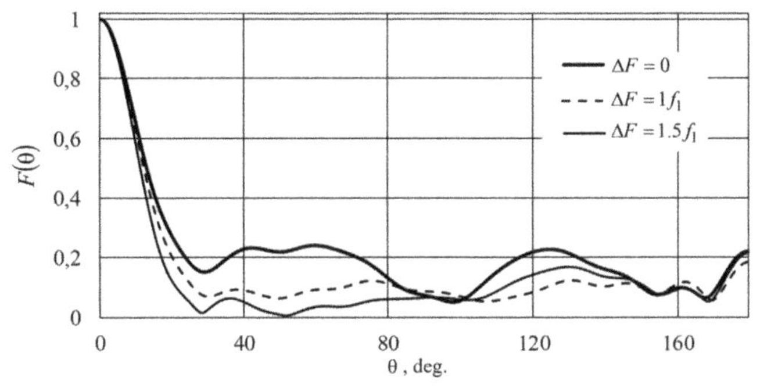

Figure 1.29. DPs of a dual-reflector Gregorian antenna featuring a secondary reflector with the aperture of $k_0c_1 = 3$ ($\theta_1 = 30°$) given various shifts ΔF of its feed from the far focus of an elliptical secondary reflector.

subsection, namely $\theta_1 = 30°$, $k_0c_1 = 4$ and $\theta_1 = 30°$, $k_0c_1 = 3$. Figures 1.20–1.22 present the computed DPs of a dual-reflector Gregorian antenna featuring the secondary dish with the aperture $k_0c_1 = 4$. Figures 1.23–1.25 present similar DPs for the antenna that featured the secondary dish with the aperture $k_0c_1 = 3$. The plots of figures 1.20–1.22 and the plots of figures 1.23–1.25 have been computed for antenna designs differing by the location of its feed with respect to its secondary dish.

As before, we consider the same options of antenna feed location with respect to the secondary dish:

- the antenna feed was placed at the far focus (point F_1) of the elliptical secondary reflector, which means that its shift with respect to that point is $\Delta F = 0$ (figure 1.17(a));
- the antenna feed was shifted from the point F_1 toward the secondary dish by the distance $\Delta F = f_1$, which means that now the feed is placed right in the middle between the two foci (figure 1.17(b));
- the antenna feed was shifted from the point toward the secondary dish by the distance $\Delta F = 1.5f_1$, which puts it closer to the near focus F_2 (figure 1.17(c)).

The DPs $F(\theta)$ presented in figures 1.20–1.25 were normalized with respect to the maximum field intensity radiated by the antenna feed. These DPs have been obtained by means of an iterative computational procedure given various numbers of iterations. The number of each curve in figures 1.20–1.25 coincides with the number of iterations used to compute each particular curve.

Analysis of the above plots shows that stable values of the equivalent surface current densities induced on the surfaces of both reflectors were already obtained at the fourth or fifth iteration. It is worth mentioning that the DPs obtained using the surface current densities computed after the first iteration (where the mutual interference between the two reflectors was not yet accounted for) differ significantly from those obtained using the surface current densities computed as a result of a few iterations.

For example, this difference between the maxima of DPs computed with and without taking into account such mutual interference of the two antenna reflectors varied within the interval of 18%–24% depending on the location of the antenna feed for an aperture radius of the secondary antenna dish of $k_0c_1 = 4$. In the case of an antenna with a secondary antenna dish aperture radius of $k_0c_1 = 3$, that same difference varied within the interval of 40%–49%.

Moreover, in the case of an antenna featuring a secondary reflector with an aperture radius of $k_0c_1 = 4$, the difference between the levels of the first sidelobe, which was computed with and without accounting for mutual interference between the two reflectors, varied by not more than 24% if the antenna feed shift was $\Delta F = 0$, not by more than 42% if the antenna feed shift was $\Delta F = f_1$, and not by more than 49% if the antenna feed shift was $\Delta F = 1.5f_1$.

Figures 1.26 and 1.27 illustrate the process of the step-by-step formation of the DPs of antennas featuring secondary reflectors with aperture radii of $k_0c_1 = 4$ and $k_0c_1 = 3$, respectively, given that the antenna feed was put at the far focus of their elliptical secondary dishes ($\Delta F = 0$).

The bold solid line corresponds to the DP of the entire antenna, the thin solid line corresponds to the DP of antenna feed, and the dotted line and the dashed line correspond to the SDs of the secondary and primary dishes, respectively.

A comparison of the DPs shown in figures 1.26 and 1.27 suggests that the relatively high level of sidelobes in the case of the antenna featuring the larger secondary dish (figure 1.26) could be caused by the fact that the secondary dish was not 100% effective in reflecting the antenna feed's field back toward the antenna's primary dish. The effect of forward wave scattering, which is pronounced in the larger secondary dish (figure 1.26), causes a wide backlobe of scattered field that has a relatively high amplitude. When combined with the field radiated by the feed, that scattering backlobe of the secondary dish causes a relatively high level of the first DP sidelobes for the entire antenna.

However, for the smaller secondary dish featured by the antenna (as in figure 1.27) the level of that scattered field backlobe of the secondary dish was 1.8 times below that in figure 1.26. The latter resulted in the lower level first sidelobes of the entire antenna.

Thus the computation method presented in this section can be helpful in optimizing the dual-reflector antenna design in the resonant waveband. As shown above, one of the possible ways of reducing the antenna DP sidelobe level in this case is to move the antenna feed closer to the secondary reflector. This would improve the screening effect of the secondary reflector with regard to the field radiated by the antenna feed.

Figures 1.28 and 1.29 are normalized with respect to their maxima DPs of the dual-reflector Gregorian antennas that featured secondary dishes with apertures of $k_0c_1 = 4$ and $k_0c_1 = 3$, respectively, given various options of their feed locations. As before, these options were characterized by different shifts ΔF of the antenna feed from the far focus of the elliptical secondary dish in the direction of the latter.

Having compared the DPs of figures 1.28 and 1.29, one can draw the obvious conclusion that the optimal configuration for the antenna that brings the level of its first sidelobe to a minimum is the one that features the antenna feed shifted from the far focus of the elliptical secondary dish by the distance $\Delta F = 1.5f_1$ toward that dish's surface.

However, if antenna designer is concerned with reducing the level of the sidelobes that are further away from the main antenna lobe, then given the antenna featuring the secondary dish with an aperture radius of $k_0c_1 = 3$ the best option for the antenna feed location is to be shifted by the distance $\Delta F = 1f_1$ from the far focus of the secondary dish. In the latter case, the antenna had the lowest level of sidelobes given the angle sector of $\theta = 105°–150°$ (figure 1.29). For instance, if we look at the relative radiation level in the direction of $\theta = 120°$ we can see that given the feed shift of $\Delta F = 1f_1$ this level is 1.7 times lower than that for the feed shift of $\Delta F = 1.5f_1$, and it is 2.7 times lower than that for the non-shifted feed. For the antenna featuring a secondary dish with an aperture radius of $k_0c_1 = 4$, the best option for it is still to have its feed shifted by the distance of $\Delta F = 1.5f_1$ from the far focus of the secondary dish.

Thus one can choose the best design options for antennas (e.g. in terms of reducing their sidelobes) based on a comparison of their DPs, which can be computed using the iterative method described above. Also, our computation results clearly demonstrate the necessity of taking into account the mutual interaction between the antenna reflectors while computing the DPs of dual-reflector antennas in the resonant waveband.

1.2 Magnetic field integral equation for solving problems of resonant electromagnetic wave scattering by perfectly conducting objects

In this section we consider a method for predicting the radar scattering characteristics of PEC objects, the sizes of which are commensurate with the wavelength of the incident wave. We refer to such scatterers as resonance-sized scatterers. Such radar scatterers can be represented by individual entities with a metallic surface or they can be integral parts of other scatterers of interest. The algorithm developed below is based on solving the Fredholm boundary IE of the second kind, or the so-called magnetic field IE (MFIE). This type of boundary IE has some advantages when computing the EM scattering characteristics of objects that are not thin electrically. The latter means that the objects occupy certain volumes rather than being flat. Real radar objects usually contain design elements that can be regarded as smooth yet their curvature radii will range from quite small to relatively large values. Thus when the MFIE is used for computing the EM responses of real objects, one needs to apply a variety of special algorithms in order to attain a desirable accuracy of IE solving.

In subsection 1.2.1 we briefly discuss the derivation of the MFIE in question. We touch upon some peculiarities of the MFIE kernel and then proceed to describing the MFIE solution algorithm in subsection 1.2.2.

The usual approach for finding solutions to the problems of EMW scattering from real radar objects is to employ the MFIE [13, 25–34], or electric field IE (EFIE) [13, 25, 29, 30, 32–36], or combined field IE (CFIE) [25, 28, 29, 32, 33, 35]. The drawbacks and advantages of each of these IEs are described in great detail in [13, 25, 29, 32, 33]. It is worth noting that IEs are usually solved by means of the method of moments (MoM) [13, 25–45], which belongs to the class of so-called projection methods.

Contemporary computation based on the MoM provides decent quality for radar scattering from the PEC objects, as well as from purely dielectric and combined objects of complex shape. However, such computations require a considerable amount of computer RAM. In addition, they depend heavily on the thorough selection of the so-called basis and test functions, the form of which is determined by the complexity of the scatterer's shape [28, 29, 31, 36]. The method involving the surface IEs drew the attention of engineers dealing with radar scattering from electrically large objects [29, 35, 41–45]. In the latter case, the multilevel fast multipole algorithm (MLFMA) [29, 44, 45] finds numerous applications. At the same time, the application of asymptotic high-frequency methods (AHFMs) [15, 29, 46–50] is devoid of non-necessary complications arising from the need to solve the IEs. Therefore, asymptotic methods would be preferable for electrically large scatterers. They will be considered in chapter 2 of this monograph.

Here in section 1.2 we employ a commonly known boundary IE of the second kind (MFIE) that guarantees us good convergence of the computation algorithm with regard to various mesh patterns of discrete nodes. Transformation of the IE into an SLE is accomplished using the interpolation method [13], which, in our opinion, has certain advantages compared to the MoM-like projection methods [13, 25–45]. The novelty of the method proposed below is in its combined approach toward: (i) an algorithm for integrating the kernel of the MFIE; (ii) a method for removing the kernel's singularity; (iii) a way to figure out the spacing between the integration nodes that will create a mesh with adaptive density, the spacing between the nodes representing the surface current density proportional to local curvature radii in the vicinity of every particular segment of the scatterer's surface; (iv) a method for eliminating ambiguous solutions to the MFIE [22, 23, 49, 51–55]. The approach proposed below allows for the rational use of computer RAM. It can also reduce the computational burden. The highest gain provided by the computational algorithm developed here is attained in the cases where the scatterer surface elements have small curvature radii and where these elements are electrically thin.

1.2.1 Deriving the magnetic field integral equation

In most cases the boundary IEs of electrodynamics are formulated by means of applying boundary conditions to an initial source representation of the EMFs that satisfy the Maxwell equations (the wave equation) and the Sommerfeld radiation condition [13, 15, 25, 29, 33, 48]. In particular, one can derive the necessary mathematical expressions using the Lorentz reciprocity theorem [13, 33],

$$\int_S \left\{(\vec{E}_2 \times \vec{H}_1) - (\vec{E}_1 \times \vec{H}_2)\right\} \cdot \vec{v}\, \mathrm{d}s = \int_V \left\{(\vec{J}_1^{\,e} \cdot \vec{E}_2) - (\vec{J}_2^{\,e} \cdot \vec{E}_1) - (\vec{J}_1^{\,m} \cdot \vec{H}_2) + (\vec{J}_2^{\,m} \cdot \vec{H}_1)\right\} \mathrm{d}v, \quad (1.69)$$

where the EMF $(\vec{E}_1, \vec{H}_1)$ would be excited by the sources which correspond to the surface densities of the equivalent electric $\vec{J}_1^{\,e}$ and magnetic $\vec{J}_1^{\,m}$ currents, and the EMF $(\vec{E}_2, \vec{H}_2)$ would be excited by the sources characterized by $\vec{J}_2^{\,e}$ and $\vec{J}_2^{\,m}$, V is an arbitrary volume bounded by the surface S, and $\vec{v}$ is the internal normal to the surface S.

To find the electric current density $\vec{J}^{\,e}$ at the surface S of a PEC object V_2 located in free space V_1, we use the method based on solving the MFIE [22, 23, 49, 51–55]. The corresponding equation can be obtained by applying Lorentz reciprocity theorem (1.69) to the sought field $(\vec{E}, \vec{H})$, and to the field of auxiliary magnetic point source residing in free space, $(\vec{E}_1^{\,m}, \vec{H}_1^{\,m})$. Considering the fact that the MFIE kernel has a singularity corresponding to the normal derivative of single-layer potential [56, 57], we can represent the MFIE in the following form [22, 23, 52, 55]:

$$\begin{aligned} 2\vec{\tau}_q^{\,0} \cdot \vec{H}^0(\vec{Q}_0) - (\vec{v}^{\,0} \times \vec{\tau}_q^{\,0}) \cdot \vec{J}^e(\vec{Q}_0)\left(1 + I_{s0}(\vec{Q}_0)\right) = \frac{2}{j\omega} \\ \int_{S-s_0} \vec{E}_1^{\,m}\left(\vec{Q} \mid \vec{Q}_0, \vec{\tau}_q^{\,0}\right) \cdot \vec{J}^e(\vec{Q})\, \mathrm{d}s. \end{aligned} \quad (1.70)$$

Here $\vec{Q}$, $\vec{Q}_0 \in S$ are the integration (source) and the observation (field) points, respectively, $\vec{H}^0$ is the magnetic vector of the incident field, $\vec{\boldsymbol{\tau}}_q^0$ $(q = 1, 2)$ are mutually orthogonal unit vectors, tangential to the surface S at point $\vec{Q}_0$ that form the right-hand system when coupled with the unit vector of the inner normal, $\vec{\boldsymbol{\nu}}^0$, at $\vec{Q}_0$, and the function $I_{s_0}(\vec{Q}_0)$ is explained in detail below.

The kernel of the MFIE (1.70) is the electric field part of the magnetic-type Green's function or, equivalently, the electric vector of the field (at point $\vec{Q}$) of the elementary magnetic Hertz dipole (with vector-moment $\vec{\boldsymbol{\tau}}_q^0$, located at $\vec{Q}_0$) residing in free space V_1,

$$\vec{E}_1^m\left(\vec{Q} \mid \vec{Q}_0, \vec{\boldsymbol{\tau}}_q^0\right) = -j\omega\left(\vec{\boldsymbol{\tau}}_q^0 \times \vec{\nabla} G_1\left(\vec{Q}, \vec{Q}_0\right)\right), \tag{1.71}$$

where

$$G_1\left(\vec{Q}, \vec{Q}_0\right) = G_1(R) = \frac{\exp(jk_1R)}{4\pi R}, \tag{1.72}$$

$R = \left| \vec{Q} - \vec{Q}_0 \right|$, and $k_1 = 2\pi/\lambda_1$ and λ_1 are the wavenumber and the wavelength in the free space, V_1, respectively.

This function has a singularity at $\vec{Q} = \vec{Q}_0$, however, this singularity is integrable [23, 29, 55–59].

IE (1.70) is the Fredholm IE of the second kind, which can be solved numerically with controlled accuracy. Here, the integration in equation (1.70) is carried out over surface S with the exception of a small neighborhood s_0 of the singular point. To evaluate the integral over s_0 we proposed the following method.

Let us consider a small area s_0 that belongs to the surface S. That particular area would be confined within the sphere with the radius $\rho_0 \ll \lambda_1$ that is centered at point $\vec{Q}_0$ as shown in figure 1.30. The scatterer's surface in this neighborhood is given by the equation $F(\vec{Q}) = 0$. Now we introduce a local Cartesian coordinate system (u_1, u_2, u_3) to be related to the s_0. Its axes $\vec{Q}_0u_1$, $\vec{Q}_0u_2$ are to be parallel to the vectors $\vec{\tau}_1^0$ and $\vec{\tau}_2^0$, which are tangential to the object's surface at point $\vec{Q}_0$, and the axis $\vec{Q}_0u_3$ is

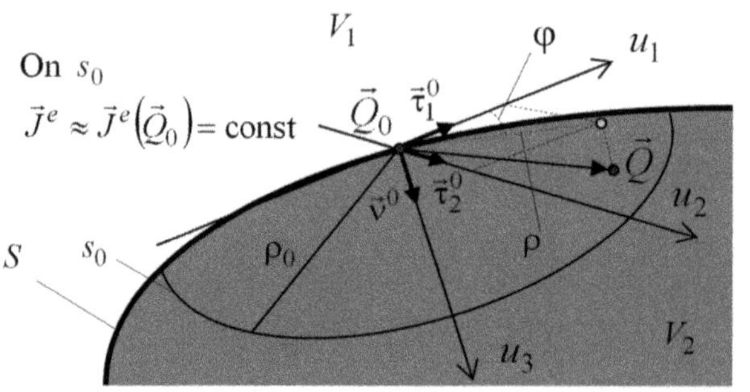

Figure 1.30. Neighborhood s_0 of the singularity at point $\vec{Q}_0$ on the object's surface.

to be directed along the vector of the inner normal $\vec{v}^0$ (figure 1.30). Then the object surface can be approximated as follows:

$$u_3 = \frac{1}{2\,|\vec{\nabla}F(\vec{Q})|_{\vec{Q}=\vec{Q}_0}}\left(\vec{M}^{\mathrm{T}}\mathbf{C}\vec{M}\right) + o\left(|\vec{M}|^2\right), \tag{1.73}$$

where $\vec{Q} = \{u_1, u_2, u_3\} = \{\rho\cos(\varphi), \rho\sin(\varphi), u_3\}$, 'T' is the transposition sign,

$$\vec{M} = \vec{Q} - \vec{Q}_0 - u_3\vec{v}^0 = u_1\vec{\tau}_1^{\,0} + u_2\vec{\tau}_2^{\,0} \tag{1.74}$$

is a vector tangential to the surface, its absolute value being $|\vec{M}| = \rho$, $o\left(|\vec{M}|^2\right)$ denotes a value of a higher infinitesimal order than $|\vec{M}|^2$, and

$$\mathbf{C} = \begin{pmatrix} c_{11} & c_{12} \\ c_{21} & c_{22} \end{pmatrix}_{\vec{Q}=\vec{Q}_0}, \qquad c_{pq} = \frac{\partial^2 F(\vec{Q})}{\partial u_p \partial u_q}. \tag{1.75}$$

In the neighborhood of point $\vec{Q}_0$, the surface $F(\vec{Q}) = 0$ can be approximately presented as follows:

$$u_3 = \rho^2 A(\varphi), \tag{1.76}$$

where

$$A(\varphi) = (c_{11}\cos^2(\varphi) + 2c_{12}\cos(\varphi)\sin(\varphi) + c_{22}\sin^2(\varphi)). \tag{1.77}$$

Then, we can introduce the element of area ds on the surface s_0 as

$$\mathrm{d}s \approx \rho\,\mathrm{d}\rho\,\mathrm{d}\varphi. \tag{1.78}$$

We can evaluate the integral over surface s_0 analytically. To that end, we need to assume that $\vec{J}^e(\vec{Q}) \approx \vec{J}^e(\vec{Q}_0)$ on surface s_0 [55, 58, 59],

$$\begin{aligned} &\int_{s_0} \vec{E}_1^m\left(\vec{Q}\,\middle|\,\vec{Q}_0, \vec{\boldsymbol{\tau}}_q^0\right)\cdot\vec{J}^e(\vec{Q})\mathrm{d}s \\ &\approx \frac{\left(\vec{\boldsymbol{v}}^0\times\vec{\boldsymbol{\tau}}_q^0\right)\cdot\vec{J}^e\left(\vec{Q}_0\right)}{2\pi}\int_0^{2\pi}\int_0^{\rho_0}\left(jk_1 - \frac{1}{\rho}\right)\frac{\exp(jk_1\rho)}{\rho}A(\varphi)\rho\mathrm{d}\rho\mathrm{d}\varphi \\ &= \left(\vec{\boldsymbol{v}}^0\times\vec{\boldsymbol{\tau}}_q^0\right)\cdot\vec{J}^e\left(\vec{Q}_0\right)I_{s_0}\left(\vec{Q}_0\right) \end{aligned} \tag{1.79}$$

where

$$I_{s_0}\left(\vec{Q}_0\right) = \frac{c_{11}+c_{22}}{4jk_1}\left[\left(jk_1\rho_0 - 2\right)\exp\left(jk_1\rho_0\right) + 2\right] \tag{1.80}$$

is the function included in equation (1.70).

Now the term $\left(\vec{v}^0\times\vec{\tau}_q^0\right)\cdot\vec{J}^e(\vec{Q})I_{s_0}(\vec{Q}_0)$ in the MFIE (1.70) is responsible for the integral contribution from the small area s_0 around the singularity. It is worth noting that if the local curvature radii of the scatterer's surface s_0 increase, the value $I_{s_0}(\vec{Q})$ will be tending to zero.

Having solved the MFIE (1.70) we can proceed to computing the magnetic field $\overline{H}$ at an arbitrary point $\vec{Q}_{sc}$ in free space, as scattered by a PEC object using the following integral representation [22, 23, 52, 55]:

$$j\omega\overline{p}^{sc}\cdot\left(\overline{H}\left(\vec{Q}_{sc}\right)-\overline{H}^{0}\left(\vec{Q}_{sc}\right)\right)=-\int_{S}\vec{E}_{1}^{m}\left(\vec{Q}\mid\vec{Q}_{sc},\overline{p}^{sc}\right)\cdot\overline{J}^{e}(\vec{Q})\,\mathrm{d}s,\qquad(1.81)$$

where $\overline{p}^{sc}$ is the unit vector pointing in the direction of the scattered EMF projection (it defines the polarization of the scattered field).

Together, the MFIE (1.70) and integral representation (1.81) form the basis for a numerical algorithm that allows computing the EM responses of PEC objects (objects with a metal surface) if the polarization, spatial, and time–frequency parameters of the probing signal are specified. The latter expressions can be employed for both monostatic and bistatic radar layouts.

1.2.2 Numerical solution to the obtained integral equation

Our original method for finding a solution to the MFIE (1.70) was described in [23] in detail. Here, we discuss some peculiarities of the computational algorithm developed there.

One can reduce the MFIE (1.70) to a matrix equation using a number of techniques [23, 55]. Our preferred choice in most cases is the use of parts of three-axial ellipsoids to represent the surfaces of the radar scatterers. The use of ellipsoids made it easier to generate a mesh of facets, which would be used for numerical integration. Sometimes a more accurate description of the object's surface can be achieved using ruled surfaces in particular [55]. The spacing between the surface current density nodes was set in an adaptive manner. We set this spacing to be proportional to the curvature radii of the facet, which meant that the facets with smaller curvature radii would be covered with current nodes more densely. Such an approach makes it possible to avoid an unnecessary increase in the dimensions of the resulting matrix equation.

In our computations we controlled the number of nodes N in which the surface density of electric current was to be evaluated (by controlling the corresponding node spacing r_c), in such manner the entire algorithm would have a property of internal convergence as controlled by margin δ_c. By such internal convergence, we mean the algorithm's property to generate a sustainable value of the object's RCS σ_N, sustainability of the RCS value being in the fact that given the number of nodes N tending to infinity the RCS would be tending to a stable value σ_{N_c}, i.e.

$$\lim_{N\to\infty}\sigma_N=\sigma_{N_c}.\qquad(1.82)$$

In addition, starting from certain value $N = N_a$, the following condition would hold:

$$\frac{|\,\sigma_{N_c}-\sigma_{N_a}\,|}{\sigma_{N_c}}100\%\leqslant\delta_c,\qquad(1.83)$$

where σ_{N_a}, σ_{N_c} are the RCS figures computed given the number of current nodes equal to N_a and N_c, respectively, in addition to $N_c > N_a$.

Below, we set the number of surface current nodes N_a in such manner that the convergence error will remain below the margin of δ_c = 1%–3%.

In particular in the VHF frequency band, one achieves a convergence error for the above algorithm of δ_c= 1%–3% for PEC scattering objects (such as cruise missiles and artillery shells) by keeping the spacing between the electric current nodes, which create a mesh over the smooth parts of these objects' surfaces, within the limits of $r_c = (0.09-0.24)\lambda_1$ [22, 23, 52]. In the case of radar scattering in the L and S frequency bands, it is possible to keep such a spacing between the current nodes, for scatterers in form of artillery shells (rockets) [53, 54], within the interval of $r_c = (0.1-0.14)\lambda_1$; and within the interval of $r_c = (0.02-0.09)\lambda_1$ for scatterers in form of metallic elements of tactical unmanned aerial vehicles (UAVs), such as their propulsion systems [55].

Figure 1.31 presents an example of mesh covering the object's surface with surface current density nodes; the object itself is a rocket fired by a generic multiple launch rocket system (MLRS).

Thus application of the mesh to surface S replaces that surface with a set of N electrically small facets, $s_n(n = \overline{1, N})$, every facet being characterized by a constant value of unknown electric current density $\vec{J}_n^e$. This makes it possible to pull these current density values out from the integral.

Now integration of the field components $\vec{\tau}_p^0 \cdot \vec{E}_1^m\left(\vec{Q} \mid \vec{Q}_0, \vec{\tau}_q^0\right)$ $(p = 1, 2;\ q = 1, 2)$ that are tangent to the surface (the latter components corresponding to point sources) over individual facets of the surface s_n can be accomplished by iterated integration. Every such iterated integral is an integral of a highly oscillating function depending on the distance $\left| \vec{Q} - \vec{Q}_0 \right|$ (equation (1.71)). Single integrals of this particular kind can be evaluated using the compound five-point Gauss formula.

Now the singularity which appears when the observation (field) and integration (source) points coincide, can be taken care of in order to enable computing the elements of the matrix equation. To that end we apply the following algorithm

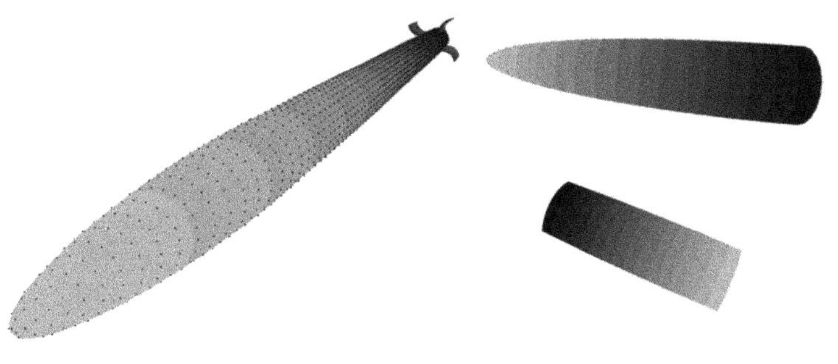

Figure 1.31. Mesh of the surface current nodes covering the surface of an MLRS rocket, the node spacing being proportional to the curvature radii of individual facets.

[22, 23, 52, 55]. Every particular facet s_n that contains the singular point is divided again into N_0 cells:

$$s_n = \sum_{g=1}^{N_0} s'_g. \tag{1.84}$$

Then we determine the distances $\rho_0 g$ between the center of facet $s_n(\vec{Q}_{n_0})$ and the centers of cells s'_g, and check the condition of $k_1\rho_{0g} > \gamma_0$. The latter parameter value $\gamma_0 = 0.01$ was established experimentally. Those elements s'_g for which this condition holds do not have any singularity and they constitute the part of s_n over which the iterated integral is evaluated numerically using a compound five-point Gauss formula for each single integration.

The elements s'_g that appear to satisfy the opposite condition of $k_1\rho_{0g} \leq \gamma_0$ are excluded from the numerical algorithm, and the integral over the surface portions they belong to is replaced with the value $I_{s_0}(\vec{Q}_{n_0})$ calculated using (1.80).

Our experience demonstrates that the evaluation of repeated integrals over facets s_n, which contain singularities, need to be carried out using the number of individual cells that satisfies the condition of $\sqrt{N_0} \geqslant 23$. The latter condition was established experimentally and it provided for the convergence of an iterated integral with an accuracy of 0.1%.

MFIE (1.70) possesses a set of ambiguous solutions or so-called 'spurious resonances' coinciding with the real-valued frequencies of natural oscillations inside the PEC object [29]. To eliminate this effect, we complement the IE (1.70) with the equations in which the field inside the object (region V_2) is assumed to be zero [22, 23, 52]:

$$j\omega\vec{\tau}_q^{\,0} \cdot \vec{H}^0(\vec{Q}_0) = \int_S \vec{E}_1^m(\vec{Q} \mid \vec{Q}_0, \vec{\tau}_q^{\,0}) \cdot \vec{J}^e(\vec{Q})\, \mathrm{d}s, \tag{1.85}$$

where $\vec{Q}_0 \in V_2\backslash S$.

Then the desired surface current density is obtained by solving the OSLE in matrix form, which is obtained after applying the mesh.

The necessary number of additional internal nodes needed to eliminate ambiguous solutions caused by natural oscillations inside the PEC object was found based on the same condition of ensuring the convergence accuracy of $\delta_c = 1\% - 3\%$ that is determined by a formula similar to (1.83). The share of auxiliary (additional) nodes must constitute 15% – 20% of the total number of nodes, at which the surface electric current density is to be evaluated.

Considering the radar scattering characteristics obtained for resonant radar objects in the VHF through S frequency bands, we have established that if the electrical sizes $a_i(i = \overline{1, 3})$ of a three-dimensional object are below the threshold of 0.6 λ_1, then ambiguous solutions to the MFIE do not appear, and the equation mentioned above can be solved by means of reducing it to an SLE. Otherwise, given that $\min(a_i) \geqslant 0.6\,\lambda_1$ (in particular this condition holds for the cruise missile fuselages irradiated in the VHF frequency band and for artillery shells (rockets) irradiated in the L, S bands), the MFIE can have ambiguous solutions. As was noted above, one can eliminate these spurious solutions using an additional equation (1.85).

Then the electric current density at the PEC object's surface can be evaluated by solving the OSLE.

Practically, we have found out that these additional points are advisable to be generated inside the object and they are to be put on an analytic continuation of the object's surface. However, a gap of about 0.2 $\lambda_{1\,\min}$ between such continuation and the surface should be provided, where $\lambda_{1\,\min}$ is the shortest wavelength in the frequency band of interest. For example, a grid of such additional nodes belonging to an analytic continuation of an MLRS rocket's surface is shown in figure 1.32.

The integral representation (1.81) of the scattered field can be transformed into the discrete form in a similar way to equation (1.70).

Comparison of the EM scattering characteristics of several benchmark models indicated a fairly high accuracy of our RCS prediction technique. The comparison involved a wide variety of methods including: (a) our original numerical algorithm (which is based on the MFIE solution); (b) other well-known mathematical methods; and (c) methods of physical modeling [23, 52].

The numerical algorithm proposed in section 1.2 for solving the MFIE with regard to EMW scattering from real radar objects has some advantages over other known methods. The above method is particularly advantageous given that the scatterers have complex shapes that involve elements with varying radii of curvature. Moreover, the algorithm developed by the authors provides for the computation of EM responses from radar objects of resonant sizes given their irradiation with EMWs at carrier frequencies varying in fairly wide ranges and given wide sectors of object attitudes toward the radar. Such EM responses are predicted with decent accuracy and without excessive computer burden.

Figure 1.32. The location of additional nodes inside the MLRS rocket.

The challenge posed by real airborne objects from an electrodynamics point of view is that the electrical sizes of their individual elements vary considerably so that some elements of the object can be considered as three-dimensional (such as fuselage, engine nacelles) whereas some can be considered as flat or electrically thin (such as wings and stabilizers). In that case, the MFIE (1.70) solution requires a considerable amount of computer RAM. In addition, the large dimensionality of the final equation coupled with the presence of flat or electrically thin elements introduces an ill-conditioned matrix of MFIE kernels, which degrades the solution accuracy. To cope with that challenge for PEC objects of complex shapes that vary considerably in size, we developed an iterative algorithm that ensures a good solution to the MFIE (1.70). This iterative algorithm is described in great detail in [22, 23]. In contrast to the known procedures [29, 44, 45] that are employed commonly, the iterative algorithm mentioned above features matrices of significantly lower dimensionality as opposed to the full-scale matrix pertaining to the entire object's surface. Our approach allowed for significantly reduced computation time while retaining a high accuracy of solution.

Additionally, it is known that the MFIE (an inhomogeneous Fredholm IE of the second kind) has better convergence of the iterative algorithm than the EFIE (an IE of the first kind) [32], which is further evidence toward choosing the MFIE for the prediction of scattering characteristics for resonant size complex-shaped scatterers.

In addition, if one introduces a field of the auxiliary point source of the EMW in the presence of an interface between the two dielectric media into equation (1.70) (the corresponding Green's function) instead of $\vec{E}_1^m$ [23, 51, 58, 59], then one can compute the surface electric current density on the surface of a PEC object buried inside the dielectric half-space. In particular, such buried objects include metallic case mines installed in the soil. In that case, the point source field representation given the interface between the two media needs to be plugged into equation (1.81) as well.

The original software was used to simulate the scattering characteristics of a number of radar objects that had resonant sizes in the VHF, L, and S frequency bands. Some results are shown and discussed in section 1.4.

1.3 The Müller integral equation set for solving problems of resonant electromagnetic wave scattering by dielectric homogeneous objects

This section is devoted to the discussion of a numerical method for predicting the EM scattering characteristics (RCS in particular) of resonance-sized dielectric homogeneous objects. Such scatterers may include the dielectric structural elements of air vehicles, dielectric mines buried in soil, and particles of precipitation (raindrops, snowflakes) as they are irradiated by radar signals on the wavelengths commensurate with the scatterers' sizes. Such scatterers may also include silver nanoparticles irradiated in the visible optical waveband, which can be considered as a scatterers filled with a homogeneous material having a complex permittivity.

The numerical computation method is also based on a solution to a MIE set [13, 31–33, 55, 58–60]. These equations are basically the boundary inhomogeneous Fredholm IEs of the second kind and they converge well after replacing the actual surfaces with their faceted versions (meshed surfaces) [32, 55].

When applied to dielectric scatterers, these boundary IEs (as was the case for PEC scatterers) are usually solved using MoM or some of its modifications [31, 32, 61]. Thus for dielectric scatterers, we are going to employ the interpolation method for obtaining a discrete version of the MIE set, which to our knowledge does have a number of advantages compared to the projection methods such as the MoM and its variations. The most distinct features of our algorithm for solving the MIE set in question include the following: (i) the algorithm for integrating the equation kernels; (ii) a method for singling out and eliminating the kernel singularities; and (iii) a way to cover the scatterer's surface with a mesh of current nodes given that their spacing would be dependent on the local curvature radii of the dielectric scatterer's surface. In addition, in this particular case the spacing between the surface current nodes is placed in additional dependence on the wavelength of the incident wave as it propagates into the scatterer's medium. As was the case for PEC scattering objects, the solutions presented here allow for a high prediction accuracy of scattering characteristics that is attained without excessive computer burden.

Further, section 1.3 presents the process of deriving the MIE set. Along the way we consider the issues with kernels involved in the equation set, as well as considering the peculiarities of the algorithm aimed at finding a numerical solution to the IE set in question. We validate the results by checking our model against a well-known numerical technique [62].

1.3.1 Deriving the Müller integral equation set

Let us consider the problem of EMW scattering by a homogeneous 3D object occupying the space V_2 that is characterized by permittivity $\varepsilon_2 = \varepsilon'_2 + j\varepsilon''_2$, the object being located in the free space V_1 that is characterized by $\varepsilon_1 = 1$ (the relative permeability is assumed to be $\mu = 1$ for all materials considered below).

For evaluating the equivalent electric $\vec{J}^e$ and magnetic $\vec{J}^m$ current densities at the surface S of a dielectric scatterer, we use the MIE set [13, 23, 31–33, 55, 58–61] that can be obtained by applying the Lorentz reciprocity theorem (1.69) to the sought field $(\vec{E}, \vec{H})$ and the fields of auxiliary point sources that reside in the domains V_1 and V_2. In subsection 1.3.1 we consider the fields (at the point $\vec{Q}$) as created by a point electric source,

$$\vec{E}^e_\alpha\left(\vec{Q} \mid \vec{Q}_0, \vec{\boldsymbol{\tau}}^0_q\right) = (\varepsilon_0\varepsilon_\alpha)^{-1}\left[\vec{\boldsymbol{\tau}}^0_q k_\alpha^2 G_\alpha(\vec{Q}_0, \vec{Q}) + \vec{\nabla}\left(\vec{\boldsymbol{\tau}}^0_q \cdot \vec{\nabla} G_\alpha(\vec{Q}_0, \vec{Q})\right)\right], \tag{1.86}$$

$$\vec{H}^e_\alpha\left(\vec{Q} \mid \vec{Q}_0, \vec{\boldsymbol{\tau}}^0_q\right) = j\omega\left(\vec{\boldsymbol{\tau}}^0_q \times \vec{\nabla} G_\alpha(\vec{Q}_0, \vec{Q})\right), \tag{1.87}$$

and by a point magnetic source,

$$\vec{E}_\alpha^m\left(\vec{Q} \mid \vec{Q}_0, \vec{\tau}_q^0\right) = -j\omega\left(\vec{\tau}_q^0 \times \vec{\nabla} G_\alpha(\vec{Q}_0, \vec{Q})\right), \tag{1.88}$$

$$\vec{H}_\alpha^m(\vec{Q}\big|\vec{Q}_0, \vec{\tau}_q^0) = \mu_0^{-1}\left[\vec{\tau}_q^0 k_\alpha^2 G_\alpha(\vec{Q}_0, \vec{Q}) + \vec{\nabla}(\vec{\tau}_q^0 \cdot \vec{\nabla} G_\alpha(\vec{Q}_0, \vec{Q})\right], \tag{1.89}$$

(i.e. by the Hertz dipoles sitting at the point $\vec{Q}_0$ that have the unit vector-moment $\vec{\tau}_q^0$) given that the dielectric constants of the object and of the host medium coincide, $\varepsilon_1 = \varepsilon_2 = \varepsilon_\alpha$.

Here ε_0, μ_0 are the permittivity and permeability of free space, respectively. It is worth noting that equations (1.86)–(1.89) include the following function

$$G_\alpha\left(\vec{Q}, \vec{Q}_0\right) = G_\alpha(R) = \frac{\exp(jk_\alpha R)}{4\pi R}, \tag{1.90}$$

where k_α is the wavenumber in the homogeneous medium characterized by dielectric constant ε_α.

Applying the Lorentz reciprocity theorem (1.69) to the sought for field $(\vec{E}, \vec{H})$ in both domains V_1, V_2 and, consequently, to the field of electric and magnetic point sources described above, one can obtain the MIE set [23, 55],

$$\begin{cases} -2\varepsilon_1\vec{\tau}_q^0 \cdot \vec{E}^0\left(\vec{Q}_0\right) - \left(\vec{\tau}_q^0 \times \vec{\nu}^0\right) \cdot \vec{J}^m\left(\vec{Q}_0\right)(\varepsilon_1 + \varepsilon_2) \\ = \dfrac{2}{j\omega}\displaystyle\int_S \left[\vec{H}_{21}^e\left(\vec{\tau}_q^0\right) \cdot \vec{J}^m(\vec{Q})\varepsilon_2 + \varepsilon_0^{-1}\vec{D}_{21}^e\left(\vec{\tau}_q^0\right) \cdot \vec{J}^e(\vec{Q})\right] \mathrm{d}s, \\ \vec{\tau}_q^0 \cdot \vec{H}^0\left(\vec{Q}_0\right) + \left(\vec{\tau}_q^0 \times \vec{\nu}^0\right) \cdot \vec{J}^e\left(\vec{Q}_0\right) \\ = \dfrac{1}{j\omega}\displaystyle\int_S \left[\vec{H}_{21}^m\left(\vec{\tau}_q^0\right) \cdot \vec{J}^m(\vec{Q})\varepsilon_2 + \vec{E}_{21}^m\left(\vec{\tau}_q^0\right) \cdot \vec{J}^e(\vec{Q})\right] \mathrm{d}s. \end{cases} \tag{1.91}$$

Here $\vec{E}^0$, $\vec{H}^0$ are the magnetic and electric vectors of the EMF incident on a dielectric scatterer, and $\vec{\nu}^0$ is the unit vector of the outer normal to its surface at the field point $\vec{Q}_0$ (the mutually orthogonal unit vectors $\vec{\tau}_q^0$ $(q = 1, 2)$ that are tangential to surface S at that particular point forming the right-hand system when coupled with the $\vec{\nu}^0$).

The kernels of the MIE set (1.91) are the combinations of the electric and magnetic Green's functions, built as differences between the fields $\left(\vec{E}_\alpha^{e(m)}, \vec{H}_\alpha^{e(m)}\right)$ pertaining to auxiliary electric (magnetic) Hertz dipoles residing in homogeneous media:

$$\vec{H}_{21}^e\left(\vec{\tau}_q^0\right) = \vec{H}_2^e\left(\vec{Q} \mid \vec{Q}_0, \vec{\tau}_q^0\right) - \frac{\varepsilon_1}{\varepsilon_2}\vec{H}_1^e\left(\vec{Q} \mid \vec{Q}_0, \vec{\tau}_q^0\right), \tag{1.92}$$

$$\vec{D}_{21}^e\left(\vec{\tau}_q^0\right) = \vec{D}_2^e\left(\vec{Q} \mid \vec{Q}_0, \vec{\tau}_q^0\right) - \vec{D}_1^e\left(\vec{Q} \mid \vec{Q}_0, \vec{\tau}_q^0\right) \tag{1.93}$$

$$\vec{D}^{e}_{1(2)}\left(\vec{Q} \mid \vec{Q}_0, \vec{\tau}^{\,0}_q\right) = \varepsilon_0\varepsilon_{1(2)}\vec{E}^{e}_{1(2)}\left(\vec{Q} \mid \vec{Q}_0, \vec{\tau}^{\,0}_q\right) \tag{1.94}$$

$$\vec{H}^{m}_{21}\left(\vec{\tau}^{\,0}_q\right) = \vec{H}^{m}_{2}\left(\vec{Q} \mid \vec{Q}_0, \vec{\tau}^{\,0}_q\right) - \vec{H}^{m}_{1}\left(\vec{Q} \mid \vec{Q}_0, \vec{\tau}^{\,0}_q\right) \tag{1.95}$$

$$\vec{E}^{m}_{21}\left(\vec{\tau}^{\,0}_q\right) = \vec{E}^{m}_{2}\left(\vec{Q} \mid \vec{Q}_0, \vec{\tau}^{\,0}_q\right) - \vec{E}^{m}_{1}\left(\vec{Q} \mid \vec{Q}_0, \vec{\tau}^{\,0}_q\right). \tag{1.96}$$

Here $\alpha = 1, 2$ correspond to either free space (domain V_1) or the space occupied by the dielectric scatterer V_2, the two latter volumes are filled with media characterized by relative permittivities ε_1 and ε_2, respectively.

The boundary integrals in (1.91) have either finite values given that $\vec{Q} \to \vec{Q}_0$ or they represent integrable singularities [23, 55, 58, 59]. The $\vec{H}^{e}_{21}$ term contains the singularity which corresponds to the normal derivative of single-layer potential [56, 57]. The rest of the components, $\vec{D}^{e}_{21}$, $\vec{H}^{m}_{21}$, $\vec{E}^{m}_{21}$ do have singularities of the order not higher than R^{-1} [23, 55, 58, 59]. Therefore, taking into account (1.78), the integrals in (1.91) take finite values if the field point $\vec{Q}_0$ belongs to the surface S. It is also worth noting that kernels of (1.91) are proportional to the difference $\varepsilon_2 - \varepsilon_1$, hence they turn to zero if $\varepsilon_2 \to \varepsilon_1$.

The MIE set (1.91), in contrast to MFIE (1.70), involves the densities of both the electric and magnetic equivalent currents, and its solution satisfies the boundary conditions (continuity of electric and magnetic field tangential components). As a result, the MIE set (1.91) has unique solutions at all real-valued frequencies [55, 60].

As soon as the densities of equivalent currents $\vec{J}^{e}$ and $\vec{J}^{m}$ are evaluated, one can compute the components of the EMF scattered by the dielectric object as measured at any given point $\vec{Q}_{sc}$ in free space V_1 using the following formulae [23, 55]:

$$j\omega\vec{p}^{\,sc} \cdot \left(\vec{E}(\vec{Q}_{sc}) - \vec{E}^{0}(\vec{Q}_{sc})\right) = \int_S [\vec{H}^{e}_{21}(\vec{p}^{\,sc}) \cdot \vec{J}^{m}(\vec{Q}) + \vec{E}^{e}_{21}(\vec{p}^{\,sc}) \cdot \vec{J}^{e}(\vec{Q})]\,\mathrm{d}s, \tag{1.97}$$

$$-j\omega\vec{p}^{\,sc} \cdot \left(\vec{H}\left(\vec{Q}_{sc}\right) - \vec{H}^{0}\left(\vec{Q}_{sc}\right)\right) = \int_S [\vec{H}^{m}_{21}(\vec{p}^{\,sc}) \cdot \vec{J}^{m}(\vec{Q}) + \vec{E}^{m}_{21}(\vec{p}^{\,sc}) \cdot \vec{J}^{e}(\vec{Q})]\,\mathrm{d}s. \tag{1.98}$$

Thus on the basis of the MIE set (1.91) coupled with representations (1.97) and (1.98) we created our numerical algorithm that allows for prediction of the EM responses of dielectric homogeneous scatterers given both monostatic and bistatic radar layouts. The above EM responses may involve various polarizations, as well as various spatial–temporal–frequency parameters of radar probing signals. The algorithm is discussed below.

1.3.2 Algorithm for finding a numerical solution to the Müller integral equation set

Let us look at the algorithm for finding a numerical solution to the MIE set (1.91) in more detail. The equation set (1.91) can be reduced to a matrix equation using a number of techniques [23, 55, 63, 64]. In order to unify this model's approach with the previous case of metallic scatterers, we adopted the same technique that uses either portions of three-axial ellipsoids to describe the surface of radar objects, or ruled surfaces that are later reduced to a mesh of facets [55]. The spacing between the mesh nodes that represents the constant values of surface current density is set to be proportional to local curvature radii on every particular facet.

A discrete representation of both IEs, (1.70) and (1.91), is carried out in the same way. As soon as the mesh is applied, the surface S can be represented as a set of N electrically small facets, $s_n(n = \overline{1, N})$. In addition, the unknown current densities $\vec{J}_n^{\,e}$, $\vec{J}_n^{\,m}$ are assumed to be constant on every particular facet. The latter means that they can be pulled outside the integral over a particular facet s_n. Now, let us break down the sought for surface densities of the equivalent currents into components along two mutually orthogonal tangential unit vectors $\vec{J}_{1(2),\,n}^{\,m} = \vec{\tau}_{1(2),\,n} \cdot \vec{J}_n^{\,m}$, $\vec{J}_{1(2),\,n}^{\,e} = \vec{\tau}_{1(2),\,n} \cdot \vec{J}_n^{\,e}$, respectively. Once simplified by introducing the index of the field (observation) point $n_0 = \overline{1, N}$, the latter MIE set (1.91) can be presented as follows:

$$\begin{cases} \sum\limits_{n=1}^{N}\left[J_{1,\,n}^{m}A_{n_0,\,n}^{11} + J_{2,\,n}^{m}A_{n_0,\,n}^{12} + J_{1,\,n}^{e}A_{n_0,\,n}^{13} + J_{2,\,n}^{e}A_{n_0,\,n}^{14}\right] = B_{1,\,n_0}, \\ \sum\limits_{n=1}^{N}\left[J_{1,\,n}^{m}A_{n_0,\,n}^{21} + J_{2,\,n}^{m}A_{n_0,\,n}^{22} + J_{1,\,n}^{e}A_{n_0,\,n}^{23} + J_{2,\,n}^{e}A_{n_0,\,n}^{24}\right] = B_{2,\,n_0}, \\ \sum\limits_{n=1}^{N}\left[J_{1,\,n}^{m}A_{n_0,\,n}^{31} + J_{2,\,n}^{m}A_{n_0,\,n}^{32} + J_{1,\,n}^{e}A_{n_0,\,n}^{33} + J_{2,\,n}^{e}A_{n_0,\,n}^{34}\right] = B_{3,\,n_0}, \\ \sum\limits_{n=1}^{N}\left[J_{1,\,n}^{m}A_{n_0,\,n}^{41} + J_{2,\,n}^{m}A_{n_0,\,n}^{42} + J_{1,\,n}^{e}A_{n_0,\,n}^{43} + J_{2,\,n}^{e}A_{n_0,\,n}^{44}\right] = B_{4,\,n_0}, \end{cases} \tag{1.99}$$

where, for $p = 1, 2$, $q = 1, 2$, we have the following:

$$A_{n_0,\,n}^{pq} = \begin{cases} (-1)^q(\varepsilon_1 + \varepsilon_2) + \dfrac{2\varepsilon_2}{j\omega}\displaystyle\int_{s_{n0}}\left(\vec{\tau}_{q,\,n_0} \cdot \vec{H}_{21}^{e}\left(\vec{Q} \mid \vec{Q}_{n_0},\ \vec{\tau}_{p,\,n_0}\right)\right)\mathrm{d}s, \\ \qquad n = n_0, \\ \dfrac{2\varepsilon_2}{j\omega}\displaystyle\int_{s_n}\left(\vec{\tau}_{q,\,n} \cdot \vec{H}_{21}^{e}\left(\vec{Q} \mid \vec{Q}_{n_0},\ \vec{\tau}_{p,\,n_0}\right)\right)\mathrm{d}s, \quad n \neq n_0; \end{cases} \tag{1.100}$$

for $p = 1, 2$, $q = 3, 4$, we have:

$$A_{n_0,\,n}^{pq} = \frac{2}{j\omega\varepsilon_0}\int_{s_n}\left(\vec{\tau}_{q-2,\,n} \cdot \vec{D}_{21}^{e}\left(\vec{Q} \mid \vec{Q}_{n_0},\ \vec{\tau}_{p,\,n_0}\right)\right)\mathrm{d}s; \tag{1.101}$$

for $p = 3, 4$, $q = 1, 2$, we have:

$$A_{n_0,\,n}^{pq} = (j\omega)^{-1}\int_{s_n}\left(\vec{\tau}_{q,\,n} \cdot \vec{H}_{21}^{m}\left(\vec{Q} \mid \vec{Q}_{n_0},\ \vec{\tau}_{p-2,\,n_0}\right)\right)\mathrm{d}s; \tag{1.102}$$

and for $p = 3, 4$, $q = 3, 4$, we have:

$$A^{pq}_{n_0,\ n} = \begin{cases} (-1)^{q-1} + (j\omega)^{-1} \int_{s_{n0}} \left(\vec{\tau}_{q-2,\ n_0} \cdot \vec{E}^{m}_{21}\left(\vec{Q} \mid \vec{Q}_{n_0},\ \vec{\tau}_{p-2,\ n_0} \right) \right) \mathrm{d}s, \\ \qquad\qquad n = n_0, \\ (j\omega)^{-1} \int_{s_n} \left(\vec{\tau}_{q-2,\ n} \cdot \vec{E}^{m}_{21}\left(\vec{Q} \mid \vec{Q}_{n_0},\ \vec{\tau}_{p-2,\ n_0} \right) \right) \mathrm{d}s,\ n \neq n_0; \end{cases} \tag{1.103}$$

$$B_{p,\ n_0} = \begin{cases} -2\varepsilon_1 \vec{\tau}_{p,\ n_0} \cdot \vec{E}^{0}(Q_{n_0}), & p = 1, 2, \\ \vec{\tau}_{p,\ n_0} \cdot \vec{H}^{0}(Q_{n_0}), & p = 3, 4. \end{cases} \tag{1.104}$$

The values $A^{pq}_{n_0,\ n}$ in expressions (1.100)–(1.103), as was the case for PEC objects, can be represented as iterated integrals, which can be evaluated by means of the compound five-point Gauss formula applied to every single integral.

Let us consider particular integrals over facets s_n that include singular points. Every singular point itself can be singled out and eliminated, thus enabling the integration in question, using the following algorithm.

As before, we break down the facet s_{n_0} containing the singularity into a collection of cells:

$$s_{n_0} = \sum_{g=1}^{N_0} s'_g. \tag{1.105}$$

Now the center of facet s_{n_0}, which is at point $\vec{Q}_{n_0}$, would be sitting at a corner of every cell s'_g adjacent to it. Then we apply expressions (1.100)–(1.103) to every single cell s'_g. To that end, we represent the values $A^{pq}_{n_0,\ n}$ as iterated integrals, as we did before. Then every single integral is evaluated using the five-point Gauss formula.

In order to evaluate the integrals over every facet s_{n_0}, one should set the number of cells $N_0 = N_{01}N_{02}$ to be fairly large. Here N_{01}, N_{02} are the numbers that represent the quantities of the segments to be used in every five-point Gauss formula that brings forth the value of every single integral included in the iterated integration. The latter is due to the fact that the auxiliary field components of both the electric and magnetic dipoles appearing in expressions (1.100)–(1.103) increase given that $\left| \vec{Q} - \vec{Q}_0 \right| \to 0$. The necessary quantity of cells N_0 that every facet would be broken down into was found by means of computer simulation.

Let us consider, for example, the results of computing the integrals (1.100)–(1.103) versus varying the number of $\sqrt{N_0} = N_{01} = N_{02}$ given that the scatterer was a three-axial ellipsoid made of a dielectric with permittivity $\varepsilon_2 = 3.17 + j1.45 \times 10^{-3}$, its semi-axes being equal to 0.016 λ_1, 0.047 λ_1, and 0.094 λ_1, for wavelength of incident wave equal to $\lambda_1 = 3.2$ cm. A mesh applied to the surface broke down the entire surface of that ellipsoid into $N = 584$ facets (which ensured the error margin of $\delta_c \leqslant 1\%$, the latter being set in the same way as for the PEC scatterer above, formula (1.83)). The number of cells to break every facet s_{n_0} down further was controlled by value $\sqrt{N_0}$, that parameter being the same for all the facets. We conducted a series of computations given that parameter $\sqrt{N_0}$ was varied from 2 through 24. Every computation round for element $A^{pq}_{n_0,\, n_0}$ was followed by the corresponding accuracy estimate:

$$\delta^{pq}_{\text{int}\, n_0}(\sqrt{N_0}) = \left| \frac{A^{pq}_{n_0,\, n_0}(\sqrt{N_0}) - A^{pq}_{n_0,\, n_0}(24)}{A^{pq}_{n_0,\, n_0}(24)} \right| 100\% \quad (n_0 = \overline{1, N}), \tag{1.106}$$

where $A^{pq}_{n_0,\, n_0}(\sqrt{N_0})$ is the value of integrals (1.100)–(1.103) computed given particular values of n_0, q, p, and N_0.

The figure 1.33 below shows the following dependence

$$\delta_{\text{int}}(\sqrt{N_0}) = \max_{n_0, p, q} \left\lfloor \delta^{pq}_{\text{int}\, n_0}(\sqrt{N_0}) \right\rfloor \tag{1.107}$$

of the maximum error pertaining to the integrals (1.100)–(1.103) computed over all the facets s_{n_0} on the surface of a dielectric ellipsoid versus the number of cells $\sqrt{N_0}$. That dependence clearly shows that given the parameter $\sqrt{N_0} \geqslant 16$ the maximum error of the surface integrals (1.100)–(1.103) computed over facets s_{n_0}, which contain the singularity, goes below one percent. In addition, that particular error does not exceed the level of 0.338% given that the parameter controlling the number of cells $\sqrt{N_0}$ is set to be equal to 20 or 24. For the parameter set to $\sqrt{N_0} > 24$, the error in question went below 0.1%.

The method proposed above for singling out and eliminating the singularity of the MIE set (1.91) as well as the method for computing the integrals over the surfaces of dielectric objects that may contain such singularities provides for accurate prediction of the scattering characteristics of dielectric objects, the method's accuracy being at

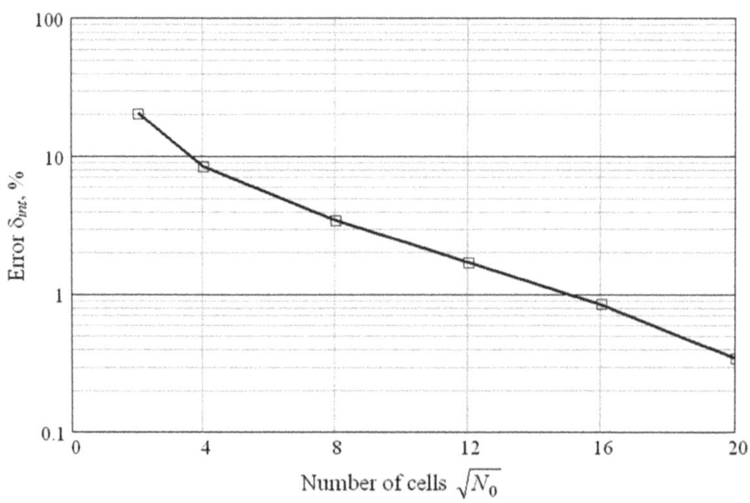

Figure 1.33. Computation error for the boundary integrals over facets s_{n_0} of a dielectric ellipsoid containing singular points versus the number of cells $\sqrt{N_0}$ that every facet was broken into.

the level of $\delta_c = 1\%–3\%$. This accuracy was maintained in cases where the EM responses of dielectric scatterers, the surface of which would include elements of greatly varying curvatures, were concerned. In particular, such scatterers might include electrically thin elements.

The book [23] contains a validation of the results obtained above. The results of the computation using the original algorithm described above are checked there against other known numerical methods for several benchmark scatterers.

In order to obtain the exact steady-state values of the RCS for the case of such benchmark scatterers, we had to determine the algorithm parameters experimentally. In particular, for the L and S frequency bands, the spacing between the current nodes that covered the surface of the UAV's dielectric fuel tank was set to be equal to $r_c = (0.02–0.3)\lambda_1$. In the case of the surface of a dielectric propeller (electrically thin element), its surface was covered by nodes spaced at $r_c = (0.01–0.08)\lambda_1$.

To confirm the advantages of the our original method, we present figure 1.34, which shows the monostatic RCS of an electrically thin dielectric ellipsoid with the relative permittivity $\varepsilon_2 = 3.17 + j1.45 \times 10^{-3}$, its semi-axes being $a_x = 0.016\lambda_1$, $a_y = 0.047\lambda_1$, and $a_z = 0.094\lambda_1$. The dashed curve shows the computation results obtained using our method (number of current nodes was $N = 710$). The solid curve on that plot presents the results obtained using FEKO software [62] (MoM, $N = 4688$). Both dependences are plotted versus angle α, which indicated the object's attitude toward the radar and belongs to the xOy plane (i.e. the radar probing along the x-axis corresponded to $\alpha = 0$). The ellipsoid's RCS was computed for two orthogonal polarizations of the incident wave, the E-plane case corresponded to the vector $\vec{H}$ being parallel to the xOy plane, and the H-plane case corresponded to vector $\vec{H}$ being perpendicular to that plane.

As a next validation step, figure 1.35 compares the errors of the ellipsoid's RCS computed for various quantities of current nodes N placed over the scatterer's surface, as well as using two different methods. The error here has the following meaning:

$$\delta(\sigma_c(N_{\max}), \sigma(N)) = \frac{|\sigma_c(N_{\max}) - \sigma(N)|}{\sigma_c(N_{\max})} 100\%, \quad (1.108)$$

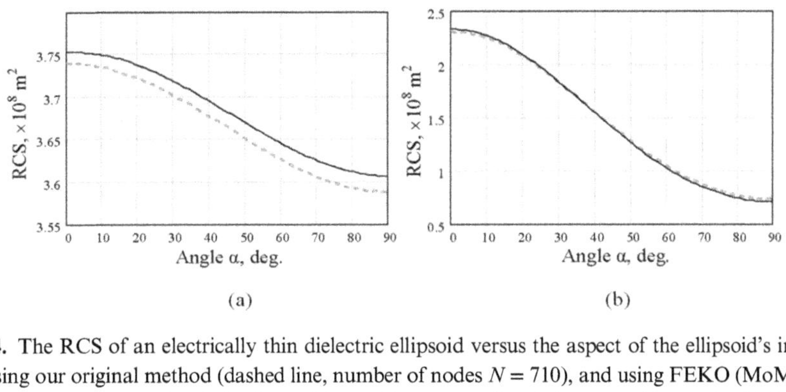

Figure 1.34. The RCS of an electrically thin dielectric ellipsoid versus the aspect of the ellipsoid's irradiation α obtained using our original method (dashed line, number of nodes $N = 710$), and using FEKO (MoM, solid line, $N = 4688$), for two orthogonal polarizations of the incident wave, the E-plane case (a) and the H-plane case (b).

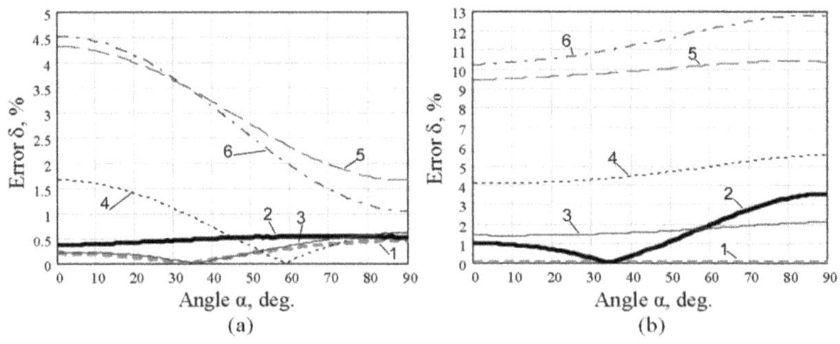

Figure 1.35. The RCS computation error versus irradiation aspect α for the case of a dielectric ellipsoid for various quantities of current nodes, and for two orthogonal polarizations of the incident wave: the E-plane case (a), and the H-plane case (b).

1—$\delta(\sigma_{cMIE}(710), \sigma_{MIE}(584))$,
2—$\delta(\sigma_{cMIE}(710), \sigma_{FEKO}(4688))$,
3—$\delta(\sigma_{FEKO}(4688), \sigma_{FEKO}(2512))$,
4—$\delta(\sigma_{cFEKO}(4688), \sigma_{FEKO}(1210))$,
5—$\delta(\sigma_{cFEKO}(4688), \sigma_{FEKO}(706))$,
6—$\delta(\sigma_{cFEKO}(4688), \sigma_{FEKO}(584))$,

where $\sigma_c(N)$ is the RCS value computed given the maximum quantity of nodes $N_{\max}$, whereas $\sigma(N)$ is the RCS computed for a reduced quantity of nodes N. The RCS computed using our original method, which is based on solving the MIE set, is denoted as σ_{MIE}, and the RCS computed by FEKO software is denoted as σ_{FEKO}. Now, for the sake of clarity, we disregard the angular dependences of the plots in figure 1.35 and put only the maximum values of the above errors into table 1.1.

Now, it is obvious that our algorithm, thanks to its good convergence, provides the error, which does not exceed 0.54%, given a smaller number of facets, $N = 584$, at both polarizations. At the same time the FEKO software can attain an error of 0.619% for the case of E-plane polarization and an error of 2.113% for the case of H-plane polarization only if the number of facets is increased to $N = 2512$.

Table 1.1. Maximum RCS computation errors obtained using two different methods for the case of an electrically thin dielectric ellipsoid.

Comparison for various N	$\delta_{\max}$, %	
	E-plane	H-plane
$\delta(\sigma_{cMIE}(710), \sigma_{MIE}(584))$	0.463	0.029
$\delta(\sigma_{cMIE}(710), \sigma_{FEKO}(4688))$	0.542	3.567
$\delta(\sigma_{FEKO}(4688), \sigma_{FEKO}(2512))$	0.619	2.113
$\delta(\sigma_{cFEKO}(4688), \sigma_{FEKO}(1210))$	1.659	5.561
$\delta(\sigma_{cFEKO}(4688), \sigma_{FEKO}(706))$	4.317	10.44
$\delta(\sigma_{cFEKO}(4688), \sigma_{FEKO}(584))$	4.514	12.80

Section 1.3 presents an original method for calculating the EM scattering characteristics of dielectric homogeneous objects. The original numerical algorithm has a number of advantages over other known methods. These advantages, as was the case with PEC scatterers, are most noticeable when the scatterers of complex shape are involved. In addition, their surfaces may include elements with significantly different curvature radii. The later curvature radii could actually be so different that they would cover electrically thin elements. The method and corresponding algorithm make it possible to predict the EM responses for a multitude of real scatterers given various polarizations, as well as various spatial and time–frequency parameters of the radar signal.

Moreover, if one employs in equations (1.91)–(1.98) the EMFs of auxiliary point sources, both electric and magnetic, given the presence of an interface between two dielectric half-spaces [23, 51, 58, 59], then the algorithm would be capable of predicting the EM responses of buried dielectric scatterers, dielectric mines located in the soil in particular.

We also employed our original software to predict the scattering characteristics of resonance-sized scatterers such as the dielectric components of a tactical UAV for their radar probing in the VHF and S frequency bands, dielectric scatterers (mines) buried in the soil for their radar probing in the VHF, UHF, and L frequency bands, electrically thin dielectric fairing for radar probing in C band, various particles of precipitation for their radar probing in the X and K_a bands, and silver nanoparticles for their irradiation in the optical frequency band (which can be considered as scatterers filled with a homogeneous material having a complex permittivity in this band). The results of computations are discussed in section 1.4.

1.4 Using an integral equation method for numerical modeling of electromagnetic scattering by metallic, dielectric, and combined resonant objects in applied problems

The numerical methods of EM scattering characteristics computation presented in sections 1.2 and 1.3, which are based upon solutions to the boundary IEs, can be applied in a number of practical cases that involve the performance assessment of various radar systems as well as the design and development of radar signal processing algorithms including those of radar target recognition. They can also be used for feasibility studies of certain aspects of nanotechnology. In this part of the book, we demonstrate the results of the application of the above methods to predict the EM responses by various radar objects of different physical character. The only thing common to all the objects examined below is that they share the same characteristic sizes that are commensurate with the wavelength of radar signal they are probed with, i.e. they are all resonance-sized scatterers.

1.4.1 Radar scattering characteristics of an unmanned aerial vehicle in the VHF and S frequency bands

UAVs (or drones) of various types are widely used in covert actions, such as reconnaissance, mapping, the tracking of moving objects, and the delivery of loads. A special place is occupied by the attack drones, equipped with precise munitions [65]. Depending on the task to be performed, these UAVs may have various electrical sizes. Micro, mini, small, and tactical UAVs have common features: as a rule, they include an insignificant amount of metallic design elements and they have comparatively small electrical dimensions in the frequency bands ranging from the VHF to the S band. These factors result in a low level of EMW scattering in the bands mentioned that complicates their radar detection and tracking [37, 39, 50, 65–74]. In addition, their relatively low speed and low altitude of flight significantly complicate their radar observation against a cluttered background.

The research and development associated with UAVs has two sides. On the one hand, drone developers strive to reduce the EM signatures of UAVs and, on the other hand, radar and signal processing engineers work on improving the efficiency of the radar detection and tracking of UAVs [37, 39, 50, 65–74]. In both cases methods of physical [37, 39, 66–73] and mathematical [37, 39, 50, 66, 67, 72] modeling of the UAV radar scattering characteristics are applied.

The mathematical methods allow, without significant material and time expenditure, the EMW scattering to be simulated from UAVs of various shapes and materials given various radar layouts. The most widely used are the AHFMs [29, 46–50] (in the case of electrically large scatterers) and methods based on solving the boundary IEs [13, 25–45, 51–55] (in the case of electrically small and resonant-sized scatterers).

There are many publications dealing with research into the radar scattering characteristics of quadcopters. In particular, in [37, 39, 66–69, 71–73] the results of simulations and measurements pertaining to the RCS of quadcopter drones in various frequency bands are presented. These works investigate the time–frequency parameters of the echo signals, including the rotor modulation spectra. For instance, paper [70] uses physical modeling and paper [50] uses mathematical modeling (using AHFM) to study the radar scattering characteristics of tactical UAVs in the X band. The scattering characteristics of small and tactical UAVs in frequencies that fall into the VHF and S bands are studied to a lesser extent. That is why, in subsection 1.4.1, we consider the radar scattering characteristics of a tactical UAV [55].

Typically, tactical UAVs have both a fuselage length and wingspan of about 2 m. The fuselage, wings, stabilizers, and fin are usually made of light-weight dielectric materials and have thin walls with a thickness of about 5 mm. Therefore, one should expect that in frequencies ranging from the VHF to S band the design elements of the UAV mentioned above would be transparent electromagnetically, i.e. they could be neglected. Then the main contributors to the EMF scattered by a drone become the following design components: the metallic (modeled as PEC) parts of its propulsion system (engine with a muffler), the metallic control unit, and the metallic payload module, and a single massive dielectric element, which is its fuel tank, consisting of a thin wall and gasoline inside. The fuel tank can be assumed to be a homogeneous dielectric object with the dielectric constant of gasoline. Additionally, he UAV's propeller can be either aluminum or dielectric. The dimensions of all the listed elements measure as dozens of centimeters. Thus a tactical UAV is a compound object of complex shape, which has both metallic and dielectric design components. For the VHF and S band frequencies, these elements are resonance-sized scatterers (i.e. commensurate with the radar wavelength). Therefore, in order to accurately predict their EM scattering characteristics, it is advisable to apply methods based on solving the IEs [55].

As long as EMW scattering from the resonance-sized metallic elements of UAV is concerned, we employ the method proposed in section 1.2, which is based upon electrodynamics computations presuming an evaluation of the electric current density excited on the surface of metallic scatterers by means of solving the MFIE.

In case of resonance-sized dielectric elements of UAV, we use instead the method proposed in section 1.3, which employs solutions to the MIE set for the surface densities of equivalent electric and magnetic currents excited on the surface of dielectric scatterers.

Additionally, the EMF scattered by a complex surface object such as a UAV drone needs to be computed while taking into account the EM interaction between its metallic and dielectric elements. The total EM response of the above UAV as a complex shape scatterer in the VHF to S wavebands can be computed as a coherent sum of fields scattered separately by its metallic and dielectric components. The latter coherent summation means that the fields in question are combined while taking into account their corresponding phase differences. Thus, using integral representations of fields scattered by metallic (1.81) and dielectric (1.98) objects, we can express the strength of the magnetic field scattered by the UAV as a whole, if that the primary field (radiated by the radar's transmitter) at the point of reception $\vec{Q}_{sc}$ is equal to zero, as follows:

$$\vec{p}^{\,sc} \cdot \vec{H}^{sc}(\vec{Q}_{sc}) = \vec{p}^{\,sc} \cdot \left(\vec{H}^{sc}_{met}(\vec{Q}_{sc}) + \vec{H}^{sc}_{diel}(\vec{Q}_{sc})\right), \tag{1.109}$$

where $\vec{H}^{sc}_{met}(\vec{Q}_{sc}) = \vec{H}_{met}(\vec{Q}_{sc}) - \vec{H}^{0}_{met}(\vec{Q}_{sc})$, $\vec{H}^{sc}_{diel}(\vec{Q}_{sc}) = \vec{H}_{diel}(\vec{Q}_{sc}) - \vec{H}^{0}_{diel}(\vec{Q}_{sc})$ are the strengths of the magnetic field scattered by the metallic and dielectric components of the UAV as measured at the reception point $\vec{Q}_{sc}$, respectively:

$$\vec{p}^{\,sc} \cdot \vec{H}^{sc}_{met}\left(\vec{Q}_{sc}\right) = -\frac{1}{j\omega}\int_{S_{met}} \vec{E}^{m}_{1}\left(\vec{Q} \mid \vec{Q}_{sc}, \vec{p}^{\,sc}\right) \cdot \vec{J}^{e}_{met}(\vec{Q})\, ds, \tag{1.110}$$

$$\vec{p}^{\,sc} \cdot \vec{H}^{sc}_{diel}\left(\vec{Q}_{sc}\right) = -\frac{1}{j\omega}\int_{S_{diel}} [\vec{H}^{m}_{21}(\vec{p}^{\,sc}) \cdot \vec{J}^{m}_{diel}(\vec{Q}) + \vec{E}^{m}_{21}(\vec{p}^{\,sc}) \cdot \vec{J}^{e}_{diel}(\vec{Q})]\, ds, \tag{1.111}$$

Here S_{met}, S_{diel} are the combined surfaces of the metallic and dielectric components, respectively, $\vec{J}^{e}_{met}$ is the electric current density excited on surfaces of metallic components of the UAV, and $\vec{J}^{e}_{diel}$, $\vec{J}^{m}_{diel}$ are the densities of the equivalent electric and magnetic currents excited on the surfaces of the dielectric components of the UAV.

One can also employ iterative methods in order to account for EM interaction of metallic and dielectric UAV components more accurately. In particular, one can see the works [75, 76] for more detail.

In general, one can account for the EM interaction between components of different nature by putting forward the corresponding set of IEs.

However, the employment of iterative methods in general, and those described in [75, 76] in particular, as well as the corresponding sets of equations that describe the current densities excited on the surfaces of real compound scatterers, involves a great deal of computer burden. The studies of EM interaction between individual components of air vehicles [22, 23, 55] have demonstrated that the method for predicting the EMF scattered by a compound UAV, which is based on solving the MFIE (1.70), the MIE set (1.91), and the integral representations (1.110) and (1.111), ensures a decent computation accuracy as long as the scattering characteristics of compound UAVs that include both metallic and dielectric components in the VHF to S frequency bands are concerned.

There may be cases when different structural components of an air vehicle belong to regions of different scattering nature (Rayleigh, resonant, and high frequency). In this case, a high enough accuracy of predicted scattering characteristics that would be obtained in a reasonable lapse of time can be ensured by a combined use of different computation methods. For instance, AHFM can be used to compute the EM responses of electrically large components of UAVs, while the method based on solving the IEs can be used to compute the EM responses of components, which, by their relative sizes, fall into the resonant scattering region. Subsequently, the EM interaction between the latter UAV's components would need to be accounted for in order to land on a final solution with regard to the field scattered by the UAV in question. In particular, this approach is described in the works [75, 76].

The most critical part of every computer model aimed at predicting the characteristics of EMW scattering by radar objects are digital three-dimensional (3D) models of the objects' surfaces. We addressed the latter problem by designing an original approach for modeling the surface of compound scatterers, this approach being particularly suitable for applying our original methods based on solutions to the IEs developed in this work. The basic idea behind our model of the compound scatterer's surface is to break it down into portions that could be presented as segments of three-axis ellipsoids. Such ellipsoids turned out to be a very good tool while modeling such UAV components as its power plant, control and payload modules, fuel tank, and propeller. Some parts of the compound scatterer surface might require other forms for their computer representation. In particular, in this section we also consider another form of scatterer's surface modeling, which employs segments of the so-called ruled surfaces. In the latter case, the surface is represented by its nodal points, neighboring nodes being connected by lines, thus creating the ruled surfaces. Subsequently, the mesh of surface current nodes is applied to such ruled surfaces.

The latter approach turns out to be particularly suitable for modeling twisted components such as propeller blades.

Having analyzed the composition of generic tactical UAVs, we have designed generalized digital models of basic UAV components for computing their scattering characteristics in the VHF, S, and X frequency bands. A model of the UAV surface [50, 55] is shown in figure 1.36. The external dielectric envelope of the UAV, which cannot be neglected in the X band, was modeled using segments of 36 three-axis ellipsoids. The dielectric envelope was assumed to be made of plastic with a dielectric constant of $\varepsilon_2 = 3.2 + j0.07$. The entire UAV envelope was assumed to be made of a single layer of plastic. However, different components of the external UAV envelope were characterized by different thicknesses of plastic. The front segment of the fuselage (the power plant's fairing), wings, and stabilizers were made of plastic that was 1 mm thick. The middle and aft sections of the fuselage that housed the fuel tank, control module, and payload module were made of plastic that was 2 mm thick.

An exploded-view drawing of the basic UAV components is shown in figure 1.37. Figures 1.38 and 1.39 show detailed geometrical features of the same basic components, all the sizes being indicated in millimeters. The fuel tank was modeled as a solid dielectric object featuring the dielectric constant of fuel $\varepsilon_2 = 2.3 + j0.03$.

The propeller was modeled as solid dielectric scatterer that was characterized by two different values of dielectric constant: $\varepsilon_2 = 2 + j0.03$ and $\varepsilon_2 = 4 + j0.03$.

It is worth noting that the characteristic dimensions of the UAV structural components were commensurate with the wavelength of the radar signal, which ranged from 10 to 30 cm.

Below we illustrate the capabilities of our computation methods, which are based on solving the boundary IEs. We concentrate on the computed RCS values for the case of a tactical UAV computer model and its individual components given various conditions of their irradiation by radar.

Figure 1.36. Model of the tactical UAV surface.

Figure 1.37. Models of the structural components of the tactical UAV.

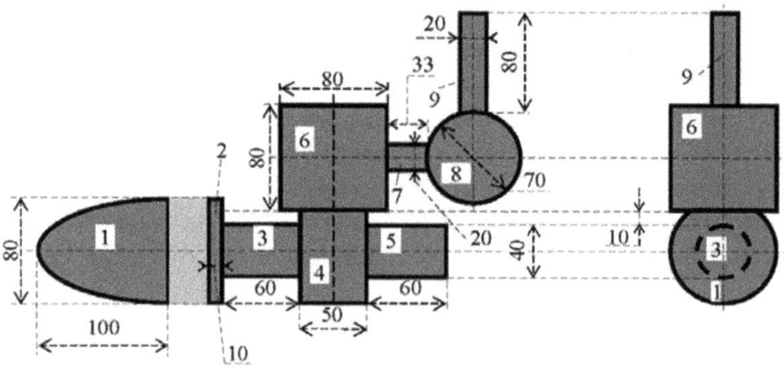

(a) (b)

Figure 1.38. PEC model of the UAV power plant: the side view (a) and the front view (b). The model comprises the following elements: 1—a part of the two-axial ellipsoid; 2—a disk; 3–5 and 7–9—cylinders; 6—cube.

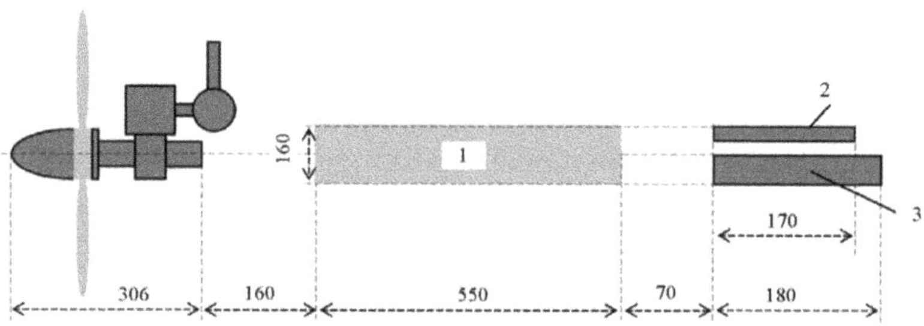

Figure 1.39. The geometry of the structural components of the UAV: 1—fuel tank (solid dielectric parallelepiped with a dielectric constant $\varepsilon_2 = 2.3 + j0.03$, its width and height being equal to 160 mm); 2—control module (PEC parallelepiped, its width being equal to 150 mm and its height being equal to 30 mm); 3—payload module (PEC parallelepiped, its width being equal to 150 mm and its height being equal to 90 mm).

The RCS computations were carried out for the following frequencies of radar signal: $f = 180$ MHz (wavelength $\lambda_1 = 1.67$ m falling into the VHF frequency band), $f = 3$ GHz ($\lambda_1 = 10$ cm, S band). For reference, subsection 1.4.1 also presents the results of the UAV RCS computation using the AHFM [50] for the frequency of illumination signal $f = 10$ GHz ($\lambda_1 = 3$ cm, X band).

As an example, we employ our original numerical methods for computing the RCS σ of a tactical UAV for a monostatic radar layout versus the following parameters:

- UAV illumination aspect angles with respect to radar both in the azimuth β (horizontal plane) and elevation ε_{uav} (vertical plane), illumination aspects in the azimuth and elevation $\beta = 0$, $\varepsilon_{\text{uav}} = 0$ corresponding to the nose-on illumination of the UAV along its longitudinal axis, and the azimuth aspect of $\beta = 90$ degrees corresponding to its side-on illumination by radar;
- illumination signal frequency f (or wavelength of the EMW λ_1);

- polarization of the illumination wave (horizontal polarization corresponding to radar illumination by the wave that has its electric field strength vector $\vec{E}$, which belongs to the horizontal (azimuth angle) plane; and vertical polarization corresponding to the vector $\vec{E}$, which is perpendicular to the direction of incident wave propagation and belongs, at the same time, to the vertical (elevation angle) plane).

All the RCS values are presented in meters squared. The increment of azimuth angle β for all the UAV RCS figures was kept equal to 0.1°. Every particular RCS corresponding to a particular azimuth aspect was computed under the condition that the value of the other aspect angle, namely, the elevation angle aspect, was set at random. These random elevation aspects were set to have a uniform distribution on the interval $\varepsilon_{\text{uav}} = -3° \pm 4°$ (negative elevation angles ε_{uav} corresponding to the UAV model illumination from below). In this manner, we simulated the effect of the UAV pitching in flight, which changes its attitude toward radar in an unpredictable manner.

The RCS figures of the UAV, the model of which is shown in figures 1.36–1.39, were obtained for illumination signal frequencies falling into the VHF and S bands using our original method based on solving boundary IEs, which was presented in sections 1.2 and 1.3 [55]. In addition, for the UAV scattering in the VHF and S bands, the thin outer shell made of dielectric, which represents the fuselage, wings, and horizontal and vertical stabilizers, is radio transparent, contributing negligibly to the total UAV's RCS. The latter means that this outer shell (envelope) was disregarded in our computations. At this stage, the electric wires inside the UAV were also excluded from consideration. Thus, given the above frequency bands almost all the EMW scattering by UAV is limited to contributions from its propulsion system (power plant + propeller), control module, payload module, and fuel tank.

In the case of the UAV EMW scattering in the X band, we employed the AHFM that was described in great detail in [50]. Given that particular waveband, we computed the field scattered by the UAV as the coherent sum of field components scattered by all its components that additionally included the EM contribution of the outer envelope, comprised of the fuselage, wings, and vertical and horizontal stabilizers.

The RCS of the propeller, considering its electric dimensions, was computed by means of solving the MIE set described in section 1.3.

The RCS of the tactical UAV model versus its azimuth aspect is shown in figures 1.40–1.42. These diagrams present the RCS figures pertaining to that particular model of tactical UAV given its irradiation in three different wavebands. As anticipated, an increase in carrier frequency of the radar signal introduces more irregularities to the dependence of the RCS on the UAV's azimuth aspect. In addition, the differences between the plots obtained given the UAV's model radar probing at two orthogonal polarizations also fade away.

Below, tables 1.2–1.4 present some statistical characteristics of the particular UAV model in the form of mean RCS (σ_{mean}) and median RCS (σ_{med}). The statistical estimates of tables 1.2–1.4 have been obtained by averaging the instantaneous RCS values on the following sectors of the azimuth aspect angles: $\beta = 0°–360°$; $\beta = 0° \pm 45°$

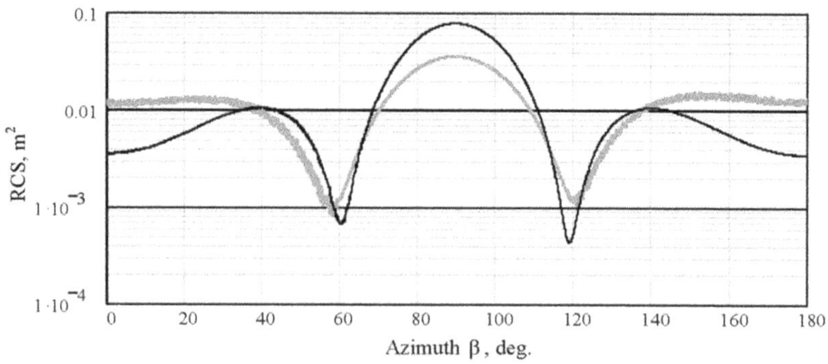

Figure 1.40. The RCS of the tactical UAV model versus its azimuth aspect irradiated at the frequency $f = 180$ MHz, for the horizontal (thin black line) and vertical (bold gray line) polarization of the radar probing signal.

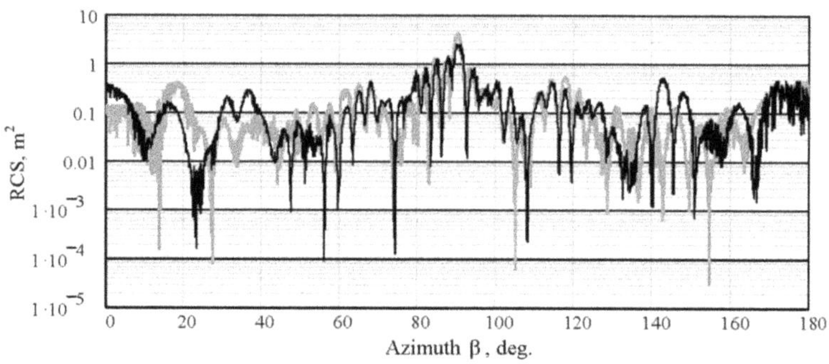

Figure 1.41. The RCS of the tactical UAV model versus its azimuth aspect irradiated at the frequency $f = 3$ GHz, for the horizontal (thin black line) and vertical (bold gray line) polarization of the radar probing signal.

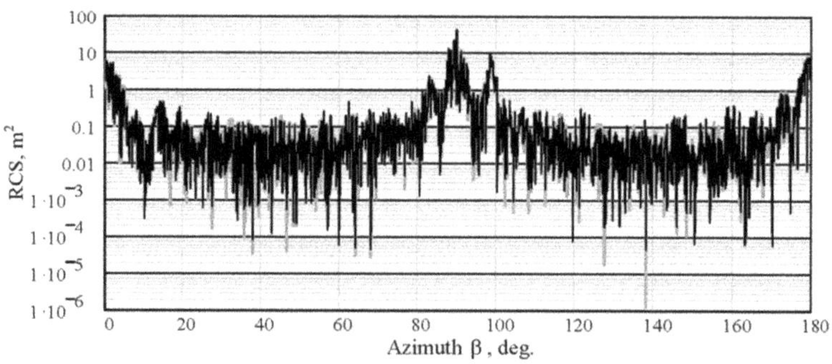

Figure 1.42. The RCS of the tactical UAV model versus its azimuth aspect irradiated at the frequency $f = 10$ GHz, for the horizontal (thin black line) and vertical (bold gray line) polarization of the radar probing signal.

(nose-on radar probing); $\beta = 45°$–$135°$ and $\beta = 225°$–$315°$ (side-on radar probing); and $\beta = 135°$–$225°$ (tail-on radar probing).

From the RCS diagrams in figures 1.40–1.42 and the values in tables 1.2–1.4, one can see that the radar echoes from the UAV in question that would be picked up by

Table 1.2. Mean and median RCSs of the tactical UAV model for its probing in different sectors of the azimuth aspects in a frequency from the VHF waveband (f = 180 MHz).

Sector of azimuth aspect	RCS, m^2			
	Horizontal polarization		Vertical polarization	
	σ_{mean}	σ_{med}	σ_{mean}	σ_{med}
Circular sector of aspects, $\beta = 0°–360°$	0.0158	0.0074	0.0126	0.0120
Nose-on aspects, $\beta = 0° \pm 45°$	0.0067	0.0064	0.0114	0.0118
Side-on aspects, $\beta = 45°–135°$, $\beta = 225°–315°$	0.0250	0.0091	0.0131	0.0071
Tail-on aspects, $\beta = 135°–225°$	0.0066	0.0061	0.0129	0.0130

Table 1.3. Mean and median RCSs of the tactical UAV model for its probing in different sectors of the azimuth aspects in a frequency from the S waveband (f = 3 GHz).

Sector of azimuth aspect	RCS, m^2			
	Horizontal polarization		Vertical polarization	
	σ_{mean}	σ_{med}	σ_{mean}	σ_{med}
Circular sector of aspects, $\beta = 0°–360°$	0.1666	0.0915	0.1708	0.0725
Nose-on aspects, $\beta = 0° \pm 45°$	0.0987	0.0762	0.0859	0.0495
Side-on aspects, $\beta = 45°–135°$, $\beta = 225°–315°$	0.2198	0.1116	0.2421	0.0999
Tail-on aspects, $\beta = 135°–225°$	0.1282	0.0808	0.1129	0.0555

Table 1.4. Mean and median RCSs of the tactical UAV model for its probing in different sectors of the azimuth aspects in a frequency from the X waveband (f = 10 GHz).

Sector of azimuth aspect	RCS, m^2			
	Horizontal polarization		Vertical polarization	
	σ_{mean}	σ_{med}	σ_{mean}	σ_{med}
Circular sector of aspects, $\beta = 0°–360°$	0.3965	0.0390	0.3928	0.0379
Nose-on aspects, $\beta = 0° \pm 45°$	0.2377	0.0341	0.2334	0.0355
Side-on aspects, $\beta = 45°–135°$, $\beta = 225°–315°$	0.5143	0.0459	0.5005	0.0450
Tail-on aspects, $\beta = 135°–225°$	0.3197	0.0270	0.3365	0.0276

the radar receiver in the S and X wavebands is about ten time as strong as the corresponding echoes in the VHF waveband.

The intensities of radar scattering by the UAV for its radar probing in the S and X wavebands are on a par with each other. However, the lower degree of irregularity of the UAV's circular RCS diagram in the S waveband implies that the UAV could be locked on its track in a more stable manner as compared to its tracking by radar operating in the X waveband.

A closer look at the UAV's sector of nose-on aspects ($\beta = 0° \pm 45°$), which is of high practical importance, reveals a couple of remarkable facts: (i) the mean values of the RCS that can be attained for the UAV in question are greater in the X rather than the S waveband; and (ii) in contrast, its median RCS values are greater in the S rather than the X waveband. Such a relation between the mean and median RCSs given the nose-on probing of the UAV in question can be attributed to a higher degree of irregularity of its RCS diagram in the X waveband.

Moreover, comparing the above figures and tables to the data presented in [15], one can conclude that the mean RCS figures of a tactical UAV given its radar probing at the nose-on aspect angles in the S waveband are about two orders of magnitude lower than those of a fighter aircraft. In the X waveband, the latter difference is even greater, putting the mean RCS of the tactical UAV to the level of 800–1400 times below that of a fighter aircraft.

Below, we consider the RCS figures of the individual UAV components.

Figures 1.43 and 1.44 present the RCS of the dominant metallic and dielectric scatterers comprising the UAV versus the azimuth aspect angle of its irradiation by radar. The RCS of the metallic UAV components includes its power plant, control unit, and payload module, their RCSs being shown by the solid black line. The RCS of the dielectric fuel tank of the UAV is shown by the dashed gray line. The term dominant scatterers implies that these metallic and dielectric components of the UAV provide the highest contribution in terms of scattering. The UAV RCS figures are presented for its probing by radar at two orthogonal polarizations. The plots of figure 1.43 were obtained for the radar probing at the frequency of $f = 180$ MHz, whereas those in figure 1.44 were obtained for the radar probing at a frequency of $f = 3$ GHz. Separate RCS plots pertaining to the metallic and dielectric UAV components give us the idea of their relative importance in terms of radar scattering.

Let us discuss in brief the RCS diagrams of the above plots.

In the case of the UAV radar probing in the VHF waveband ($\lambda_1 = 1.67$ m), all the above components are electrically small scatterers. The RCS of dielectric fuel tank is about ten time as low as the RCS of all the dominant metallic components.

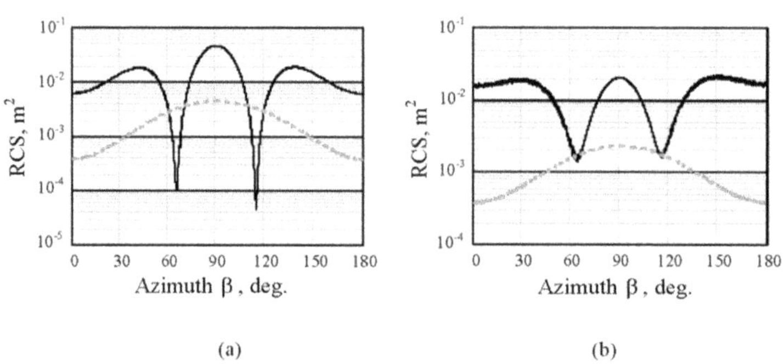

Figure 1.43. The RCS of the UAV's dominant components versus the UAV's azimuth aspect for its probing at the frequency $f = 180$ MHz, (a) corresponding to the horizontal polarization case and (b) corresponding to the vertical polarization case.

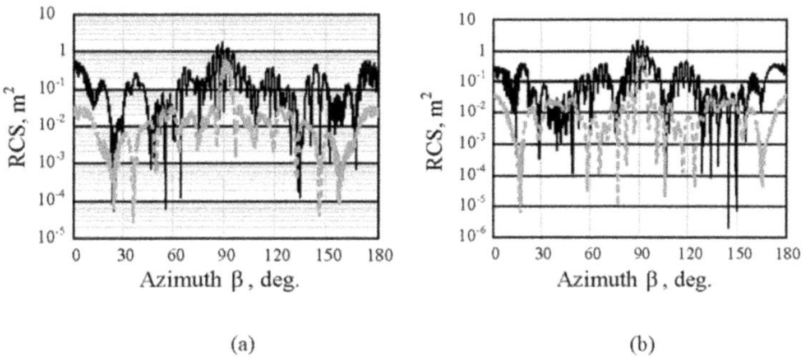

Figure 1.44. The RCS of the UAV's dominant components versus the UAV's azimuth aspect for its probing at the frequency $f = 3$ GHz, (a) corresponding to the horizontal polarization case and (b) corresponding to the vertical polarization case.

The RCS diagram of metallic components is characterized by three maxima. The latter means that their scattering behavior is similar to that of a 'dipole'.

The shape of the dielectric fuel tank (parallelepiped) produces its SD where the maximum scattering is attained for radar probing in the direction perpendicular to its wall.

In the case of the UAV radar probing in the S waveband ($\lambda_1 = 10$ cm), all the above components of the UAV are resonance-sized scatterers. The RCS figures of the metallic components and of dielectric fuel tank are commensurate with each other. In this particular situation, both the metallic and dielectric components provide a similar scattering contribution to the UAV. All the RCS diagrams of figure 1.44 tend to have their maxima for an azimuth aspect of the UAV irradiation equal to $\beta = 90°$. The RCS of the metallic components for the azimuth aspects close to the straight nose-on irradiation of the UAV are a bit lower than that for its side-on irradiation. In the case of the dielectric fuel tank, the difference between its local RCS maxima given its nose-on and side-on radar probing is significantly greater.

In order to study the nature of radar scattering by the UAV's propeller in the S and X wavebands, we adopted two different models. Both models had the same propeller diameter; they had two blades of the same characteristic dimensions. At the same time, the geometry of the blades was different. The geometry of the first propeller model was much simpler than that of the second propeller model. The blades of propeller model #2, unlike those of propeller #1, were twisted along the line connecting the ends of the two blades (the blade's generatrix) and were thinner electrically. Comparing the scattering characteristics of the two kinds of propeller, one can study the impact of propeller shape on the RCS for various aspect angles of their irradiation by radar.

The model of propeller #1 is shown in figure 1.45. It comprises a central hub shaped as cylinder with a foundation diameter equal to $d_1 = 4.66$ cm and with a height equal to $d_2 = 2$ cm (figure 1.45(a)). The surface of the blade is generated by segments of four ellipsoids, their leading faces (with respect to the UAV's flight direction) being convex and their trailing faces being flat. The maximal thickness of the blade is equal to $d_3 = 2$ cm, and its maximal width is equal to $d_4 = 4.6$ cm (figure 1.45(b)). Each blade has a rotational pitch angle equal to 10°. The diameter

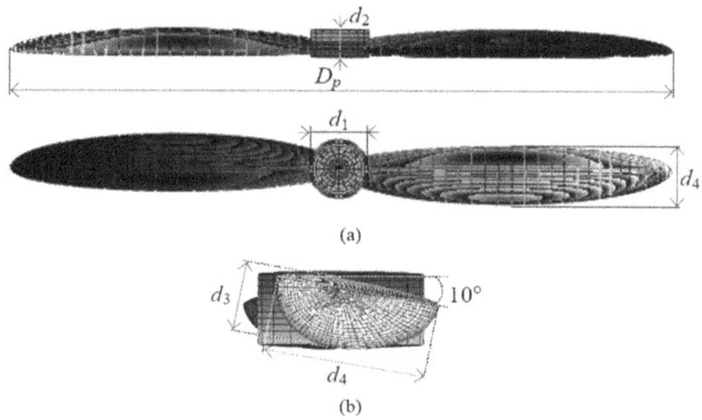

Figure 1.45. Propeller model #1.

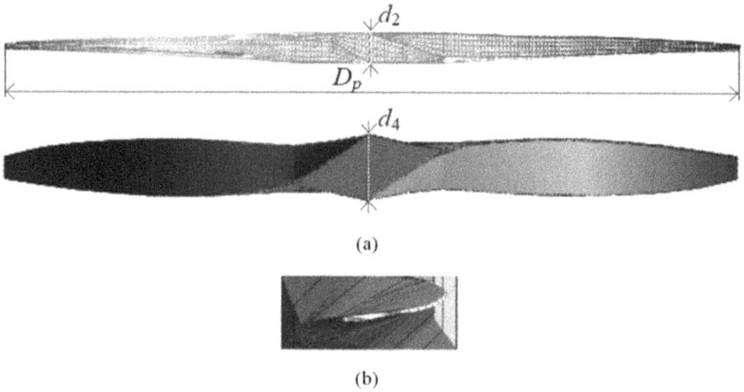

Figure 1.46. Propeller model #2.

of the propeller is equal to D_p = 52 cm. The dielectric constant of the material that the propeller was made of was assumed to be equal to $\varepsilon_2 = 2 + j0.03$.

The model of propeller #2 is shown in figure 1.46. The surface of each blade is approximated using a ruled surfaces. The maximal thickness of each blade is equal to d_2 = 2 cm, and its maximal width is equal to d_4 = 4.6 cm. The diameter of the propeller is equal to D_p = 52 cm (figure 1.46(a)). Each blade has a twist (figure 1.46(b)). The RCS of the propeller #2 model is evaluated for two different values of its material dielectric constant: $\varepsilon_2 = 2 + j0.03$ and $\varepsilon_2 = 4 + j0.03$.

In addition, to account for the propeller rotation, we repeated the process of the RCS, σ, computation for various angles of its rotation, as shown in figure 1.47. Angle α, which defines the current orientation of the propeller blades in the plane of their rotation, was assumed to have running discrete values in the interval from 0° through 180° with an increment of 0.1°. In addition, the angle $\alpha = 0$ corresponds to the propeller standing vertically. The angle θ, which defined the attitude (the tilt) of the propeller rotation plane with respect to the direction of the incident EMW and which was read off the axis of propeller rotation, was simulated to be taking values of either 0° or 30° (figure 1.47).

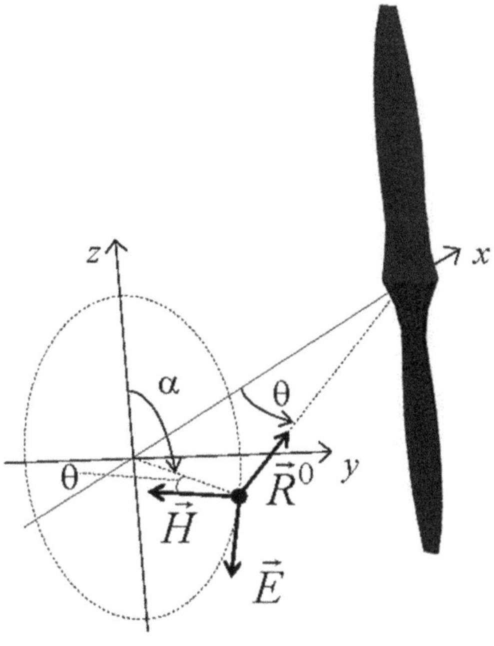

Figure 1.47. The orientation of the rotating propeller with respect to the field point, and with respect to the electric and magnetic strength vectors.

The RCSs of both propeller models have been evaluated for two wavelengths of the radar probing signal: $\lambda_1 = 10$ cm (carrier frequency $f = 3$ GHz) and $\lambda_1 = 3$ cm ($f = 10$ GHz). The characteristic dimensions of both propellers were commensurate with the wavelength of the incident EMW, i.e. they both represented resonance-sized scatterers. Therefore, we used an electrodynamics computation method based on solving the MIE set, as described in section 1.3, to evaluate their RCS figures.

The results of the propeller model RCS computations are presented below for the case of a monostatic radar layout and for radar probing at two orthogonal polarizations defined as follows:

- Horizontal polarization means that vector $\vec{E}$ belongs to the plane of propeller rotation and makes the angle of 90° with the axis Oz (and propeller itself) for an angle of propeller rotation of $\alpha = 0$, and makes an angle of 0° for an angle of propeller rotation of $\alpha = 90°$;
- Vertical polarization means that vector $\vec{E}$ belongs to the plane created by the propeller rotation axis and the EMW incidence direction $\vec{R}^0$, vector $\vec{E}$ being, at the same time, perpendicular to the vector $\vec{R}^0$.

Figures 1.48–1.50 present the RCSs of propeller #1. Figure 1.48 shows that a particular RCS versus the propeller rotation angle α for an angle of the rotation plane tilt equal to $\theta = 0$ for the case of its irradiation by radar at the frequency $f = 3$ GHz ($\lambda_1 = 10$ cm). The mean RCS value in the latter case is $\sigma_{\text{mean}} = 0.052$ m^2. The lateral sizes (widths) of blades are smaller than the incident signal wavelength. Therefore, the

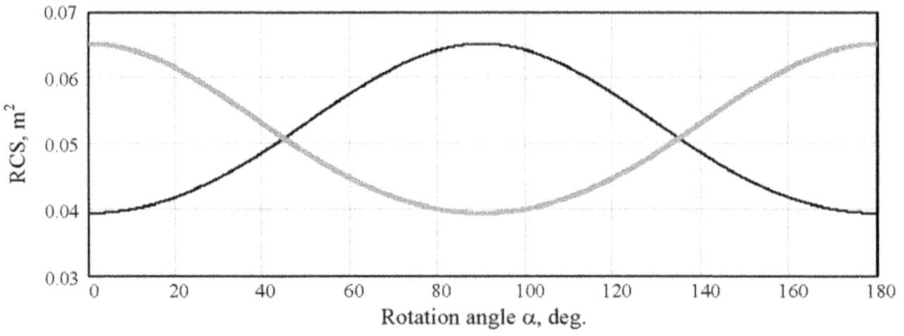

Figure 1.48. The RCS of propeller #1 for radar probing along its axis of rotation at $f = 3$ GHz, the thin black line corresponding to horizontal polarization and the bold gray line corresponding to the vertical polarization of the incident EMW.

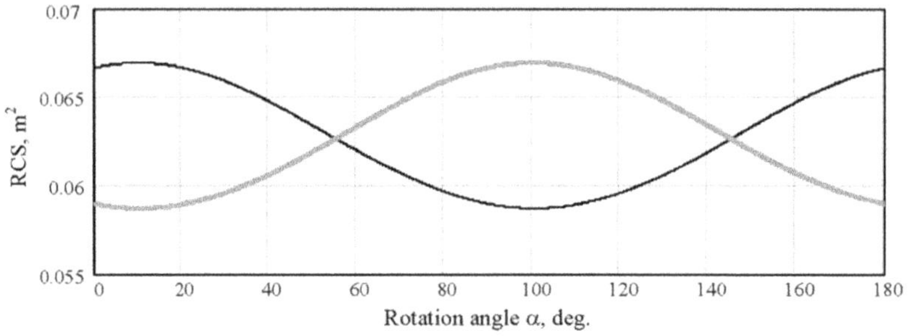

Figure 1.49. The RCS of propeller #1 versus its angle of rotation α for radar probing along its axis of rotation at frequency $f = 10$ GHz, the thin black line corresponding to horizontal polarization and the bold gray line corresponding to vertical polarization of the incident wave.

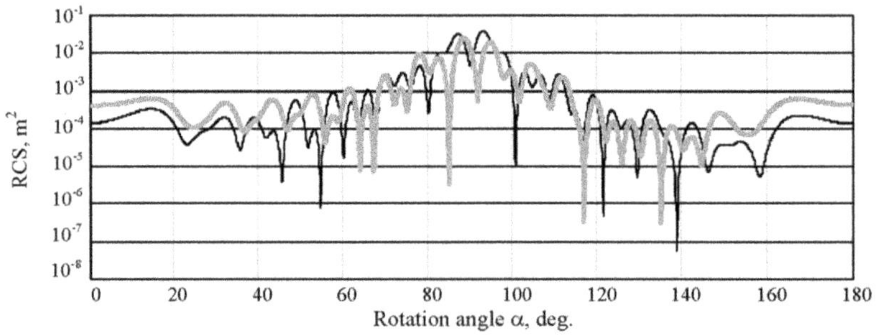

Figure 1.50. The RCS of propeller #1 versus its angle of rotation α given its radar probing at the angle $\theta = 30°$ with respect to the propeller's axis of rotation at a frequency $f = 10$ GHz, the thin black line corresponding to horizontal polarization and the bold gray line corresponding to vertical polarization of the incident wave.

maximum of the propeller RCS is observed when the electric strength vector lines up (becomes co-linear) with the line that connects the ends of two its blades.

Figure 1.49 presents similar dependences for the case of propeller #1 irradiation by radar at a frequency $f = 10$ GHz ($\lambda_1 = 3$ cm). The angle of rotation plane tilt was assumed to be equal to $\theta = 0$ in figure 1.49, and it was assumed to be equal to $\theta = 30°$

in figure 1.50. The mean RCS value of propeller #1 for a wavelength of incident wave equal to $\lambda_1 = 3$ cm and for a rotation plane tilt angle equal to $\theta = 0$ is $\sigma_{mean} = 0.063$ m^2. In this particular case, the width of the propeller #1 blades exceeds the wavelength of the incident EMW and the maximum of its RCS becomes shifted by an angle of 10° with respect to the case where the electric strength vector is co-linear with the line connecting the ends of the two blades.

When the plane of propeller rotation becomes tilted by an angle $\theta = 30°$ with respect to the direction of incidence (figure 1.50), the dependence of the RCS versus the propeller rotation angle α attains a more complicated structure. The total RCS figure becomes significantly reduced compared to the case of propeller irradiation along the axis of propeller rotation. Now, the mean RCS values constitute the value $\sigma_{mean} = 2.053 \times 10^{-3}$ m^2 given the propeller's radar probing at horizontal polarization, and the value $\sigma_{mean} = 1.479 \times 10^{-3}$ m^2 given its radar probing at vertical polarization.

Next, the figures 1.51 and 1.52 present the RCS of propeller #2 versus its rotation angle α given that it was made of a material with dielectric constant equal to $\varepsilon_2 = 2 + j0.03$; whereas figures 1.53 and 1.54 correspond to a propeller made of a material with a dielectric constant equal to $\varepsilon_2 = 4 + j0.03$. The RCS plots were computed such that the tilt angle of the propeller's rotation plane θ was assumed to be equal to 0°, 5°, 10°, 20°, and 30° (figure 1.51 presents an additional RCS curve obtained for the tilt angle $\theta = 2°$). For reference, figure 1.55 plots the RCS figures of propeller #1 (dashed lines) and propeller #2 (solid lines) for their radar probing along the axis of propeller rotation.

It also can be seen that when propeller #2 is irradiated by radar at the frequency $f = 3$ GHz ($\lambda_1 = 10$ cm), then the lateral width of the blade is less than the wavelength of the incident EMW. Therefore, given the EMW incidence along the propeller's axis of rotation ($\theta = 0$, figures 1.51 and 1.53) the scattering behavior of

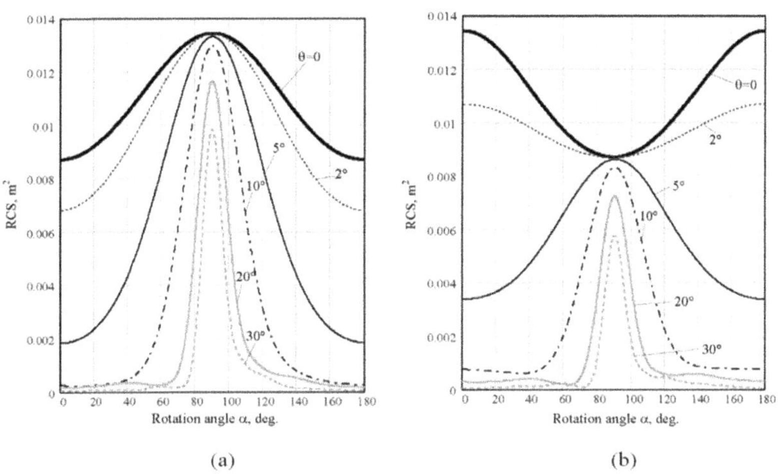

Figure 1.51. The RCS of propeller #2 ($\varepsilon_2 = 2 + j0.03$) probed by radar at the frequency $f = 3$ GHz for horizontal (a) and vertical (b) polarization of the incident wave.

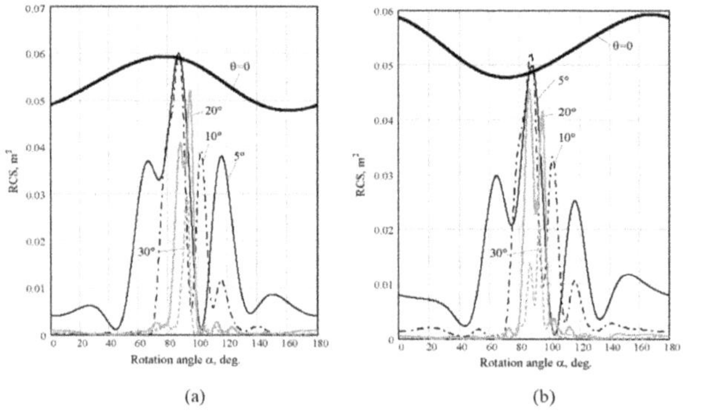

Figure 1.52. The RCS of propeller #2 ($\varepsilon_2 = 2 + j0.03$) probed by radar at the frequency $f = 10$ GHz for horizontal (a) and vertical (b) polarization of the incident wave.

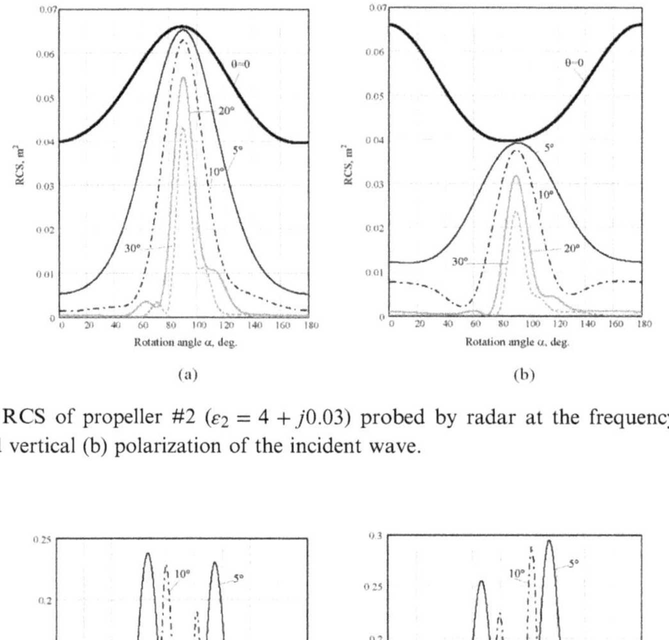

Figure 1.53. The RCS of propeller #2 ($\varepsilon_2 = 4 + j0.03$) probed by radar at the frequency $f = 3$ GHz for horizontal (a) and vertical (b) polarization of the incident wave.

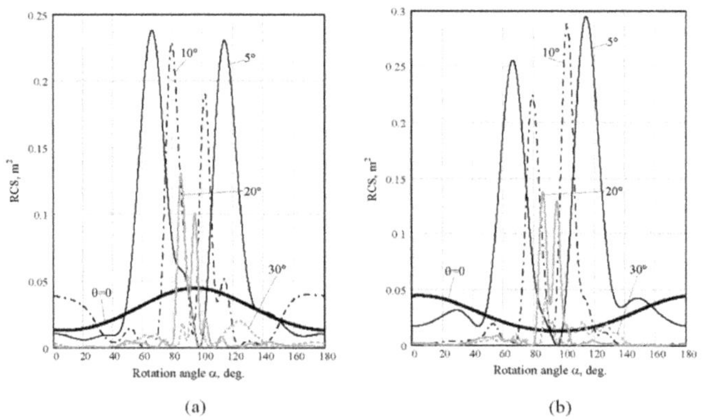

Figure 1.54. The RCS of propeller #2 ($\varepsilon_2 = 4 + j0.03$) probed by radar at the frequency $f = 10$ GHz for horizontal (a) and vertical (b) polarization of the incident wave.

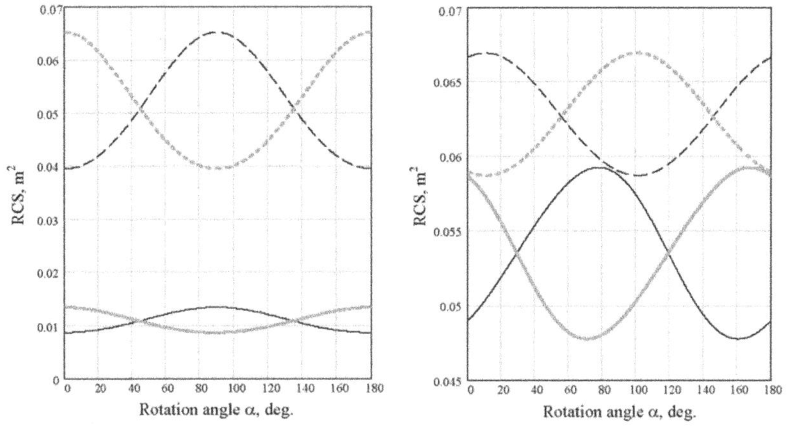

Figure 1.55. The RCS of propeller #1 (dashed lines) and propeller #2 (solid lines) for the case of their material's dielectric constant for their radar probing along the axes of rotation at the frequencies $f = 3$ GHz (a) and $f = 10$ GHz (b). The thin black lines correspond to horizontal polarization and the bold gray to vertical polarization.

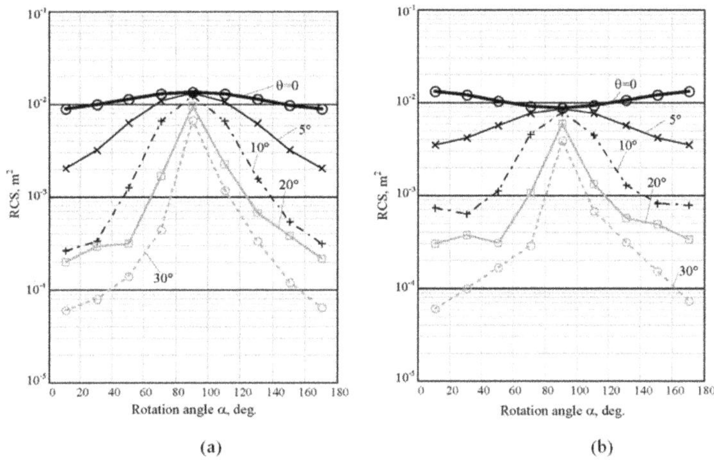

Figure 1.56. The RCS of propeller #2 ($\varepsilon_2 = 2 + j0.03$) probed by radar at the frequency $f = 3$ GHz that was obtained by averaging instantaneous RCSs over the 20° sectors of propeller rotation angle α for horizontal (a) and vertical (b) polarization of the incident wave.

the propeller is similar to that of the dipole, and its RCS attains its maximum value when the electric strength vector lines up (becomes co-linear) with the line connecting the ends of the two blades (the blade's generatrix).

Figures 1.56–1.59 present the RCS figures of propeller #2 that were obtained by averaging its instantaneous RCSs over sectors of rotation angles α that were 20° wide. The RCS plots of figures 1.56–1.59 were computed given that the tilt angle of the propeller's rotation plane θ was assumed to be equal to 0°, 5°, 10°, 20°, and 30°. Tables 1.5 and 1.6 present the RCS figures of propeller #1 and #2, respectively, that were obtained by averaging its instantaneous RCSs over the sector of rotation angles

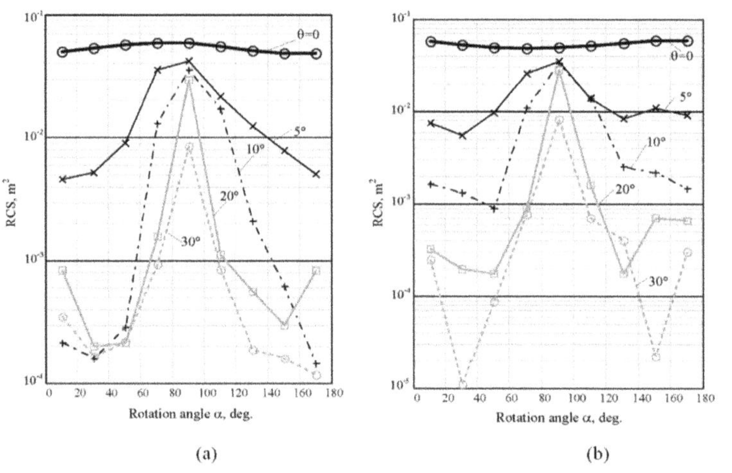

Figure 1.57. The RCS of propeller #2 ($\varepsilon_2 = 2 + j0.03$) probed by radar at the frequency $f = 10$ GHz that was obtained by averaging instantaneous RCSs over the 20° sectors of propeller rotation angle α for horizontal (a) and vertical (b) polarization of the incident wave.

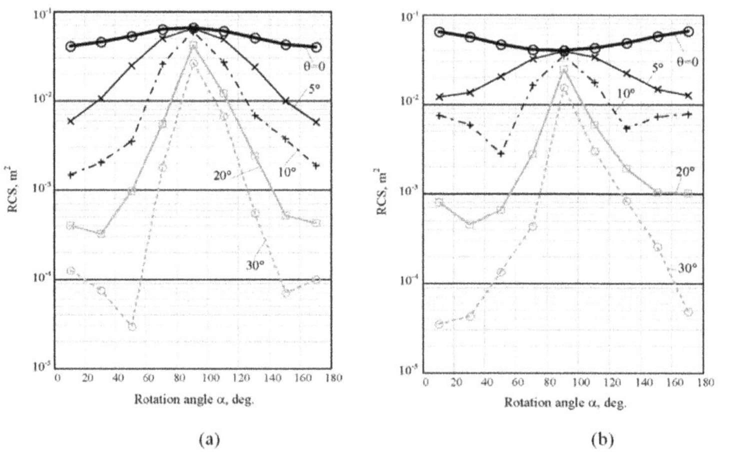

Figure 1.58. The RCS of propeller #2 ($\varepsilon_2 = 4 + j0.03$) probed by radar at the frequency $f = 3$ GHz that was obtained by averaging instantaneous RCSs over the 20° sectors of propeller rotation angle α for horizontal (a) and vertical (b) polarization of the incident wave.

α from 0° to 180°. For reference, table 1.7 presents the total RCS figures of the tactical UAV in question, which were obtained by averaging its instantaneous RCSs over the sector of ±45° of nose-on azimuth aspects.

When the propeller is probed by a radar at the frequency $f = 10$ GHz ($\lambda_1 = 3$ cm) then the lateral size of its blade exceeds the wavelength of the incident wave, which means that propeller is electrically large in two dimensions (the third dimension, which is the blade's thickness does not count). Thus, for propeller irradiation along its rotation axis, one obtains its scattering maxima that appear when the vector of

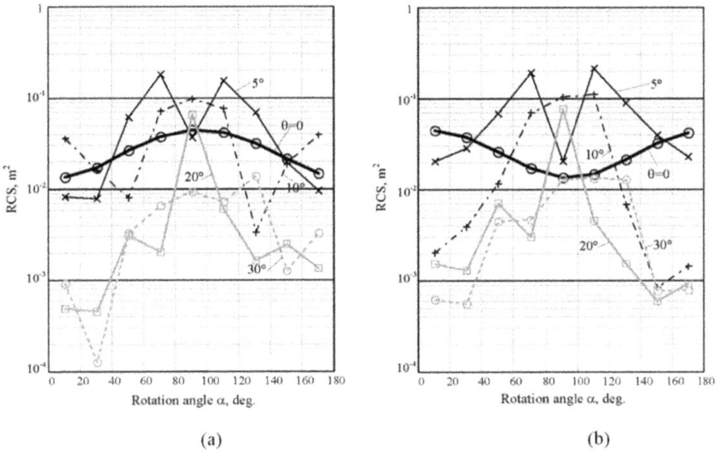

Figure 1.59. The RCS of propeller #2 ($\varepsilon_2 = 4 + j0.03$) probed by radar at the frequency $f = 10$ GHz that was obtained by averaging instantaneous RCSs over the 20° sectors of propeller rotation angle α for horizontal (a) and vertical (b) polarization of the incident wave.

Table 1.5. The RCS of propeller #1 obtained by averaging instantaneous RCSs over the sector of propeller rotation angles $\alpha = [0°, 180°]$.

Dielectric constant, ε_2	Wavelength, cm	Polarization	Averaged RCS, σ_{mean}, m^2	
			Tilt angle θ, degree	
			0	30
2+j0.03	10	Horizontal	0.052	—
		Vertical		—
	3	Horizontal	0.063	2.05×10^{-3}
		Vertical		1.48×10^{-3}

electric strength $\vec{E}$ makes an angle of about 12°–18° with the propeller's generatrix (figures 1.52, 1.54, and 1.55).

Plots of figure 1.55(a) show that for their radar probing at the wavelength $\lambda_1 = 10$ cm (the lateral size of the blades is less than the wavelength) the RCS of both propeller models attains its maximum when the vector $\vec{E}$ becomes co-linear with the propeller generatrix. At the same time, given their radar probing at the wavelength $\lambda_1 = 3$ cm (figure 1.55(b), the lateral size of the blades is commensurate with the wavelength) the RCS maximum shifts in the direction of propeller rotation with respect to the generatrix. It is evident from the plots of figure 1.55 and the data of tables 1.5 and 1.6 that for propeller #1 radar probing at the wavelength $\lambda_1 = 10$ cm its mean RCS is 4.7 times as high as the mean RCS of propeller #2. The latter could be explained by the blades of propeller #1 having a greater area compared to those of propeller #2. When both propellers are probed by radar at the wavelength $\lambda_1 = 3$ cm the difference between their means RCSs is not as large, the mean RCS of propeller #1 being just 1.17 times the mean RCS of propeller #2.

Table 1.6. The RCS of propeller #2 obtained by averaging instantaneous RCSs over the sector of propeller rotation angles $\alpha = [0°, 180°]$.

Dielectric constant, ε_2	Wavelength, cm	Polarization	Averaged RCS σ_{mean}, m^2				
			Tilt angle θ, degree				
			0	5	10	20	30
2+ j0.03	10	Horizontal	0.011	6.36×10^{-3}	3.29×10^{-3}	1.72×10^{-3}	1.01×10^{-3}
		Vertical		5.57×10^{-3}	2.46×10^{-3}	1.18×10^{-3}	6.37×10^{-4}
	3	Horizontal	0.054	0.016	7.67×10^{-3}	3.87×10^{-3}	1.27×10^{-3}
		Vertical		0.014	7.50×10^{-3}	3.67×10^{-3}	1.18×10^{-3}
4+j0.03	10	Horizontal	0.051	0.027	0.015	7.17×10^{-3}	3.98×10^{-3}
		Vertical		0.025	0.012	4.36×10^{-3}	2.22×10^{-3}
	3	Horizontal	0.027	0.061	0.040	9.31×10^{-3}	5.05×10^{-3}
		Vertical		0.076	0.034	0.011	5.62×10^{-3}

Table 1.7. Total RCS of UAV obtained by averaging instantaneous RCSs over the sector of ±45° of nose-on azimuth aspect angles.

Wavelength, cm	Averaged RCS σ_{mean}, m^2	
	Horizontal polarization	Vertical polarization
10	0.099	0.086
3	0.238	0.233

As soon as the field point shifts away from the propeller rotation axis by the tilt angle θ, then the propeller's RCS decreases quickly. The only exception from that rule comes with the RCS of propeller #2 that is made of a dielectric material with a dielectric constant $\varepsilon_2 = 4 + j0.03$ for its radar probing at the wavelength $\lambda_1 = 3$ cm (figure 1.54). In the latter case, the RCS of propeller #2 dependence versus its angle of rotation α is characterized by sharp peaks, at which the RCS values exceed the corresponding levels attained when that particular propeller is irradiated along its axis of rotation.

It is also worth mentioning that the mean RCS of the propeller models considered above in the VHF waveband is about only 10^{-7} m^2, which can be neglected considering the RCS figures of the other UAV components in that particular waveband.

Finally, analyzing the plots of figures 1.48–1.59 and the data presented in tables 1.5–1.7 we can draw the following conclusions:

1. The main factors affecting the intensity of radar scattering from propellers include the surface area of their blades, the propeller's orientation relative to the vector of electric field strength, and the dielectric constant of the material the propeller is made of. Different shapes and sizes of the propeller blades could also affect the RCS of the propeller in particular and the UAV as a whole.

2. As a rule, when the field point shifts away from the axis of propeller rotation (or from the straight nose-on aspect of the UAV irradiation by radar) the RCS of the propeller decreases pretty fast.
3. However, as far as the UAV scattering in the S and X wavebands is concerned, then our data regarding the propeller RCS figures and in particular their comparison with the RCS figures of the UAV as a whole (table 1.7) clearly indicate that the radar returns from the propeller can make up a significant portion of the UAV radar echoes. In addition, given the complex nature of EM interaction between the rotating propeller and the other UAV scattering components, one is compelled to take the scattering by propellers into consideration so that the spectra of propeller modulation would be also available to radar engineers for study.

The parameters of these spectra can also be used to improve the quality of the radar detection, tracking, and recognition of tactical UAVs against a background of clutter created by strong ground returns.

1.4.2 The high-resolution radar range profiles of artillery shells

The basic feature of counter-battery radar operation is the reception and processing of signals scattered by artillery shells, which are low-altitude objects of relatively small electrical size, against a background of powerful returns from the underlying surface. Notwithstanding such a challenging environment, such radars are required to provide high measurement accuracy of the three coordinates of shells while being capable of non-cooperatively identifying the type of artillery system that fired every particular shell. Intensive studies aimed at increasing the effectiveness of counter-battery radars and improving the methods of their use are currently being conducted in countries that manufacture such radars [53, 77, 78]. One of the ways to improve the shell range measurement accuracy and to improve the selection of echoes from artillery shells from the cluttered returns created by the underlying surface is to employ WB or UWB probing signals. In addition, the application of WB (or UWB) signals provides for non-cooperative identification of the shell type by their high-resolution radar range profiles (HRRPs). This fact is quite important, since using only the shell's trajectory parameters one cannot always ensure the high quality of their identification [53]. Algorithms of air target non-cooperative identification are described in [46]. For further development and improvement of these methods as applied to artillery shells of different types, it is important to have information about the radar scattering characteristics of the objects in question. Considering a wide variety of conditions related to the radar probing of these objects, necessary information can be obtained using computer modeling.

In subsection 1.4.2 we demonstrate the WB (UWB) responses of a 122 mm rocket of MLRS and a 152 mm artillery shell in the L and VHF frequency bands [53, 54]. Since the objects considered have resonant sizes in the frequency band concerned, the responses are obtained using the calculation technique described earlier in section 1.2, which is based on solving the MFIE. The main goal of our work is to investigate the HRRPs of the two objects mentioned above. The results of our simulation are discussed below.

In our work we study the radar responses of an MLRS rocket (diameter $d = 0.122$ m, length $L = 2.87$ m, stabilizer span 0.25 m), and of an artillery shell ($d = 0.152$ m, $L = 0.709$ m). The radar objects in question and models of their surfaces are presented in figure 1.60.

The HRRPs correspond to a radar probing signal with a rectangular amplitude–frequency spectrum and a phase–frequency spectrum equal to zero in the given band of $[f_{\min}, f_{\max}]$. The calculation was carried out given the bandwidth of sounding signal Δf equal to 0.3 GHz and to 0.1 GHz. For modeling, we used the WB (UWB) probing signals featuring the parameters shown in table 1.8.

The frequency responses given fixed object aspect angles were obtained using a numerical technique based on solving the MFIE, the frequency increment being equal to 10 MHz. The responses that were smoothed out by means of a Hamming window were then used to compute the impulse responses (HRRPs) of the objects by applying the Fourier transform to the above frequency responses. In our particular case, the range resolution constituted the values $\delta_r = 0.71$ m ($\Delta f = 0.3$ GHz), and $\delta_r = 2.13$ m ($\Delta f = 0.1$ GHz), respectively.

The HRRPs presented in figures 1.61–1.64 correspond to vertical polarization, monostatic reception (electric vector $\vec{E}$ was perpendicular to both the object axis and the direction of the probing/reception), a fixed tilt angle equal to zero, and an azimuth aspect angle β equal to 0°, 30°, and 60°. The amplitudes of the computed responses were normalized to the maximum of the signal scattered by the rocket in

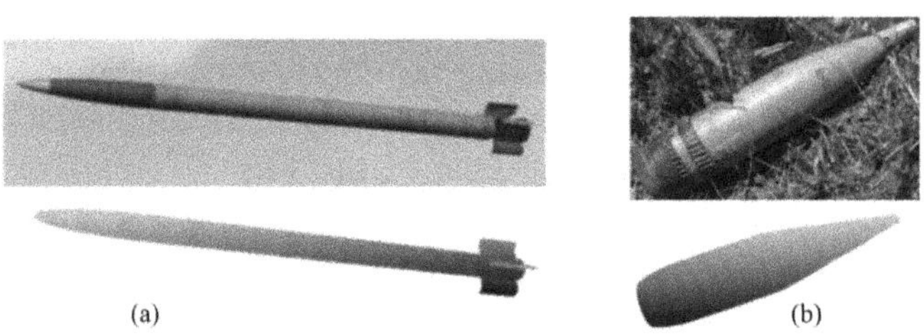

Figure 1.60. Models of radar objects: MLRS rocket (a) and artillery shell (b).

Table 1.8. Parameters of the radar probing signals used to obtain the HRRPs.

	Frequency, GHz		Wavelength, m	
Bandwidth Δf, GHz	$f_{\min}$	$f_{\max}$	$\lambda_{l\ \max}$	$\lambda_{l\ \min}$
L frequency band				
0.3	1	1.3	0.3	0.231
0.1	1.1	1.2	0.273	0.25
VHF frequency band				
0.3	0.1	0.4	3	0.75
0.1	0.2	0.3	1.5	1

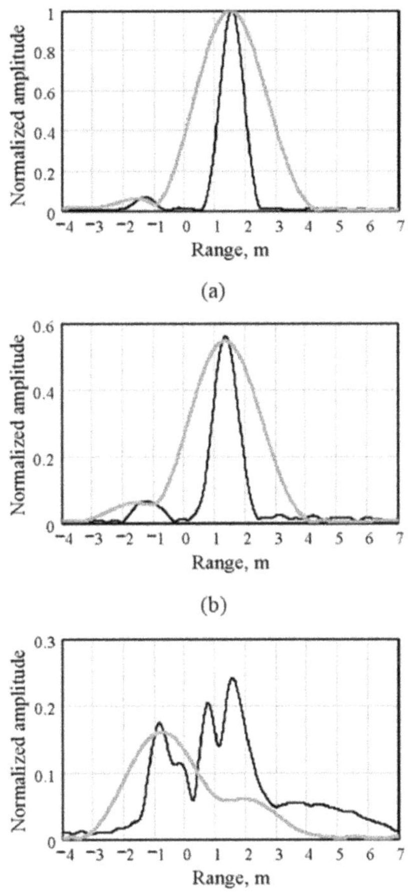

Figure 1.61. HRRPs of a rocket (d = 0.122 m, L = 2.87 m) of an MLRS. Radar probing signal of the L frequency band: (a) β = 0°; (b) β = 30°; (c) β = 60°.

the L band given angle $\beta = 0$. The responses for Δf = 0.3 GHz are shown by black lines, and for Δf = 0.1 GHz are shown by gray lines.

HRRPs of the rocket in the L frequency band are shown in figure 1.61. In this case, the range resolution δ_r (of 0.71 m and 2.13 m, respectively) did not exceed the rocket length ($L = 2.87$ m). As a result, given the aspect $\beta = 0$, the HRRPs consist of two distinct peaks, the distance between them being equal to the rocket length. As soon as the probing direction β deviates from the axial one by angle of 30° and 60°, the distance between the two end peaks decreases due to the reduced path-length difference between the signals scattered by the front and rear parts of the object (as seen from the reception point). In addition, in the latter case the echo signals are characterized by a more complicated shape.

When a UWB probing signal with a bandwidth Δf = 0.3 GHz in the VHF frequency band is used (figure 1.62), the HRRPs also have two distinct responses

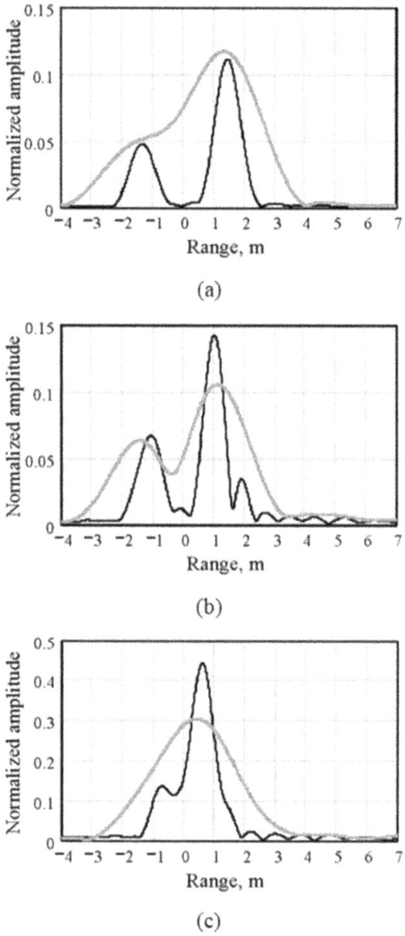

Figure 1.62. HRRPs of a rocket ($d = 0.122$ m, $L = 2.87$ m) of an MLRS. Radar probing signal of the VHF frequency band: (a) $\beta = 0°$; (b) $\beta = 30°$; (c) $\beta = 60°$.

echoed by the front and rear parts of the rocket. However, their shapes differ from the profiles shown in figure 1.61 that correspond to the L band. These results demonstrate the different mechanisms of secondary radiation for the two frequency bands in question. The difference in mechanisms is explained by the different electrical dimensions of the scatterer.

The profiles in figure 1.62 show that, in the VHF frequency band, when $\Delta f = 0.1$ GHz, the local responses echoed from the front and rear ends of the scatterer are also visible but not so clearly as was in given in the L band. The latter effect is observed despite the fact that the range resolution $\delta_r = 2.13$ m is less than the rocket length.

This fact can be explained by the existence of waves which propagated along the scatterer's surface. These waves are caused by the resonant effects of EM scattering. The complex natural resonances (CNRs) associated with this phenomenon are

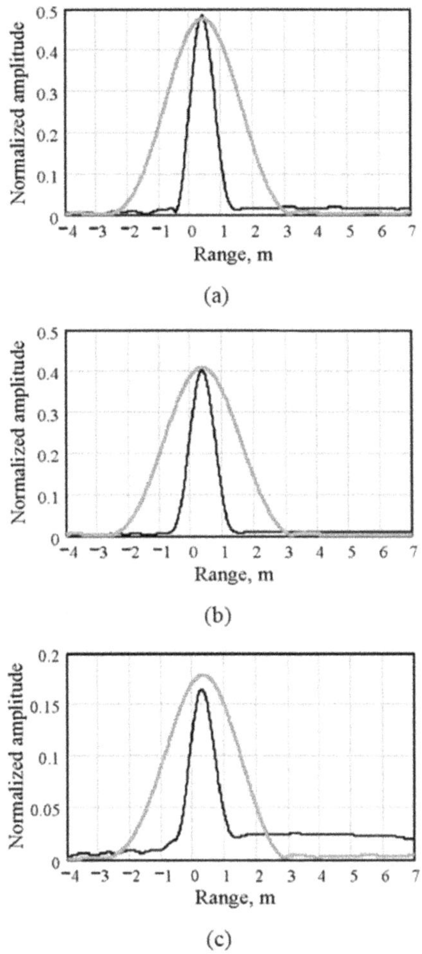

Figure 1.63. HRRPs of an artillery shell ($d = 0.152$ m, $L = 0.709$ m). Radar probing signal of the L frequency band: (a) $\beta = 0°$; (b) $\beta = 30°$; (c) $\beta = 60°$.

independent of the target aspect angle and they can be used as stable signatures for non-cooperative radar recognition of resonant objects [23, 53, 54, 79–84].

The HRRPs of an artillery shell ($L = 0.709$ m) for their probing in the L and VHF bands are presented in figures 1.63 and 1.64, respectively. In this case, the distinct peaks echoed from the front and rear ends are not observed for either $\Delta f = 0.1$ GHz ($\delta_r = 2.13$ m) or for $\Delta f = 0.3$ GHz ($\delta_r = 0.71$ m).

The radar signal scattered by an artillery shell represents basically the echo of a point target, which represents the 'non-distorted' probing signal, the latter being the Fourier transform of the WB (UWB) signal having a rectangular amplitude–frequency spectrum.

The results presented show that when one employs the WB (UWB) radar probing signal with $\Delta f = (0.1–0.3)$ GHz, one can also use the resulting echoes for non-cooperative identification that would be based on analysis of their shape. If the

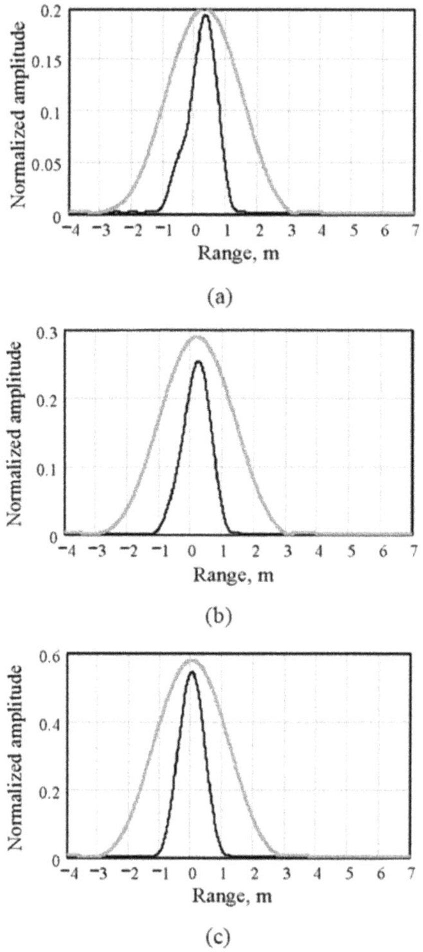

Figure 1.64. HRRPs of an artillery shell (d = 0.152 m, L = 0.709 m). Radar probing signal of the VHF frequency band: (a) $\beta = 0°$; (b) $\beta = 30°$; (c) $\beta = 60°$.

received echo contains the two distinct peaks scattered by the object's ends, then this suggests an obvious decision to be made that the object is longer than the range resolution δ_r cell (the MLRS rocket). Otherwise, if the received echo is an undistorted single peak response, this suggests that the object has a length that is below the range resolution cell δ_r (the artillery shell). Thus, the type of weapon can be identified—artillery or MLRS. Moreover, a decision can be made for an azimuth aspect angle β varying from 0 through more than 60°, and there would be no necessity to check the received echo against a set of references obtained for various aspect angles. Therefore, the proposed approach does not require a large database of radar profiles obtained for various aspect angles of objects of interest, thus saving considerable time, which is critical for counter-battery radars.

Analyzing the simulation results presented in figures 1.61–1.64, we can conclude that the maximum amplitude of the responses of artillery shells for their probing in the L frequency band and for azimuth aspects β of 0° and 30° exceeds that of the responses obtained for their radar probing in the VHF band. If the shell's aspect $\beta = 60°$, then its maximum amplitude in the L band is less than that in the VHF band. The more the signal is spread in time, the smaller its amplitude. In general, radar signals with spectra located in the L band do have some advantages as far as the radar probing of artillery shells is concerned [53].

When the non-cooperative radar recognition of objects is required, then a WB (UWB) short pulse or stepped-frequency signal (with pulse compression) can be employed for radar probing.

Let us consider the physical implications of the results obtained above. Current advances of radar technique provide for wide employment of UWB signals, the spectrum bandwidth Δf of which is commensurable with its central frequency f_0. This brings the so-called bandwidth ratio $K = \Delta f / f_0$ close to unit. Notwithstanding some advantages, the application of UWB signals is limited to rather short ranges. UWB radars are used widely for probing objects buried in the ground (including anti-tank and anti-personnel mines). The feasibility of employing UWB signals for the radar probing of cruise missiles and ground objects is also being studied. The employment of UWB signals in radar places fairly strict demands on the parameter stability of the transmitter, receiver, and AS, which need to hold across a wide frequency range. The greater the broadband coefficient the harder it is to maintain the specified frequency performance of these devices. Moreover, a radar scan across angular coordinates becomes more challenging as the bandwidth ratio becomes too great. Therefore, the HRRPs of artillery shells considered here for the VHF frequency band using probing signals with a bandwidth of 0.3 GHz ($K = 1.2$) are largely of pure academic interest. However, we present them as a reference for the sake of better understanding the EMW scattering mechanisms. At the same time, the HRRPs obtained by more narrowband signals in the VHF band that had the bandwidth $\Delta f = 0.1$ GHz (K = 0.4) have more practical importance. The latter observation makes even more sense in the L frequency band (the values of K being equal to 0.26 and 0.08, see to table 1.8), where the higher values of the carrier frequency reduce the bandwidth ratio even further, thus providing for wider practical applications of the aforementioned HRRPs.

A UWB high-resolution response of a radar object can be considered as a combination of radar returns from its individual segments. The coherent integration of such returns is usually a more challenging task than the coherent integration of single peak narrowband radar echoes. That is why the true UWB signal could be used in a separate radar mode devoted to radar recognition, while the usual tasks of radar detection and tracking could be solved using more narrowband signals.

Another major issue in practical radar applications is the necessity to ensure fairly high signal-to-noise values. Our qualitative analysis of HRRPs obtained with respect to an MLRS rocket clearly shows that those HRRPs usually comprise two major peaks, one being stronger and the other being considerably weaker (see figures 1.61(a) and (b)). Under condition of insufficient signal-to-noise ratio, such a lower amplitude of one of the peaks can degrade the performance of the radar recognition algorithm. Therefore, a quick look at the plots in figures 1.61–1.64 show that the most advisable option for implementing the radar recognition of artillery shells by their HRRPs in particular would be to set the bandwidth of the radar probing signal equal to $\Delta f = 0.3$ GHz.

1.4.3 The radar detection and identification of metallic and dielectric mines buried in the ground

Anti-tank and anti-personnel mines buried in the ground constitute a serious menace not only to the military, but to civilians as well. In the last few decades a whole range of landmine searching devices (systems) have been developed. These systems are based on different physical principles and potentially have significantly greater capabilities compared to the standard mine detectors used in the armies of different countries. However, the main disadvantages of the known systems used for mine detection are a low probability of correct detection and a high probability of false alarms.

Ground penetrating radar (GPR) [85–91] possesses a number of advantages when solving the problem under consideration, specifically from the standpoint of the information carried by the echo signal. However, when implementing classical radar detection methods to detect a landmine, one will face a number of challenges:

i. The landmine response is masked by powerful reflection from the ground surface. To extract a useful signal against a background of interfering signal, it is reasonable to use probing signals that provide the high-range resolution with respect to depth; the latter option can be complemented with either oblique radar probing or with a radar layout where its transmitting and receiving antennas sit directly on the surface.
ii. The ground is a heterogeneous dispersive medium with attenuation. Its electro-physical parameters (dielectric constant and/or specific conductance) that affect the shape and parameters of the echo signal depend significantly on the soil type, as well as its humidity and density. Hence GPR should incorporate a device for estimating the soil parameters.
iii. The low contrast (in terms of the permittivities of plastic mines compared to the soil in its immediate vicinity) makes the radar echoes from such mines very weak. When detecting such objects, it is highly desirable to envision a change of the probing signal's frequency spectrum (FS) in order to provide for a maximum amplitude of echoes reflected from low-contrast subsurface objects (SSO).
iv. Landmines are observed on a background of heterogeneities (litter) in the ground, both natural and artificial (cartridge cases, parts of vehicles and weapons, stones and other metal and dielectric objects), which leads to an increased probability of false alarms. To reduce the false alarm rate, the mine detectors need to be equipped with additional algorithms for the input signal processing.

One way of improving the landmine detector performance is to account for the *a priori* information regarding the scattering properties of various landmines while implementing the input signal processing. The aim of this subsection is to simulate numerically the scattering properties of mines and other SSOs in various conditions, to generalize them and to develop a method of radar detection and identification for different types of mines buried in the ground.

The characteristic geometric dimensions of widely used landmines (15–35 cm) are comparable to the wavelengths of the probing signals, which vary from 3 m to 10 cm and less. The latter wavelengths correspond to optimal ground penetrating capability down to the depths at which the mine would be planted (up to 10 cm for anti-personnel mines, and up to 70 cm for anti-tank mines). Considering that landmines are resonant scatterers, numerical methods based on solving boundary IEs (described in sections 1.2 and 1.3) are used to calculate their EM responses. In this case, the EMFs of auxiliary dipoles were used as the kernels of equations (1.70) and (1.91) that accounted for the presence of an interface between the dielectric half-spaces [23, 51, 58, 59]. The CNRs of the objects of interest [23, 53, 54, 79–84] are used as characteristic features used for their identification.

During numerical modeling we used the subsurface probing model depicted in figure 1.65. The AS of the GPR was located in half-space V_1 that had relative permittivity $\varepsilon_1 = 1$ at the height h_a. During calculations, we assumed that the transceiving AS was using two low-gain antennas (simulated by magnetic dipoles) located at a distance B_a from each other, the antennas having the same height above the ground at their location points A_1 and A_2, and having magnetic vector-moments $\vec{p}^{a1}$ and $\vec{p}^{a2}$, respectively (figure 1.65). The half-space V_2 (with relative permittivity $\varepsilon_2 = \varepsilon'_2 + j\varepsilon''_2$) includes the object V_3 that is bounded by surface S. The object is filled with material, the relative permittivity of which is $\varepsilon_3 = \varepsilon'_3 + j\varepsilon''_3$. All the regions V_1, V_2, and V_3 are homogeneous, and the complex permittivities ε_2, ε_3 are the functions of frequency f. A metallic object is considered to be a PEC object ($\varepsilon''_3 \to \infty$). Relative permeability is $\mu = 1$ in all the considered regions. The problem consists in finding the field, $(\vec{E}, \vec{H})$, scattered by object V_3 at an arbitrary point of half-space V_1.

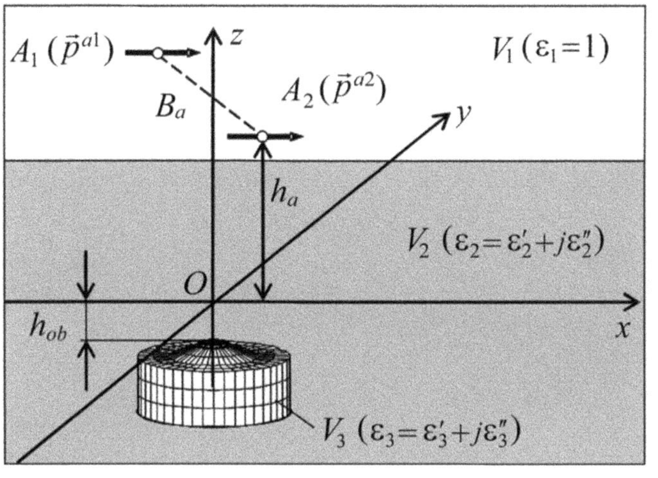

Figure 1.65. The radar probing model of a subsurface object.

The models of anti-tank anti-track mine M15 (push action in a metal housing, made in the USA) and anti-personnel landmine DM-11 (in a plastic housing, made in Germany), which are used in this section as simulation examples aimed at radar detection and identification of landmines, are shown in figure 1.66. Below we consider the decision regarding the landmine detection and identification to be the same decision, the mine is to be considered as detected only when the identification tag has been pinned on it. In computations, we assumed the surface of M15 to be PEC while the volume inside by DM-11 was assumed to be filled with homogeneous material with relative permittivity (dielectric constant) $\varepsilon_3 = 2.2$.

The method considered below is based on using CNRs as the feature for distinguishing between various objects of resonant sizes. The properties of CNRs were studied with respect to both SSOs [87, 92] and the objects in free space [79, 84, 92–95]. In the references just mentioned it was shown that the CNRs are determined by the geometric and electro-physical parameters of objects and they occur independently of radiation conditions. The mentioned properties of the CNRs provide a basis for using these parameters as identification features for various types of landmines.

The method uses the following experimentally proven fact: due to the limited nature of the FS of real radar signals, the major contribution into the impulse responses of objects of resonant dimensions is provided by a finite number M of decaying oscillations. Such oscillations correspond to the pairs of poles (α, ω) of the object's response in the complex frequency $S = \alpha + j\omega$ domain. Considering this, the object's impulse response can be presented as [96, 97]

$$r(t) = \sum_{m=1}^{M} A_m \exp(S_m t), \tag{1.112}$$

where A_m are complex coefficients to be found; $S_m = \alpha_m + \mathrm{i}\omega_m$, which can be presented in discrete form by introducing the samples of initial data

$$r(t_n) = r(n) = \sum_{m=1}^{M} A_m \exp(S_m t_n) = \sum_{m=1}^{M} A_m \exp(S_m (n-1)\Delta t), \tag{1.113}$$

where $n = \overline{1, N}$; N denotes the number of time samples in the response, Δt is the sampling interval. Thus after receiving and sampling the object's EM response, we can obtain a set of N non-linear equations with $2M$ unknowns A_m and S_m. If $N \geqslant 2M$ and given that the response is uniformly sampled, then our target

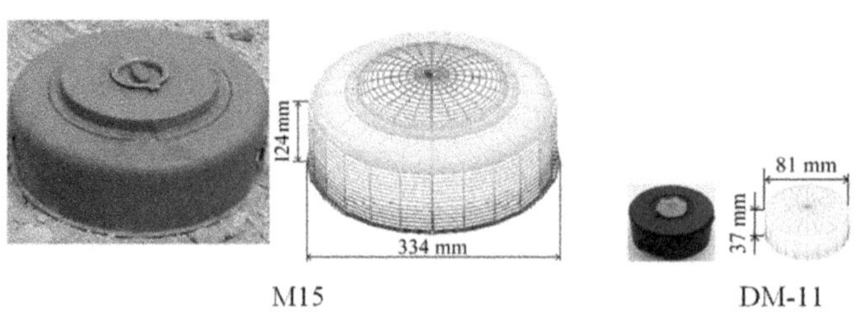

Figure 1.66. Models of anti-tank and anti-personnel mine surfaces.

parameters can be found using the Prony method [96, 97]. The method is based on the fact that signal samples in formula (1.113) satisfy a homogeneous Mth order differential equation (commonly called the Prony differential equation [96, 97]) that has the following form

$$\sum_{m=0}^{M} \gamma_m r(n+m) = 0, \tag{1.114}$$

where γ_m are differential equation coefficients, $n = 0,1,\ldots, N - M$.

To identify the SSOs, we employ the correlation algorithm [87, 98]. This algorithm makes implicit use of the Prony differential equation coefficients γ_m as features for the object identification, which can be calculated using CNRs of the object of interest S_m. The latter are determined in advance using physical or mathematical modeling and are saved to a database of target objects, the structure of which is discussed below.

In order to determine the parameters of the algorithm suggested above and to assess the potential capabilities of CNRs as a mine identification feature, we employed the model of scattering described above to calculate the radar characteristics of landmines and other SSOs that actually litter the site given various conditions of their probing. We based our computations on the set-up shown in figure 1.65. In this subsection, we consider an example library of reference EM responses of the M15 and DM-11 landmines (figure 1.66). The library basically includes the frequency response poles of landmines. The library also includes some benchmark objects (the litter) shaped as ellipsoids and spheres of various dimensions and filling media. For future reference, we call these objects foreign bodies since they would be planted into the natural environment and they would also have salient features similar to those of the EM responses of interest. The foreign objects we simulated are as follows:

- Object 1—a dielectric sphere of 10 cm radius that had relative permittivity equal to that of stone $\varepsilon_3 = 8.3$.
- Object 2—a dielectric sphere of 5 cm radius ($\varepsilon_3 = 8.3$).
- Object 3—a dielectric ellipsoid ($\varepsilon_3 = 8.3$) with semi-axes $a_x = 7.5$ cm, $a_y = 7.5$ cm, $a_z = 3$ cm (the ellipsoid's semi-axes were oriented along the coordinate axes x, y, z of figure 1.65).
- Object 4—a dielectric ellipsoid ($\varepsilon_3 = 8.3$, $a_x = 15$ cm, $a_y = 15$ cm, $a_z = 6$ cm).
- Object 5—a PEC sphere of 5 cm radius.
- Object 6—a PEC sphere of 10 cm radius.
- Object 7—a PEC ellipsoid ($a_x = 7.5$ cm, $a_y = 7.5$ cm, $a_z = 3$ cm).
- Object 8—a PEC ellipsoid ($a_x = 15$ cm, $a_y = 15$ cm, $a_z = 6$ cm).
- Object 9—a dielectric ellipsoid with a dielectric constant of the DM-11 mine ($\varepsilon_3 = 2.2$, $a_x = 15$ cm, $a_y = 15$ cm, $a_z = 6$ cm).
- Object 10—a dielectric ellipsoid ($\varepsilon_3 = 2.2$, $a_x = 7.5$ cm, $a_y = 7.5$ cm, $a_z = 3$ cm).
- Object 11—a dielectric ellipsoid ($\varepsilon_3 = 8.3$, $a_x = 4$ cm, $a_y = 4$ cm, $a_z = 1.7$ cm).

For the sake of the simulation, we assumed that at any given time there was only one SSO (either landmine or foreign object) within the operating range of the GPR and the radar echo contained only the object's EM response without any interference imparted by the interface between the two media. Practically speaking, the returns from the interface between the two media were assumed to be canceled. Detection and identification were accomplished based on the results of single radar probing into the ground, however, several radar returns were being measured simultaneously and averaged at different receiver locations in the object's vicinity.

Before actually running the landmine detection/identification algorithm, we performed some preparatory work. It included the generation of a library of reference EM responses from the objects of interest as well as adjustment of the parameters of the identification algorithm.

The salient features of the reference EM responses were extracted from the CNRs of the SSOs listed above given that noise interference was absent. We also assumed that all the objects were buried in the ground at a depth of 6 cm.

The impulse responses of SSOs $r(t)$ were used as inputs for the landmine detection/identification algorithm, which can be observed directly by a GPR employing a UWB probing signal [85, 87, 88, 90], or they can be synthesized while processing the linear-frequency modulated or the so-called stepped-frequency signals [86, 88, 89]. We extracted the so-called 'late-time' component from the impulse response, which is a decaying portion that appears after the probing signal stops illuminating the object of interest. Figure 1.67 shows the impulse responses of the M15 and DM-11 landmines which are planted in the ground. The responses were computed using the expressions (1.70) and (1.81) for mines made of metal, and using expressions (1.91), (1.97), and (1.98) for the mines made of dielectric material. The amplitudes are presented in relative units (r.u.). The corresponding 'late-time' components are marked in the plots of figure 1.67. The impulse responses presented in figure 1.67 were obtained under the following conditions. Magnetic dipoles with equal vector-moments ($\vec{p}^{a1} = \vec{p}^{a2}$) were oriented in parallel to the air–ground interface (figure 1.65), they were placed at the same height h_a = 0.5 m, and they were used as a model of transceiving AS. The distance between centers of antennas was B_a = 0.5 m. Antennas were considered to be decoupled. The geometric center of the antenna was sitting right above the center of the landmine. The probing signal had a uniform frequency spectrum $G(f) = 1$ and the phase spectrum $\varphi(f) = 0$ within the frequency range from $f_{\min}$ = 100 MHz through $f_{\max}$ = 1000 MHz for the case of the M15 landmine and through $f_{\max}$ = 1500 MHz for the case of the DM-11 landmine.

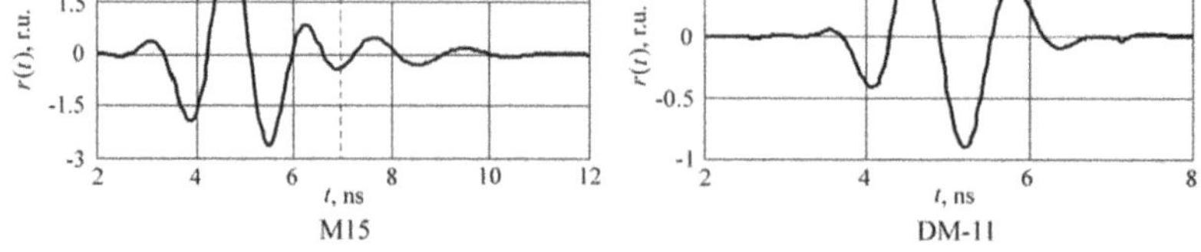

Figure 1.67. The impulse responses of landmines buried in the ground.

The mines were buried at the depth $h_{\text{ob}} = 6$ cm into the gray loam that was assumed to have a density of 1.2 g cm^{-3} given its moisture content of 10%. The dependences of the real and imaginary parts of the soil's dielectric constant versus frequency for this kind of soil are shown in figure 1.68 [99].

The 'late-time' response component is caused by the interaction of two signals, one scattered from the front (with respect to the radar) of the object's surface and the second generated by the so-called 'crawling' wave for an object made of metal, or by the wave scattered from the rear surface of the object for an object that was dielectric. When the characteristic object dimensions become commensurate with the probing signal's wavelengths (as represented by its frequency spectrum) in the medium where the object was planted and provided that the lag time resolution of the probing signal is sufficiently high, then the signal generated by that 'crawling' wave (or the wave backscattered from the object's back surface) reaches the object's front by the time the probing signal has already ended. In this particular case, the aforementioned interaction between the two wave components creates a 'late-time' response that is primarily shaped by the phase shifted 'crawling' wave (or the wave scattered back from the object's rear). The latter fact makes the geometric dimensions and filling material sources of major impact upon the late-time response. Obviously the way the 'late-time' component is extracted depends on the type of the object of interest (metallic or dielectric).

For the sake of convenience, from now on we refer to this 'late-time' component of the object's response as the measured response, and we denote it as $r_m(t)$. When the EM response is checked against the reference metallic object, then the 'late-time' component is found by truncating the response's front by the time corresponding to the probing pulse duration. On the other hand, when the response is checked against the reference dielectric object, then it needs to be truncated by the time it takes the probing signal to propagate to the object's rear and back.

The considerations provided above suggest the direct necessity of estimating the parameters of the ground in which the object of interest is planted. The latter is due to the strong influence of the probing signal's wavelength on the dielectric constant of soil and, consequently, on the electric dimensions the SSO has in that soil.

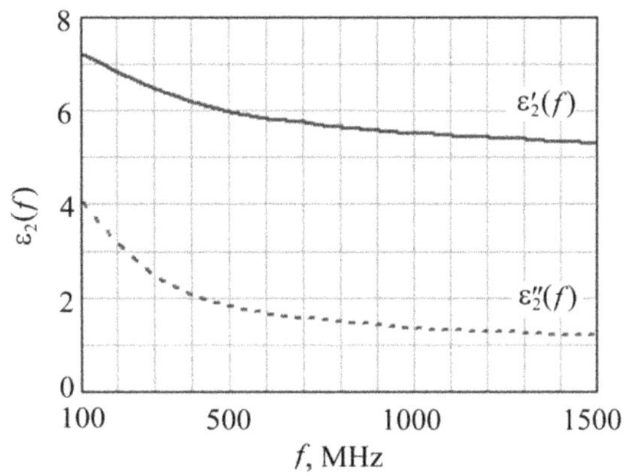

Figure 1.68. Relative permittivity of gray loam.

The mentioned dependence also affects the selection of the probing signal bandwidth tailored for the identification of certain metallic or dielectric objects of various geometries.

As soon as the parameters of the soil are evaluated and the measured response is extracted from the input signal, the detection and identification algorithm can be broken down into three stages [87, 98].

1. *Prediction.* Based on the coefficients of the reference response that appear in the Prony differential equation and on the 'late-time' component of object's response, the so-called predicted response is determined:

$$r_c(n + k_0) = \sum_{\substack{k=0 \\ k \neq k_0}}^{N} \left\{ -\frac{\gamma_k}{\gamma_{k_0}} r_m(n + k) \right\}, \tag{1.115}$$

where γ_k is the kth coefficient of the differential equation, $\gamma_{k_0} = \max\{\gamma\}$.

Now, comparing expressions (1.115) and (1.114), one could assume that as long as the measured response corresponds to that of the reference object, then the functions r_m and r_c would be identical, which takes us to the next stage of the algorithm.

2. *Evaluation* of the mutual correlation coefficient between the measured and predicted responses, ρ:

$$\rho = \frac{\sum_{n=1}^{N} 2 r_c(n) r_m(n)}{\sum_{n=1}^{N} \left[r_c^2(n) + r_m^2(n) \right]}. \tag{1.116}$$

3. *Decision making.* At this stage, the value of ρ is compared to some threshold value ρ_0. If $\rho < \rho_0$, then the decision is to be made that there is a foreign object present within the operation range of GPR. If $\rho \geqslant \rho_0$, then one concludes that the object, which corresponds to a particular set of reference coefficients appearing in the differential equation, is found. The threshold value is set based on the condition of setting the desired conditional probability of correct identification (CPCI) rather than to ensure some predefined rate of false alarms.

While simulating the landmine responses, we noticed that parameters of the poles pertaining to individual frequency responses depended on the sampling interval Δt of the input response. Thus, in order to determine an optimal value for the Δt, we introduced quality indicators that allowed us to quantify the difference between the impulse response represented by exponential parameters of the model (1.113) and the corresponding impulse responses obtained using electrodynamic models (1.70) and (1.81) for metallic landmines, and the models (1.91), (1.97), and (1.98) for the dielectric landmines (or the impulse responses obtained by measurement). Such indicators included the sums of squared residuals between the measured impulse responses and those predicted on the basis of the poles of their frequency responses. The indicators also depend on the maximum value of absolute deviation between these values as well as the correlation coefficient between them. Our analysis shows that the best value for the sampling interval Δt would be around $\Delta t_{opt} = (2.5 f_{max})^{-1}$–$(5 f_{max})^{-1}$, where f_{max} denotes the upper frequency boundary of the probing signal's spectrum. Setting Δt both above and below that range leads to poor performance of the pole selection procedure.

For the suggested method of detection/identification to be effective, one needs to ensure that the resonant frequency range of all the objects of interest would overlap with the frequency range of GPR operation. In addition, it would also be advisable to fit the GPR with the capability of adjusting its operating frequency range to cover a wide range of reference objects, in particular if their relative sizes and/or the material they are made of vary considerably. The smaller the object size, the higher the probing frequencies to be used for its identification.

Analysis of our simulated impulse responses with regard to anti-tank mines planted in the ground demonstrated that the mine detection/identification algorithm achieves its best performance when the probing signal spectrum frequencies go as high as 1000 MHz. Further increase of the maximum spectrum frequency leads to an increase of the specular component in the impulse response, whereas its 'late-time' component becomes weaker. The latter degrades the algorithm described above. Our analysis of impulse responses with regard to the anti-personnel DM-11 landmine revealed the reasons for increasing the maximum spectrum frequency up to 1500 MHz in a GPR aimed at detection/identification of such mines.

Analysis of the obtained results proved the reasonableness of the assumption made earlier. Namely, that the imaginary parts of the object's frequency response poles do not practically depend on the location of the GPR AS. When the AS was placed with an offset of 0.5 m from the SSO center in the horizontal plane, the above parameter values varied by only around 4% as far as the objects mentioned above are concerned. In addition, we observed a similar result (a variation of about 5%) as we tried to gradually increase the AS height. The simulation results proved, however, that once the GPR antenna is placed at a height below 0.2 m, then the imaginary parts of the frequency response poles pertaining to objects of interest had significant variance and they could not be used for landmine identification. The latter effect is caused by the probing wave front, which is not locally planar (with respect to the SSO aperture) when it reaches the ground surface and, in particular, when it reaches the object itself. Hence the minimum height of antenna placement is determined by the lateral dimension of the target objects.

As a next step, we evaluated the performance of the landmine identification algorithm described above using probabilistic quality indices. These indices were evaluated given that landmines were observed against a background of foreign objects.

The receiver's input was simulated as an additive mixture of the SSO impulse response (obtained under the same observation conditions that had been adopted earlier to generate the database of reference EM responses) and realizations of Gaussian noise with zero mean and standard deviation σ. For every SSO concerned, we generated 50 input signals differing by individual noise realizations and the signal-to-noise ratio $q = u/\sigma$, where u denotes the maximum amplitude of the pulse reflected by the mine after its processing in the receiver. The signal-to-noise ratio q was varied from 10 dB through 40 dB. As a result, the correlation coefficients were calculated for each SSO using formula (1.116). The value of every correlation coefficient was obtained by averaging over 50 realizations of the input signal. We also evaluated the empirical distributions of the latter coefficients since they also represented random variables. The detection/identification threshold of correlation coefficient ρ_0 used by the landmine detection/identification algorithm was determined by the required CPCI P_{lim}.

Table 1.9 presents such threshold values that we obtained for the landmines in question for various values of CPCI P_{lim}. All the threshold values are presented for the signal-to-noise ratio $q = 30$ dB.

The identification algorithm performance was characterized by the following probability estimates:

- CPCI $\hat{M}_{\text{lim}} = N_1/N_2$, where N_1 is the quantity of correctly identified input signals that corresponded to the target object (one of the two landmines), i.e. the number of times the correlation coefficient ρ, computed by formula (1.116), went above the corresponding threshold value ρ_0 given that the target landmine response was present in the input signal, N_2 stands for the total number of input signal realizations containing EM responses of the target landmine.
- Conditional probability of false identification (CPFI) $\hat{M}_{\text{f}} = N_3/N_4$, where N_3 denotes the number of instances that the object was identified as the target landmine given that the input signal contained a response from one of the foreign objects, N_4 stands for the total number of input signals containing responses from a particular foreign object. This probability basically indicated the share of every particular foreign object in the total false alarm rate.

Tables 1.10 and 1.11 present both the estimated values of CPFI $\hat{M}_{\text{f}}$ versus the required level of CPCI P_{lim} for different values of signal-to-noise ratio q. All the numbers are presented for the case where a single measurement of EM response was carried out for the M15 (table 1.10) and DM-11 (table 1.11) landmines (AS hovering about the center of SSO). Some foreign objects with negligible impact on the false

Table 1.9. Threshold values of the correlation coefficients.

P_{lim}	ρ_0	
	M15	DM-11
0.9	0.7	0.679
0.8	0.715	0.680
0.7	0.724	0.682
0.6	0.731	0.683
0.5	0.737	0.684

Table 1.10. Identification results for the M15 landmine.

	q				
	20				30
	P_{lim}				
	0.6	0.7	0.8	0.9	0.9
Foreign object	$\hat{M}_{\text{f}}$				
1	0.125	0.188	0.25	0.313	0
2	0.056	0.056	0.056	0.167	0
3	0.111	0.111	0.278	0.333	0
4	0.056	0.056	0.056	0.278	0
5	0.091	0.182	0.182	0.273	0
6	0	0	0.091	0.273	0
7	0.182	0.182	0.182	0.273	0
8	0.273	0.364	0.545	0.636	0.273
9	0.091	0.182	0.273	0.273	0
10	0.091	0.091	0.091	0.182	0

Table 1.11. Identification results for the DM-11 landmine.

	q				
	10				30
	P_{lim}				
	0.6	0.7	0.8	0.9	0.9
Foreign object	$\hat{M}_{\text{f}}$				
2	0	0.056	0.056	0.056	0
5	0.909	0.909	0.909	0.909	0.364
6	1	1	1	1	0
9	0.455	0.455	0.545	0.545	0.182
10	0	0	0	0.091	0
11	0	0.056	0.056	0.167	0

alarm rate of the landmine detection/identification algorithm were left out for the sake of clarity.

It can be seen from table 1.9 that in order to increase the correct detection/identification rate P_{lim} one would need to lower the threshold ρ_0. However, it is obvious from table 1.10 that the higher probability of correct landmine detection comes at the price of an increase in the false alarm rate. Yet in the case of the M15 landmine it was possible to achieve the desired level of the correct detection/identification rate and, at the same time, get rid of false alarms due to almost all foreign objects except one (foreign object #8 in table 1.10). The latter was achieved by increasing the signal-to-noise ratio from 20 dB to 30 dB. Moreover, through analysis of the individual shares of various foreign objects in the total false alarm rate with respect to the metallic M15 landmine, we found that most false decisions came from object #8, which is a PEC ellipsoid of similar size. The rest of the foreign objects accounted for a significantly lower number of false alarms. The latter result proved the fact that the CNRs depended on the geometric dimensions of the scatterer.

It is worth noting that variations in the signal-to-noise ratio within the range of 10–40 dB did not result in any significant decrease of the correct landmine detection/identification rate. In addition, the detection/identification of a scatterer with such a salient EM response as the M15 landmine does not require a special noise cancellation technique, and a single measurement is usually enough to determine the presence of such an object.

On the other hand, in our simulation involving the DM-11 landmine we found out that even the high signal-to-noise ratio q = 30 dB did not guarantee the algorithm would achieve its desired performance in terms of correct landmine detection/identification rate, which only went as high as $\hat{M}_{\text{lim}} \approx 0.7$ in this case. Given the lower signal-to-noise ratios (q = 10–20 dB), the latter rate did not go above the level of $\hat{M}_{\text{lim}} = 0.545$. The detection/identification algorithm would also have a higher rate of false alarms, as seen from table 1.11, even if the signal-to-noise ratio was fairly high. The latter is because the plastic DM-11 landmine's response is more similar to those of the foreign objects (it possesses fewer salient features).

However, as soon as we employed arithmetic averaging of the decisions obtained for several received signals obtained while shifting the AS positions in the horizontal plane (the AS center was being moved with respect to the center of the SSO by the values ranging from 0 to 0.5 m), we were capable of achieving the desired performance of the landmine detection/identification algorithm. However, in the case of a lower contrast object, such as the DM-11 landmine for instance, some additional measures would be needed to decrease the distortions of its EM response. Arithmetic averaging of multiple consecutive decisions aimed at reducing the impact of noise, and every new decision being obtained using an input signal received from a slightly shifted position of the GPR AS in the horizontal plane, also helped to increase the overall performance of the detection/identification algorithm.

Our simulation regarding the identification of landmines EM responses that were planted at unknown depth in the ground has demonstrated that the correlation coefficient did not go high enough to reach its threshold value ρ_0. For example, given the M15 landmine that was planted at the depth h_{ob} = 10 cm, the correlation coefficient happened to be confined within the range ρ = 0.56–0.83, while the threshold was as high as $\rho_0 = 0.9$. Thus, in order to provide for a high performance mine detection/identification algorithm, the mine planting depth needs to be estimated. The corresponding database needs to include the reference EM responses of landmines for different planting depths. In addition, our data indicated that in order to achieve the CPCI $P_{\text{lim}} > 0.9$, the landmine planting depth needs to be estimated with an accuracy of at least 2 cm.

From simple physical considerations, we see a direct necessity to estimate the electrical parameters of the soil rather precisely in order to achieve a high performance of the landmine detection and identification algorithm. The latter requirement becomes particularly important in the case of metallic scatterers because their CNRs are determined by the character of the EMW propagation around the object's surface in its immediate vicinity.

In order to estimate the desired accuracy with which the dielectric permittivity of the soil is to be estimated, we simulated the above algorithm's operation in the case that the electric parameters of the soil differed from those adopted for the generation of the library (database) of reference EM responses of landmines. The following results were obtained during the detection/identification of the M15 landmine.

While simulating single radar probings into the ground with the signal-to-noise ratio q varied from 10 dB through 40 dB and a relative measurement error of the soil's permittivity of $\Delta_{\varepsilon_2} = 5\%$, we were capable of achieving an estimated value of the CPCI $\hat{M}_{\text{lim}}$ for the landmines equal to 0.364. When the relative measurement error increased to $\Delta_{\varepsilon_2} = 10\%$, the algorithm performance dropped to $\hat{M}_{\text{lim}} = 0.273$. Given the relative measurement error Δ_{ε_2} increased to 20%, we ended up with a very poor algorithm performance of $\hat{M}_{\text{lim}} = 0.09$.

Combining the radar information obtained for several different positions of the GPR AS in the horizontal plane and employing arithmetic averaging, we managed to achieve errorless landmine detection/identification.

As far as dielectric landmines are concerned, the CNRs of which are primarily determined by the parameters of the waves propagating through the object's medium, the measurement accuracy of the soil's permittivity is not so critical. We checked this out in our simulation. For example, given a relative accuracy of the soil's permittivity measurement Δ_{ε_2} equal to 20%, we obtained almost the same landmine identification performance as that obtained using the exactly known electric parameters of the soil.

However, considering the results of our study, we can state that the required accuracy of measuring the electro-physical parameters of the soil the landmine is planted in would be rather high. However, contemporary methods and available devices do provide sufficient relative measurement accuracy of the soil's permittivity of about 20% [89].

Thus the database needed for practical implementation of the above method for detecting and identifying landmines planted in the ground should contain the salient features of the landmines of interest. One of the possible ways to extract such features is to use the coefficients of the Prony differential equation that could be computed using the CNRs of the target objects. The database would need to include the aforementioned coefficients given various values of the landmine planting depth ranging from 10 cm through 70 cm with the increment of 2 cm, as well as different values of the soil permittivity. The relative measurement accuracy of the soil permittivity must not be worse than 20% across the frequencies within the probing signal's spectrum. The CNRs of landmines are to be determined in advance while accounting for the functions $\varepsilon_2(f)$ that correspond to different types of soils, as well as their moisture content (humidity) and density [99]. The dependence of CNRs on the depth of landmine planting, as well as on the type of soil and its moisture content, complicates the practical implementation of the suggested method.

The method for the landmine detection and identification described above in combination with the available mathematical model of SSO scattering, which allows for computation of the landmine CNRs in various conditions, makes practical implementation of landmine detection/identification possible. The suggested digital algorithm designed for that purpose utilizes the correlation coefficient between the measured EM responses of the buried objects and the reference responses contained in the database, the latter corresponding to the estimated planting depth and soil type. Final decisions regarding the detection/identification are to be made by comparing the above correlation coefficient with some predefined threshold value. If the threshold is exceeded, then the decision is made that there is a landmine of a certain type within the operation range of the GPR.

1.4.4 Electromagnetic wave scattering by meteorological particles of various shapes

Radar sounding can give a lot of important information about atmospheric phenomena [100–105]. Currently, weather radar are used to find the speed and direction of movement of meteorological formations, to detect atmospheric fronts as well as the boundaries of clouds and precipitation, to detect turbulent and convective atmospheric formations, and to find the phase composition of meteorological formations [103]. Solving these problems requires a great deal of data about the radar scattering performance of rain and snow particles [104, 105]. As long as the radar wavelength is in the range from a few millimeters to a few centimeters, these particles of precipitation become, basically, dielectric objects of resonant size. The works [104, 105] describe the well-known numerical methods that make it possible to calculate the RCS of spherical and ellipsoidal raindrops. Usually, precipitation includes particles of different types, shapes, and sizes. In addition, particles of the same type can have different size ratios [104–108]. The wide intervals that the parameters of these particles occupy can make it problematic to find their scattering characteristics, and, consequently, complicates the evaluation of the microphysical parameters of hydrometeors. The later issue applies in particular to the problem of quantifying the moisture content.

The scattering performance of hydrometeors has not been studied enough in the known literature. The lack of statistical information about the radar properties of hydrometeors, as well as the incomplete data obtained by remote radar sensing, makes the mathematical simulation of radar scattering by hydrometeors of various types very important.

Considering the variety of geometric shapes, and taking into account the fact that the electrical sizes of precipitation particles (raindrops, snow crystals) put them into the resonant scattering region, we can use the surface IEs presented in the frequency domain to compute the scattering characteristics of such scatterers. The development of the necessary numerical methods is based on solving the IEs representing the scattering by dielectric objects, and they are the subject of many works [13, 23, 25, 31–33, 60, 61].

In section 1.3 we propose a numerical method for solving the MIE set [23, 55]. In this subsection we also discuss some features of the algorithm we designed for solving that particular IE set pertaining to models of hydrometeors. As we show below, our algorithm is aimed at predicting the scattering characteristics of dielectric objects, yielding stable solutions while having a smaller number of current nodes compared to other known methods. The latter advantage becomes particularly noticeable when calculating the scattering characteristics of electrically thin dielectric objects. This makes it possible to expand the method's applicability domain to include a variety of shapes and sizes of scatterers that were not available before and to reduce the computational burden. Below, we present the RCS of ellipsoidal raindrops and snow crystals in the form of hexagons that were computed using the proposed method. The results of the computations feature wavelengths of radar signal equal to 3.2 cm and 0.8 cm, as well as various aspects of the scatterer radar probing at two orthogonal polarizations.

Raindrops and snow crystals can be considered as homogeneous dielectric objects (with complex relative permittivity $\varepsilon_2 = \varepsilon'_2 + j\varepsilon''_2$) located in free space (with relative permittivity ε_1). We also assume that the raindrops were of ellipsoidal shape (figure 1.69) and that the crystals of snow were hexagon shaped (figure 1.70).

Throughout this subsection, we assume that all the scatterers of concern have relative magnetic permeability $\mu = 1$. In addition, all the EMFs are assumed to have the time dependence expressed by the term $\exp(-j\omega t)$.

Our computation method provides sufficiently high computation accuracy as far as the densities of equivalent electric and magnetic currents on the surface of a dielectric object and the field scattered by it are concerned. Coupled with the method of applying a mesh of current nodes that takes into account the curvature of different surface parts of a dielectric scatterer, the developed numerical algorithm provides

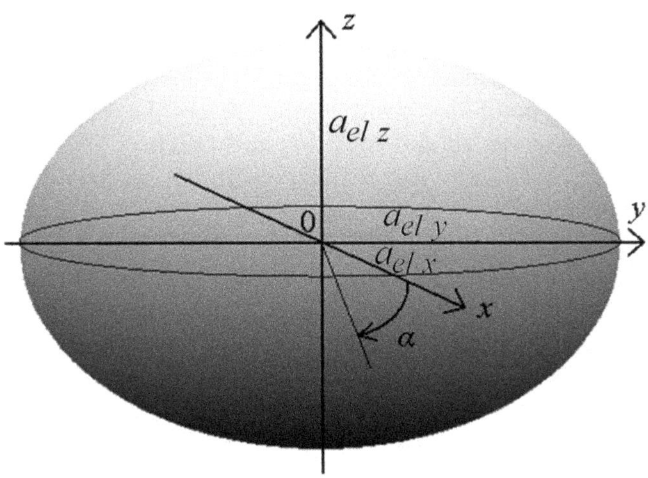

Figure 1.69. Model of a raindrop shaped as triaxial ellipsoid.

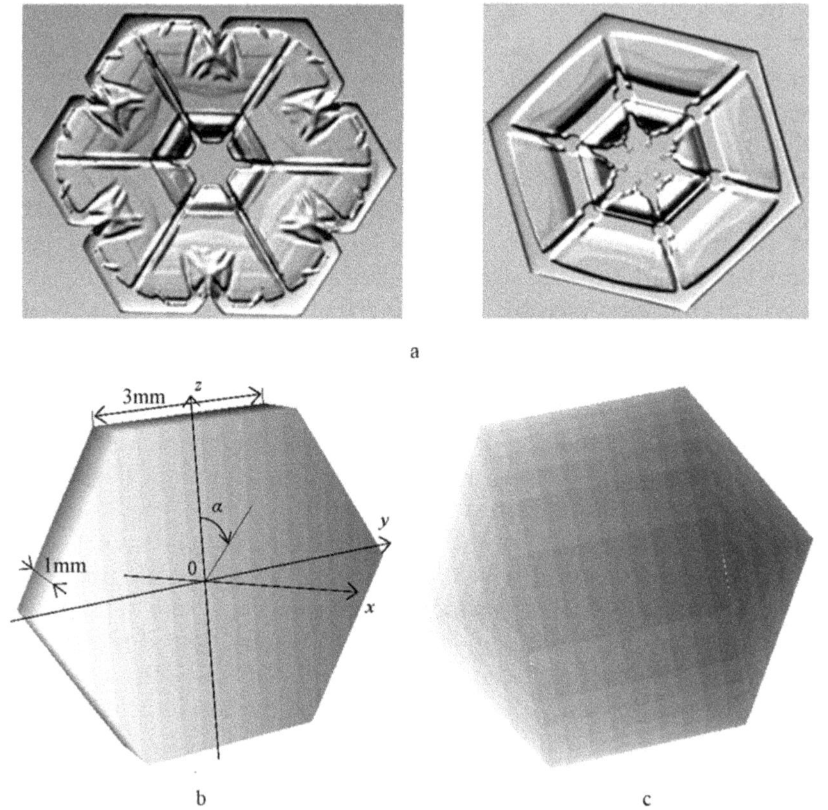

Figure 1.70. Snow crystals: (a) actual hexagon shapes of snow crystals; (b) model of a hexagon shaped snow crystal; and (c) model of a hexagon shaped snow crystal with smooth edges.

for rational use of computer memory. The latter results in reduced computation time and in wider nomenclature of the scatterer shapes. The algorithm advantages are particularly noticeable with electrically thin scatterers.

The above advantages are illustrated in figure 1.35 and table 1.1 in subsection 1.3.2.

Below, we apply our method to compute the RCS of the most common ellipsoidal raindrops [104, 105] and hexagonal snow crystals [108].

In our models, we assume the hydrometeors to be homogeneous dielectric scatterers. An ellipsoid of revolution with its major semi-axes $a_{elx} = a_{ely}$ and its minor semi-axis a_{elz} was employed as a raindrop model (figure 1.69). The semi-axes of the ellipsoid (raindrop model) were related to the so-called average cubic radius of the raindrop (radius of the equivalent sphere) r_3, and they were computed using the well-known formulas from [104]. Below, we present the results of calculations for two raindrops with $r_3 = 1.5$ mm ($a_{elx} = a_{ely} = 1.587$ mm, $a_{elz} = 1.34$ mm) and $r_3 = 3$ mm ($a_{elx} = a_{ely} = 3.449$ mm, $a_{elz} = 2.27$ mm). In our computations, we used the values of complex relative permittivity equal to that of water. They were calculated by the known formulas [107] as $\varepsilon_2 = 62.44 + j31.93$ at $\lambda_1 = 3.2$ cm (for $f = 9.375$ GHz); and $\varepsilon_2 = 17.91 + j27.83$ at $\lambda_1 = 0.8$ cm (for $f = 37.5$ GHz).

The study models of snow crystals are shown in figure 1.70. The first model is a hexagon that is 1 mm thick, its edge length being 3 mm (figure 1.70(b)). The second model is the same hexagon but it has its edges rounded (figure 1.70(c)). According to [109], the relative permittivity of snow crystals is $\varepsilon_2 = 3.17 + j\,1.45 \times 10^{-3}$ at the wavelength $\lambda_1 = 3.2$ cm.

The aspect angle α corresponding to monostatic radar probing belongs to the xOz plane and it is counted off from the Ox axis in the case of raindrops and from the Oz axis in the case of snow crystals, as shown in figures 1.69 and 1.70, respectively.

RCS are presented for the case of horizontal ($\vec{H}^0$, $\vec{H}$ belonged to the plane passing through the illumination direction vector and being perpendicular to the xOy plane) and of vertical ($\vec{H}^0$, $\vec{H}$ were parallel to the xOy plane) polarizations.

Figure 1.71 presents the computed plots of monostatic RCS versus illumination aspect for the case of ellipsoidal raindrop models. Considering the fact that actual raindrops are to be illuminated at random aspect α, we also present the statistical characteristics of their RCS in table 1.12. The table contains the maximum σ_{max}, minimum σ_{min}, average σ_{av}, and median σ_{med} RCS values of raindrops. Let us briefly analyze these results.

The electrical size of the smaller drop ($a_{\text{el}x} = a_{\text{el}y} = 1.587$ mm, $a_{\text{el}z} = 1.34$ mm) given the illumination signal wavelength $\lambda_1 = 3.2$ cm is equal to $2a_{\text{el}x}/\lambda_1 = 0.1$. At this

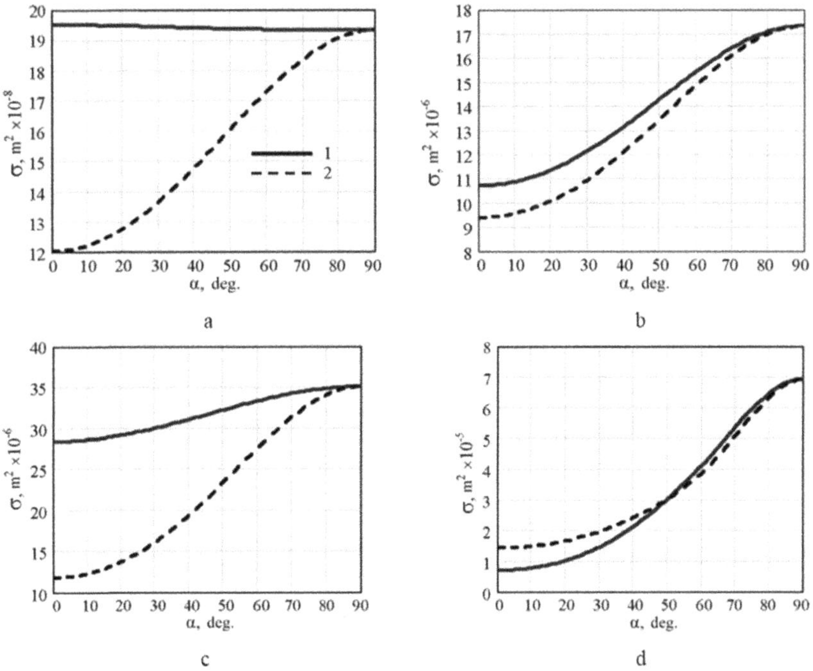

Figure 1.71. Monostatic RCS of ellipsoidal raindrops: (a) and (b) correspond to the smaller raindrop of $r_3 = 1.5$ mm; (c) and (d) correspond to the larger raindrops of $r_3 = 3$ mm; (a) and (c) correspond to illumination signal wavelength $\lambda_1 = 3.2$ cm; (b) and (d) correspond to illumination signal wavelength $\lambda_1 = 0.8$ cm; line #1 presents the case of horizontal polarization; line #2 presents the case of vertical polarization.

Table 1.12. RCS of ellipsoidal raindrops.

Raindrop	λ_1, cm	Polarization	σ, m^2 × 10^{-5}			
			σ_{max}	σ_{min}	σ_{av}	σ_{med}
r_3 = 1.5 mm, ellipsoid,	3.2	hor.	0.0195	0.0193	0.0194	0.0194
$a_{elx} = a_{ely}$ = 1.587 mm,		vert.	0.0193	0.0120	0.0156	0.0154
a_{elz} = 1.34 mm	0.8	hor.	1.73	1.07	1.39	1.37
		vert.	1.73	0.940	1.31	1.28
r_3 = 3.0 mm, ellipsoid,	3.2	hor.	3.51	2.84	3.17	3.16
$a_{elx} = a_{ely}$ = 3.449 mm,		vert.	3.51	1.18	2.24	2.14
a_{elz} = 2.27 mm	0.8	hor.	6.92	0.699	3.14	2.55
		vert.	6.91	1.43	3.34	2.70

wavelength the RCS of a raindrop, depending on its orientation, deviates by less than 1% from the value corresponding to its illumination along the minor semi-axis (figure 1.69) in the case of horizontal polarization and by less than 38% for the case of vertical polarization (figure 1.71(a), table 1.12).

At the same time, the well-known formula for calculating the RCS of a dielectric sphere, which is small electrically, that was obtained using the Mie series [46],

$$\sigma = \pi^5 \frac{(2r_3)^6}{\lambda_1^4} \left| \frac{\varepsilon_2 - 1}{\varepsilon_2 + 2} \right|^2, \tag{1.117}$$

yields the RCS of the equivalent sphere (r_3 = 1.5 mm) that is equal to $\sigma = 1.972 \times 10^{-7}$ Z m^2. This RCS coincides (within an accuracy of 1%–2%) with the RCS computed by us for a raindrop given its illumination at horizontal polarization. However in the case of vertical polarization, the actual RCS depends on the illumination aspect and it differs significantly from the result predicted by (1.117). The latter indicates that even for small electrical sizes of such radar scatterers, the exact calculation methods and in particular those based on solving the IE should be employed to evaluate the RCS of raindrops.

The plots above show that the RCS of an ellipsoidal raindrop with the equivalent radius r_3 = 3 mm ($a_{elx} = a_{ely}$ = 3.449 mm, a_{elz} = 2.27 mm, $2a_{elx}/\lambda_1 = 0.22$) has a stronger dependence on the aspect of its illumination α (figure 1.71(c), table 1.12). In the case of the shorter wavelength of λ_1 = 0.8 cm, the raindrops become bigger electrically, which makes the dependence of their RCS on the aspect α even more noticeable (figures 1.71(b) and (d), table 1.12).

Figure 1.72 and table 1.13 present the monostatic RCS that was computed for the case of snow crystals given their illumination at the wavelength of λ_1 = 3.2 cm.

For the sake of comparison, table 1.13 also contains the statistical characteristics of scattering by a dielectric triaxial ellipsoid (a_{elx} = 1 mm, a_{ely} = 3 mm, a_{elz} = 1.5 mm) and by cylinder with a base radius of a_c = 3 mm and a height of d_c = 1 mm (the axis of the cylinder coinciding with the Ox axis). The ellipsoid and

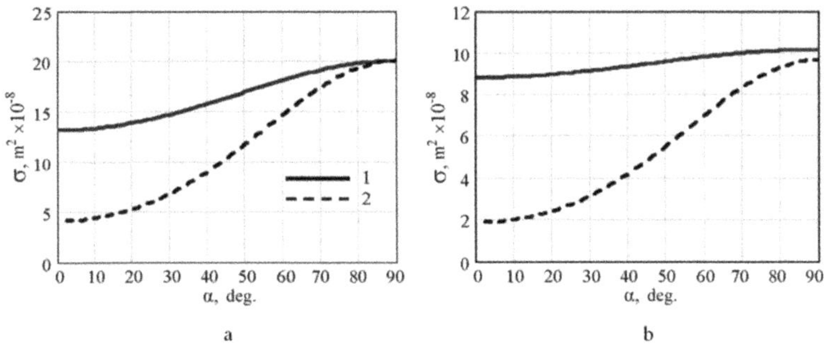

Figure 1.72. Monostatic RCS of snow crystals at the wavelength of $\lambda_1 = 3.2$ cm: (a) hexagon, (b) hexagon with smooth edges; line #1 corresponds to horizontal polarization; line #2 corresponds to vertical polarization.

Table 1.13. RCS of dielectric scatterers with relative permittivity of snow at the wavelength of $\lambda_1 = 3.2$ cm.

		σ, m$^2 \times 10^{-7}$			
Scatterer	Polarization	σ_{max}	σ_{min}	σ_{av}	σ_{med}
Snow crystal (hexagon)	hor.	2.01	1.32	1.65	1.64
	vert.	2.01	0.413	1.12	1.03
Snow crystal (smoothed hexagon)	hor.	1.02	0.881	0.949	0.949
	vert.	0.968	0.187	0.532	0.486
Ellipsoid ($a_{elx} = 1$ mm, $a_{ely} = 3$ mm,	hor.	0.374	0.359	0.366	0.366
$a_{elz} = 1.5$ mm)	vert.	0.231	0.0732	0.146	0.144
Cylinder ($a_{cyl} = 3$ mm, $d_{cyl} = 1$ mm)	hor.	2.97	2.09	2.52	2.50
	vert.	2.97	0.538	1.61	1.46

the cylinder have relative permittivities corresponding to that of snow crystals ($\varepsilon_2 = 3.17 + j1.45 \times 10^{-3}$).

Comparison of the RCS plots shown in figures 1.72(a) and (b) (hexagon and smoothed hexagon) shows that, despite the small electrical sizes, the edges of the snowflake make a fairly significant contribution to its RCS. The smoothed edges lead to a decrease in the intensity of the snowflake backscattering.

Finally, the plots of figures 1.71 and 1.72 and the data of tables 1.12 and 1.13 clearly show that as far as the radar signals at wavelengths of 3.2 cm and 0.8 cm are concerned, the RCS of hydrometeors strongly depends on their shape. Therefore, accurate prediction of radar scattering by hydrometeors should take into account the variety of their shape and EM parameters.

The numerical method presented above enables studying the scattering characteristics of hydrometeors of resonant sizes that can have various shapes. Such characteristics are of great practical importance for engineers dealing with the design of algorithms aimed at processing echo signals in weather radar, as well as algorithms aimed at finding the microphysical parameters of meteorological formations.

1.4.5 Electromagnetic scattering and radiation characteristics of an antenna inside a dielectric radome of resonant size

In this subsection we consider a calculation technique that could be used for reliable simulation of the EM radiation and scattering by antenna placed inside a dielectric radome of resonance size. Our technique is based on solving the set of boundary MIEs. The peculiarities of the proposed algorithm are discussed briefly. Calculation results for a model of a resonance-sized dielectric radome are demonstrated.

The ASs of objects in the air such as aircraft and missiles that are equipped with homing heads are usually placed under the nose dielectric fairings (radomes). On the one hand, the requirements for structural strength and radio transparency are fairly stringent for such radomes. On the other hand, the antennas inside radomes should not significantly increase the total RCS of their platforms. Therefore, qualitative and quantitative data about the scattering and radiation performance of antennas inside radomes are a subject of acute interest for specialists involved in the development of advanced airborne radar systems. At the same time, the scattering characteristics of antennas inside radomes are also necessary for the assessment of the RCS figures of airborne objects. These data can also be used for developing algorithms for aircraft radar detection.

To address the problems mentioned above we employed computer simulation. Common airborne radar systems are equipped with dielectric radomes that are electrically large, in which case the AHFMs are commonly applied [110–114] (a more detailed list of publications can be found in the sources mentioned). However, some airborne objects such as special purpose drones and decoy missiles employ smaller radomes, the geometrical dimensions of which are commensurate with the radar signal wavelength. It is obvious then that the AHFMs cannot provide the necessary accuracy of computation. A possible solution is the use of numerical methods based on solving the boundary IEs, which have important advantages in modeling the EM scattering characteristics of resonance-sized dielectric objects [23, 31–33, 41, 51, 59–61, 63, 64]. Unfortunately, quantitative data regarding the radiation and scattering performance of ASs inside dielectric radomes of resonance size are practically absent in the literature known to the authors.

In section 1.3 above, we described a technique for modeling the EMW scattering from homogeneous dielectric objects based on solving the MIE set [23, 55].

The sketch of an antenna placed inside a homogeneous dielectric radome V_2 (with surface S, and relative permittivity $\varepsilon_2 = \varepsilon'_2 + j\varepsilon''_2$) located in free space V_1 (with $\varepsilon_1 = 1$) is shown in figure 1.73. The permittivity ε_2 is a function of the wavelength, and the relative permeability is $\mu = 1$ in both V_1 and V_2. The radiating antenna was represented by the two elementary electric dipoles with vector-moments $\vec{p}^{e01}$ and $\vec{p}^{e02}$ spaced along the y-axis by a quarter of the sounding wavelength λ_1. The task is to obtain the field $(\vec{E}, \vec{H})$ in free space V_1 that was radiated by antenna placed inside the dielectric radome V_2. In addition, the calculation of the field $(\vec{E}, \vec{H})$ scattered by

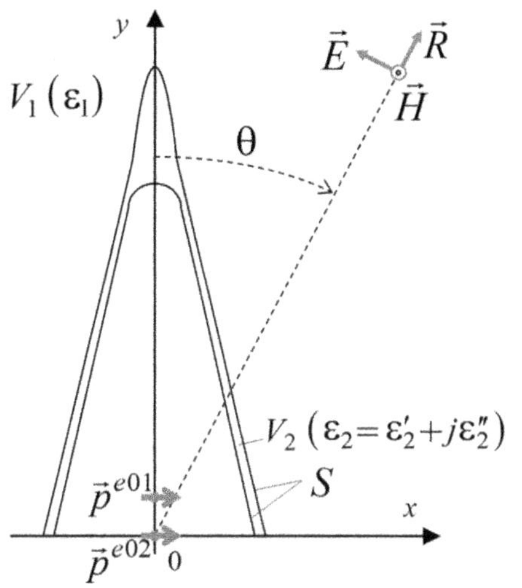

Figure 1.73. Cross-section of the model geometry with regard to the EMF ($\vec{E}$, $\vec{H}$) in free space V_1 that was radiated by antenna formed by two electric dipoles located inside a homogeneous dielectric radome V_2 defined by surface *S*.

the radome at any point in free space, given that a plane wave was incident on it, can also be considered. Thus, the numerical method must yield the DPs of the transmitting or receiving antenna if it was placed inside the radome or it must output the RCS of such an AS.

On the basis of equations (1.91), (1.97), and (1.98), a numerical algorithm was developed, which allows calculation of the components of the field scattered by a 3D dielectric radome of arbitrary shape, given various polarization as well as spatial and time–frequency parameters of the radar probing signal. In this subsection, we demonstrate some calculation results with regard to a resonance-sized nose dielectric radome.

Two elementary electric dipoles characterized by vector-moments $\vec{p}^{e01}$ and $\vec{p}^{e02}$ were used as radiating antenna. The centers of dipoles were located at the y-axis at a distance of $\lambda/4$ from each other, as shown in figure 1.73. As long as the free space field is concerned, the dipoles generated the fields $\left(\vec{E}^{01}, \vec{H}^{01}\right)$ and $\left(\vec{E}^{02}, \vec{H}^{02}\right)$, respectively. The total field was represented as follows:

$$\vec{E}^{0} = \vec{E}^{01} + \vec{E}^{02}\exp(-j\pi/2), \tag{1.118}$$

$$\vec{H}^{0} = \vec{H}^{01} + \vec{H}^{02}\exp(-j\pi/2), \tag{1.119}$$

In computations, we considered a 3D axis-symmetric homogeneous dielectric radome that had a relative dielectric permittivity $\varepsilon_2 = 7.3 + j0.0219$ at the wavelength $\lambda_1 = 7.5$ cm, and $\varepsilon_2 = 7.3 + j0.02044$ at $\lambda_1 = 7$ cm. A cross-section of the radome's geometry is shown in figure 1.73.

The inner surface of the radome is formed by the lateral surface of the truncated cone with bases $r_{\text{in}1} = 0.71$ cm and $r_{\text{in}2} = 2.697$ cm, and with height $L_{\text{in}} = 9.8056$ cm.

A hemisphere with a radius of 0.71 cm caps the upper base. The external surface of the radome is also formed by the lateral face of a similar truncated cone with the following parameters: $r_{ex1} = 0.5954$ cm, $r_{ex2} = 2.977$ cm, $L_{ex} = 11.2$ cm. A two-axial ellipsoid with semi-axes $a_x = a_z = 2.1898$ cm and $a_y = 74.322$ cm, its center being at the point with coordinates $x_0 = z_0 = 0$, $y_0 = -60.322$ cm, forms the upper side of radome's external surface. Thus, the radome had the following parameters: height $L_{rad} = 14$ cm, diameter of the base $d_{rad} = 5.954$ cm, and wall thickness $\Delta_{rad} \approx 2.8$ mm.

Given the above model of the radome, we generated a mesh of $N = 780$ current nodes to cover its entire surface for radar probing wavelengths of 7.5 and 7 cm. The maximum difference between our estimates of the EMF scattered by the radome in the most important sector of angles θ that were obtained for a different number of nodes of $N = 780$ and of $N = 607$ did not exceed 1%. A further increase in N produced a decrease in the value of the maximum error, which was in full agreement with the theoretical proof of convergence, as guaranteed by the Fredholm nature of the MIE set.

Normalized radiation patterns (RPs) $F(\theta) = |\vec{E}(\theta)|$ of the considered antenna inside a dielectric radome (figure 1.73) are presented in figures 1.74 and 1.75 for $\lambda_1 = 7.5$ cm, $L_{rad} = 1.867\ \lambda_1$, $d_{rad} = 0.794\ \lambda_1$, and $\Delta_{rad} \approx 0.037\ \lambda_1$, and in figures 1.76, 1.77 for $\lambda_1 = 7$ cm, $L_{rad} = 2\ \lambda_1$, $d_{rad} = 0.851\ \lambda_1$, and $\Delta_{rad} \approx 0.04\ \lambda_1$. The angle θ varied in the xy-plane, and $\theta = 0$ corresponded to the direction along the radome axis.

RPs in the E-plane ($\vec{p}^{\,e01}$ and $\vec{p}^{\,e02}$ are parallel to the x-axis; the $\vec{E}$ component of the scattered EMF belongs to the xy-plane and is orthogonal to the propagation direction $\vec{R}$, see figure 1.73) and in the H-plane ($\vec{p}^{\,e01}$, $\vec{p}^{\,e02}$ are also parallel to the x-axis; the $\vec{E}$ vector is orthogonal to both the xy-plane and to $\vec{R}$) are presented in figures 1.74, 1.76, and figures 1.75, 1.77, respectively.

The RPs of an antenna in free space without any radome are shown by the thin black lines, and the RPs of the AS with a radome are shown by the bold gray lines.

The RPs obtained in the E-plane reveal the collimation effect of the radome on the radiated field in the axial direction. The impact of the radome on the RP in the H-plane manifests in the appearance of a strong sidelobe (in the direction $\theta = 73°$ in figure 1.75; and in the direction $\theta = 71.5°$ in figure 1.77). In general, for a radiation

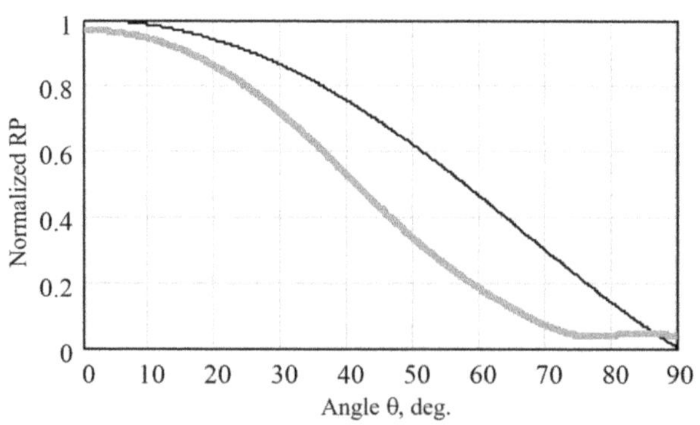

Figure 1.74. The RP of the considered AS at $\lambda_1 = 7.5$ cm in the E-plane.

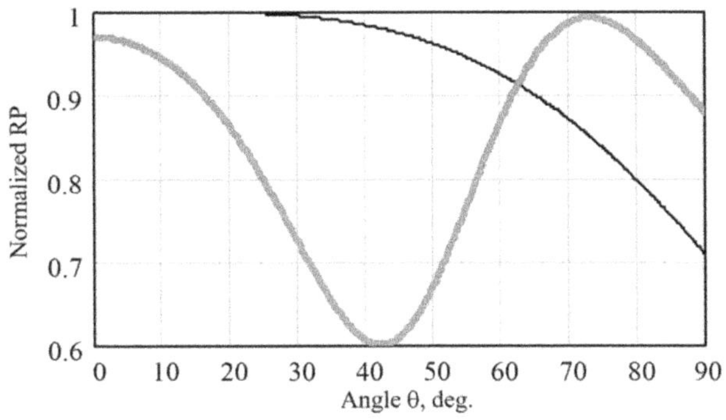

Figure 1.75. The RP of the considered AS at $\lambda_1 = 7.5$ cm in the H-plane.

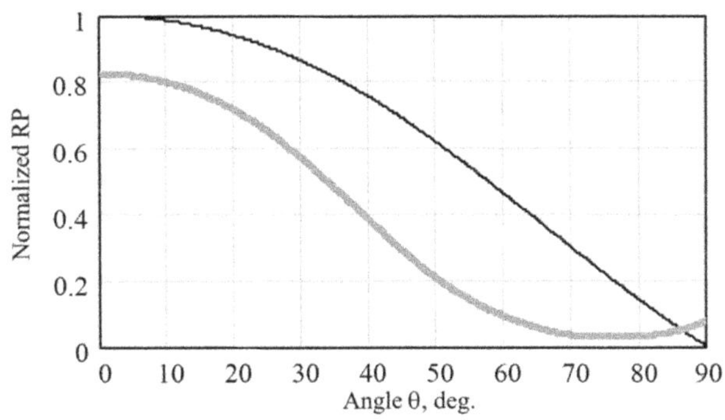

Figure 1.76. The RP of the considered AS at $\lambda_1 = 7$ cm in the E-plane.

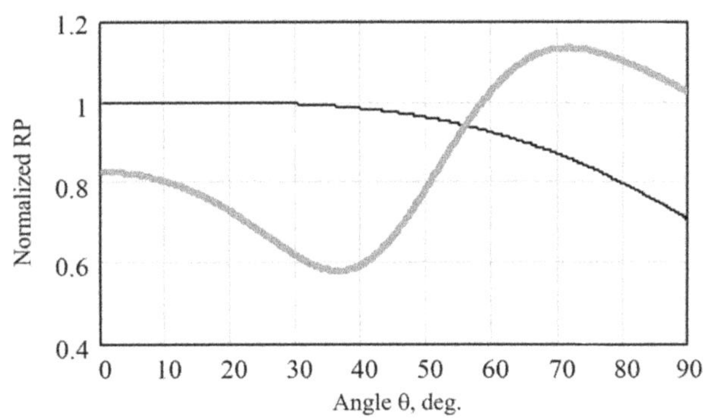

Figure 1.77. The RP of the considered AS at $\lambda_1 = 7$ cm in the H-plane.

wavelength of $\lambda_1 = 7.5$ cm, the drop in the radiation performance caused by the radome (within the sector of angles $\theta = \pm 40°$) did not exceed 3.16 dB in the E-plane, and 4.22 dB in the H-plane.

For a wavelength of $\lambda_1 = 7$ cm, the maximum degradation of radiation performance was below the values of 5.97 dB and 4.33 dB, respectively.

The presented results show that the radome was quite well matched to the wavelength of $\lambda_1 = 7.5$ cm. If the wavelength is changed to 7 cm (a frequency change of 7%), the matching becomes worse. However, the difference was not significant.

The plots of monostatic RCS pertaining to the considered dielectric radome alone (without antenna) as a function of the radar probing aspect θ are presented in figure 1.78 for the E-plane ($\vec{E}^0$, $\vec{E}$ belong to the xy-plane, orthogonal to the propagation direction, thin black line) and for the H-plane ($\vec{E}^0$, $\vec{E}$ are orthogonal to both the xy-plane and the propagation direction, bold gray line) for a radar probing wavelength of $\lambda_1 = 7.5$ cm. It can be seen that this particular radome had a very small RCS, which did not exceed 5.29×10^{-3} m^2. Thus it would not make a considerable contribution to the total RCS of the airborne object.

The results presented above clearly demonstrate that our algorithm based on solving the MIE set can be employed successfully for predicting the scattering characteristics of various 3D dielectric objects of resonance size that are of arbitrary shapes. In addition, when complemented with a MFIE, this algorithm can be used for predicting the scattering characteristics of complex objects that have both dielectric and metallic components.

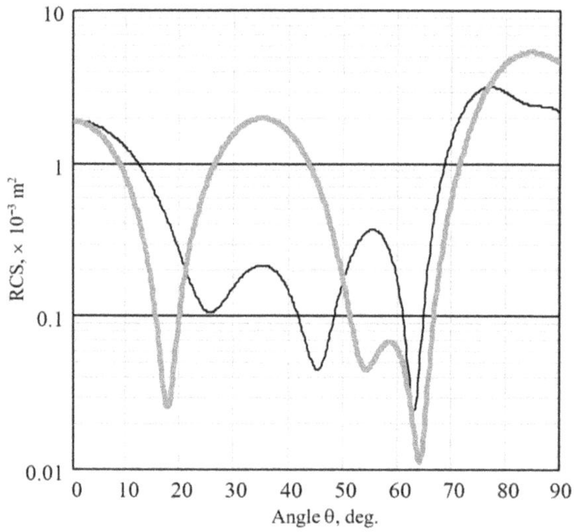

Figure 1.78. Monostatic RCS pertaining to the dielectric radome alone versus the aspect angle of radar probing.

1.4.6 Electromagnetic wave scattering by nanoparticles in the optical band

Some content in this section has been reproduced with permission from [125]. Copyright 2018 IEEE.

In this subsection, a mathematically convergent method for predicting the 3D EMW scattering by metal nanoparticles is presented for the optical wavelength band. The method is based on solving the MIE set. The formulas are derived for the components of the optical theorem, namely: the total scattering cross-section (TSCS) and extinction cross-section (ECS). Sample results of these characteristics computed using the above MIE sets are shown for various angles of incidence and various polarizations of a planar EMW that is incident on a nanoparticle characterized by two different shapes and electric sizes. These results demonstrate that the resonances in the surface-plasmon (SP) modes occur at certain wavelengths.

Nanoparticles and their arrays are considered as enabling technologies in many important applied problems associated with the use of optical systems. In particular, the resonance effects in the scattering and absorption of light by noble-metal nanosized particles are being studied intensively. These resonances make it possible to obtain additional advantages in the performance of optical antennas, sensors, and absorbers, to increase the intensity of fluorescence signals in the optical band, etc. Here, numerical modeling is used widely to reduce the time and effort needed in the design and development in the nanoscale where technologies are very costly.

In the visible optical band (wavelengths from 300 nm to 900 nm), a metal nanoparticle, which has dimensions measured in tens to hundreds of nanometers, can be considered as a scatterer filled with a homogeneous material with a complex dielectric permittivity that depends on the wavelength and has a negative real part. Therefore, in the scattering analysis, one can apply such a powerful instrument as the MIE set, originally developed to analyze the scattering by conventional dielectric scatterers [23].

Previously, considerable attention was paid to the application of MIEs for solving 2D wave scattering problems. This enabled the dependable prediction of the scattering and absorption characteristics of thin metal and dielectric strips and wires of various cross-sections [63, 115–117]. In addition, eigenvalue problems adapted to the analysis of the modal thresholds of 2D microcavity lasers were also studied with the aid of MIEs in [118, 119]. Simultaneously, papers touching on the use of this powerful approach in 3D EMW scattering are scarce [23, 61, 120, 121].

In this subsection we apply the 3D MIE method, which was coupled with the original boundary meshing scheme developed earlier by us to study dielectric scatterers of arbitrary shape [23], to the case of a plasmonic metal scatterer. Since the metals are lossy in the visible band, we derive expressions for TSCS and ECS as well as determining the absorption cross-section (ACS) using the optical theorem [122, 123]. The sample results of the calculation of TSCS and ACS are presented for the case of two silver nanoparticles shaped as ellipsoids-of-revolution versus the wavelength for different sizes and shapes. We also present some results regarding different angles of planar EMW incidence and various wave polarizations.

The geometry and notations adopted below for the case of plane wave scattering by a silver nanoparticle V_2 (simulated here by a homogeneous ellipsoid of revolution with semi-axes $a_x = a_y$, a_z, surface S, and relative permittivity $\varepsilon_2 = \varepsilon'_2 + j\varepsilon''_2$) located in free space V_1 (with $\varepsilon_1 = 1$) is shown in figure 1.79. The permittivity ε_2 is a function of the wavelength, and the relative permeability is $\mu = 1$ both in V_1 and V_2 (nonmagnetic materials). The incident field is denoted as $\left(\vec{E}^0, \vec{H}^0\right)$. The objective is to obtain the scattered field $(\vec{E}^{sc}, \vec{H}^{sc})$ using the classical EM boundary-value problem for 3D Maxwell equations coupled with associated boundary and radiation conditions. Our eventual goal is to analyze the scattering characteristics of such

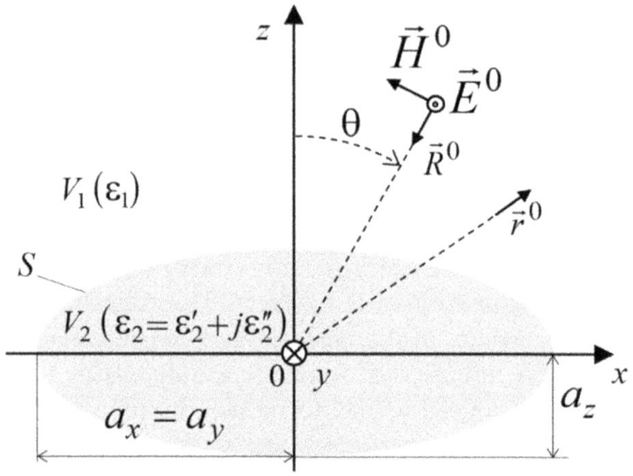

Figure 1.79. Cross-section of the problem's geometry: plane wave scattering by a particle shaped as ellipsoid of rotation around the z-axis.

nanoparticles made of silver versus the wavelength and the angle θ of wave incidence for two orthogonal polarizations.

Equivalent electric $\vec{J}^e$ and magnetic $\vec{J}^m$ current densities can be obtained by solving the MIE set (1.91), as was proposed in section 1.3.

As soon as $\vec{J}^e$ and $\vec{J}^m$ are found from a solution to the MIE set (1.91) in the far zone, the electric field $\vec{E}^{sc}$ scattered by nanoparticles in the direction of unit vector $\vec{r}^0$ can be found using the known representation [15, 33]

$$\begin{aligned}\vec{p}\cdot\vec{E}^{sc}(\vec{r}^0) &= jk_1\exp\left[jk_1 r\right]/(4\pi r)\\ &\times\int_S [\vec{J}^e(\vec{Q})\cdot(\vec{p}-\vec{r}^0(\vec{p}\cdot\vec{r}^0))\zeta_0+\vec{J}^m(\vec{Q})\cdot(\vec{p}\times\vec{r}^0)]\\ &\times\exp\,[jk_1(\vec{r}^0\cdot\vec{Q})]\,\mathrm{d}s_Q,\end{aligned} \tag{1.120}$$

where k_1 is the free space wavenumber, ζ_0 is the free space impedance, and r is the distance to the point of observation (field point).

Having introduced two mutually orthogonal unit vectors $\vec{p}_1$ and $\vec{p}_2$ tangential to the sphere of large radius r, we can rewrite representation for the object TSCS as follows [125]:

$$\begin{aligned}\sigma_{sc} &= \int_\Omega\left\{\left|\vec{p}_1\cdot\vec{E}^{sc}\right|^2+\left|\vec{p}_2\cdot\vec{E}^{sc}\right|^2\right\}r^2\sin(\theta)\,\mathrm{d}\theta\,\mathrm{d}\varphi\\ &=\frac{1}{k_1^2}\int_\Omega\left\{|F_1(\theta,\varphi)|^2+|F_2(\theta,\varphi)|^2\right\}\sin(\theta)\,\mathrm{d}\theta\,\mathrm{d}\varphi,\end{aligned} \tag{1.121}$$

where

$$F_\alpha(\theta, \varphi) = -\frac{k_1^2}{4\pi}\int_S \Big[\vec{J}^e(\vec{Q}) \cdot \big(\vec{p}_\alpha - \vec{r}^0(\vec{p}_\alpha \cdot \vec{r}^0)\big)\ \zeta_0 \\ + \vec{J}^m(\vec{Q}) \cdot \big(\vec{p}_\alpha \times \vec{r}^0\big)\Big] \exp\Big[jk_1\big(\vec{r}^0 \cdot \vec{Q}\big)\Big]\, ds_Q, \quad (\alpha = 1, 2). \tag{1.122}$$

For simplicity, the dependence of vectors $\vec{p}_\alpha$ and $\vec{r}^0$ on the angles θ, φ in (1.122) is not shown.

To obtain an equation for the ECS in the 3D case, we used the considerations presented in [122]. As result, the following integral representation was obtained:

$$\sigma_{\text{ext}} = -\frac{4\pi}{k_1^2} \operatorname{Re}\Big\{(\vec{p}_0 \cdot \vec{p}_1) F_1(\theta, \varphi)\,|_{\vec{R}^0} + (\vec{p}_0 \cdot \vec{p}_2) F_2(\theta, \varphi)\,|_{\vec{R}^0}\Big\}, \tag{1.123}$$

where the unit vector $\vec{p}_0$ indicates the orientation of the electric vector $\vec{E}^0$ of the incident plane wave.

In expression (1.123) the functions $F_\alpha(\theta, \varphi)$ are to be calculated in the direction of the incident plane wave $\vec{R}^0$.

According to the optical theorem [122, 123], after we obtain the values σ_{sc} and σ_{ext}, we can compute the ACS as

$$\sigma_{\text{abs}} = \sigma_{\text{ext}} - \sigma_{\text{sc}}. \tag{1.124}$$

In our computations, we used experimental measurements of the complex dielectric function of silver that were tabulated in [124], which we complemented with a spline interpolation algorithm to obtain arbitrary wavelengths. Now, according to these data, the dielectric constant ε'_2 of silver varies from 1 to 40 as the wavelength changes from 300 nm to 900 nm.

Figures 1.80 and 1.81 present the plots of normalized TSCS and ACS spectra for the case of two silver nanoparticles that differed slightly in shape and size. They are modeled as ellipsoids. The first had the semi-axes $a_x = a_y = a = 75$ nm, $a_z = 25$ nm, and the other had semi-axes $a = 100$ nm and $a_z = 20$ nm. The normalized TSCS and ACS have been computed using the numerical algorithm developed above that was based on a solution to the MIE set with subsequent computation of the functions representing the σ_{sc}, σ_{ext}, and σ_{abs} using (1.121)–(1.124). The plots of figures 1.80 and 1.81 have been obtained to a relative accuracy of not greater than 0.3% across the entire visible band, the accuracy being controlled by the sampling rate. We found that the number of elementary mesh cells required to represent the ellipsoid's boundary in order to provide for such accuracy was 708. The latter value can seem to be unexpectedly large given that we are dealing with a 1/10 wavelength size particle. However, in this case the IEs depend not only on the scatterer's size (say, its largest diameter) but also on its curvature. The latter dependence manifests via the terms that contain the source-function derivatives.

In our calculations, we assumed that the angle of the plane EMW incidence took the values of $\theta = 0$ (which means that $\vec{R}^0$ is to be parallel to the z-axis, or broadside illumination, see figures 1.79), 30°, 60° and 90°. The particle was illuminated by the

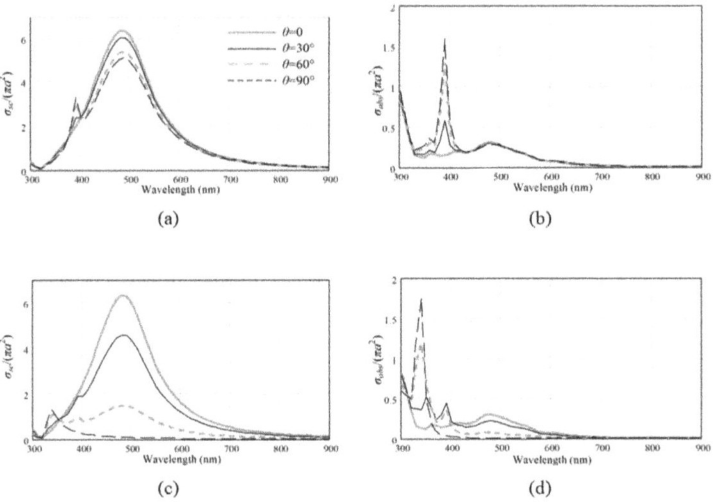

Figure 1.80. Normalized TSCS (a) and (c) and ACS (b) and (d) for a silver nanoparticle shaped as an ellipsoid with semi-axes $a_x = a_y = a = 75$ nm and $a_z = 25$ nm versus the wavelength for various incidence angles θ of the horizontally (a) and (b) and of vertically (c) and (d) polarized incident plane EMW.

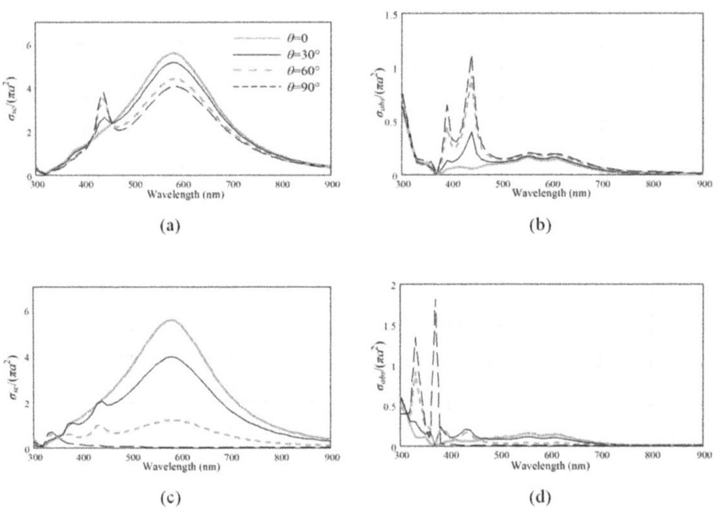

Figure 1.81. Normalized TSCS (a) and (c) and ACS (b) and (d) for a silver nanoparticle shaped as an ellipsoid with semi-axes $a_x = a_y = a = 100$ nm and $a_z = 20$ nm versus the wavelength given various incidence angles θ of the horizontally (a) and (b) and vertically (c) and (d) polarized plane incident EMW.

waves of horizontal ($\vec{E}^0$ being parallel to the y-axis, and $\vec{H}^0$ belonging to the $x0z$ plane, figure 1.79) and of vertical ($\vec{H}^0$ being parallel to the y-axis, and $\vec{E}^0$ belonging to the $x0z$ plane) polarization. Plots representing the scattering and absorption were obtained for the wavelengths from 300 to 900 nm, with an increment of 10 nm.

The results show that for an angle of wave incidence $\theta = 0$ the plots of TSCS contain a single hump, which is due to the principal SP mode. The hump's maximum was observed at the wavelength of 490 nm for the smaller particle, and at 580 nm for the larger particle. The maximum value of TSCS decreases as the deviation of the angle of incidence goes away from the broadside. In addition, as angle θ increases, additional humps due to other SP modes appear at the shorter wavelengths. The spectra of ACS demonstrate the same dynamics except for the fact that the higher-order SP modes (appearing if $\theta \neq 0$) feature more pronounced resonances (for the absorption) than the principal one. In the ultraviolet part of the optical band, ACS goes up fairly quickly because of increased losses in the bulk silver.

As anticipated, the SP-mode resonances shift to the shorter wavelengths as the particle itself or its visible cross-section become smaller. The presence or absence of specific resonances due to the higher-order SP modes depends on the angle of incidence and polarization of the incident wave.

Thus the method for numerical simulation of 3D EM scattering characteristics of metal nanoparticles of arbitrary shape is based on solving the MIEs. The method allows TSCS and ACS to be predicted for silver nanoparticles as functions of the wavelength in the visible band. It accounts for various incidence angles and polarizations of the plane incident wave. The resulting plots reveal the fact that SP resonances of ellipsoidal silver nanoparticles depend on the incidence angle and polarization. The algorithm presented above can be applied to the analysis of resonance scattering and absorption by various plasmonic nanoparticles.

References

[1] Sukharevsky I O, Zalevsky G S, Nechitaylo S V and Sukharevsky O I 2010 Electromagnetic wave scattering by a circular PEC plate of finite thickness *Electromagn. Waves Electron. Syst.* **15** 42–7

[2] Zalevsky G, Nechitaylo S, Sukharevsky O and Sukharevsky I 2010 EM wave scattering by perfectly conducting disc of finite thickness *Proc. of the 13th Int. Conf. on Mathematical Methods in Electromagnetic Theory (Kyiv, 6–8 September)* pp 316–9

[3] Sukharevsky I O, Nechitaylo S V, Ivanchenko D D and Melezhik P N 2012 Electromagnetic wave scattering by a curved screen of finite thickness *Telecommun. Radio Eng.* **71** 413–22

[4] Kyurkchan A G and Smirnova N I 2009 The auxiliary current and the null field methods *Electromagn. Waves Electron. Syst.* **14** 4–12

[5] Kyurkchan A G and Smirnova N I 2009 Solution of wave diffraction problems by the null-field method *Acoust. Phys.* **55** 691–7

[6] Stratton J A 2007 *Electromagnetic Theory* (Hoboken, NJ: Wiley)

[7] Silver S (ed) 1949 *Microwave Antenna Theory and Design* (New York: McGraw-Hill)

[8] Leontovich M A 1948 About approximate boundary conditions for an electromagnetic field on the surface of well-conducting bodies *Radiowave Propagation Research. Collection 2* (Moscow: USSR Academy of Sciences) pp 5–12 (in Russian)

[9] Appolonskiy S M and Yerofeenko V T 1993 Generalization of Leontovich boundary conditions for non-stationary fields *Electr. J.* **12** 64–7 (in Russian)

[10] Ramo S, Whinnery J R and Duzer T V 1994 *Fields and Waves in Communication Electronics* 3rd edn (Hoboken, NJ: Wiley)

[11] Nechitaylo S V, Sukharevsky I O, Ivanchenko D D and Melezhik P N 2010 Diffraction by a curved arbitrary screen of finite thickness *Proc. of the 13th Int. Conf. on Mathematical Methods in Electromagnetic Theory (Kyiv, 6–8 September)* pp 358–61

[12] Vinogradov S S, Smith P D and Vinogradova E D 2002 *Canonical Problems in Scattering and Potential Theory. Part 2: Acoustic and Electromagnetic Diffraction by Canonical Structures* (London/Boca Raton, FL: Chapman and Hall/CRC)

[13] Vasiliev E N 1987 *Excitation of Bodies of Revolution* (Moscow: Radio and Communication) (in Russian)

[14] Bakhvalov N S, Zhidkov N P and Kobel'kov G M 2008 *Numerical Methods* (Moscow: BINOM. Laboratoriya znanii) (in Russian)

[15] Sukharevsky O I, Vasylets V O, Nechitaylo S V and Orlenko V M 2014 *Electromagnetic Wave Scattering by Aerial and Ground Radar Objects* ed O I Sukharevsky (Boca Raton/London: CRC Press/Taylor and Francis) p 334

[16] Ivanchenko D D and Sukharevsky I O 2010 Backscattering measurements for metallic unclosed spherical screens *Telecommun. Radio Eng.* **69** 423–8

[17] Nechitaylo S, Orlenko V, Sukharevsky O and Vasylets V 2018 Electromagnetic wave scattering by a screen of finite thickness and conductivity *Telecommun. Radio Eng.* **77** 1409–21

[18] Sukharevsky O, Nechitaylo S, Orlenko V and Vasilets V 2018 Frequency response of impedance screens of finite thickness in the resonance band *Proc. of the 9th Int. Conf. on Ultrawideband and Ultrashort Impulse Signals (Odessa, 4–7 September)* pp 28–33

[19] Goldstein L D and Zernov N V 1971 *Electromagnetic Fields and Waves* (Moscow: Soviet Radio)

[20] Lazorenko O V and Chernogor L F 2009 *Ultrawideband Signals and Processes: Monograph* (Kharkov: VN Karazin Kharkov National University)

[21] Milligan T A 2005 *Modern Antenna Design* (Hoboken, NJ: Wiley)

[22] Zalevsky G S and Sukharevsky O I 2014 Calculation of scattering characteristics of aerial radar objects of resonant sizes based on iterative algorithm *Radioelectron. Commun. Syst.* **57** 244–53

[23] Sukharevsky O I, Zalevsky G S and Vasilets V A 2016 Modeling of ultrawideband (UWB) impulse scattering by aerial and subsurface resonant objects based on integral equation solving *Advanced Ultrawideband Radar: Signals, Targets, and Applications* ed J D Taylor (Boca Raton/London: CRC Press/Taylor and Francis) ch 5 pp 195–235

[24] Milligan T A 2005 *Modern Antenna Design/by Thomas A Milligan* 2nd edn (Hoboken, NJ: Wiley)

[25] Mittra R 1973 *Computer Techniques for Electromagnetics* (Oxford: Pergamon)

[26] Rius J M, Úbeda E and Parron J 2001 On the testing of the magnetic field integral equation with RWG basis functions in method of moments *IEEE Trans. Antennas Propag.* **49** 1550–3

[27] Gürel L and Ergül Ö 2005 Singularity of the magnetic field integral equation and its extraction *IEEE Antennas Wirel. Propag. Lett.* **4** 229–32

[28] Ergül Ö and Gürel L 2007 Linear-linear basis functions for MLFMA solutions of magnetic field and combined field integral equations *IEEE Trans. Antennas Propag.* **55** 1103–10

[29] Gibson W C 2008 *The Method of Moments in Electromagnetics* (Boca Raton/London: Chapman and Hall/Taylor and Francis)

[30] Ylä-Oijala P, Taskinen M and Järvenpää S 2009 Advanced surface integral equation methods in computational electromagnetics *Proc. of Int. Conf. on Electromagnetics in Advanced Applications (Turin, 14–18 September)* pp 369–72

[31] Ubeda E, Tamayo J M and Rius J M 2011 Taylor-orthogonal basis functions for the discretization in method of moments of second kind integral equations in the scattering analysis of perfectly conducting or dielectric objects *Prog. Electromagn. Res.* **119** 85–105

[32] Yan S, Jin J-M and Nie Z 2011 Improving the accuracy of the second-kind Fredholm integral equations by using the Buffa-Christiansen functions *IEEE Trans. Antennas Propag.* **59** 1299–310

[33] Volakis J L and Sertel K 2012 *Integral Equation Methods for Electromagnetics* (Raleigh, NC: SciTech)

[34] Ubeda E, Tamayo J M, Rius J M and Heldring A 2013 Stable discretization of the electric-magnetic field integral equation with the Taylor-orthogonal basis functions *IEEE Trans. Antennas Propag.* **61** 1484–90

[35] Eibert T F 2007 Some scattering results computed by surface-integral-equation and hybrid finite-element–boundary-integral techniques, accelerated by the multilevel fast multipole method *IEEE Antennas Propag. Mag.* **49** 61–9

[36] Liu Y, Guo X-Y and Zhou H-J 2018 A memory saving augmented EFIE with modified basis functions for low-frequency problems *IEEE Trans. Antennas Propag.* **66** 1359–65

[37] Schröder A, Renker M and Aulenbacher U 2015 Numerical and experimental radar cross section analysis of the quadrocopter DJI Phantom 2 *Proc. of IEEE Radar Conf. (Johannesburg, October)* pp 463–8

[38] Ylä-Oijala P *et al* 2018 Surface integral equation-based characteristic mode formulation for penetrable bodies *IEEE Trans. Antennas Propag.* **66** 3532–9

[39] Li T *et al* 2019 Optimized radar waveform parameter design for small drone detection based on echo modeling and experimental analysis *IEEE Access.* **7** 101527–38

[40] Wang J *et al* 2020 Simulation of 3-D electromagnetic scattering and inverse scattering by arbitrary anisotropic dielectric objects embedded in layered arbitrary anisotropic media *IEEE Trans. Antennas Propag.* **68** 6473–8

[41] Kong B-B *et al* 2018 A discontinuous Galerkin surface integral equation method for scattering from multiscale homogeneous objects *IEEE Trans. Antennas Propag.* **66** 1937–46

[42] Liu Y *et al* 2018 A wavelet-enhanced PWTD-accelerated time-domain integral equation solver for analysis of transient scattering from electrically large conducting objects *IEEE Trans. Antennas Propag.* **66** 2458–70

[43] Tasic M S and Kolundzija B M 2018 Method of moment weighted domain decomposition method for scattering from large platforms *IEEE Trans. Antennas Propag.* **66** 3577–89

[44] Kalfa M, Ergül Ö and Ertürk V B 2018 Error control of multiple-precision MLFMA *IEEE Trans. Antennas Propag.* **66** 5651–6

[45] Zhang Q *et al* 2020 Multilevel fast multipole algorithm enhanced characteristic mode analysis for half-space platform *IEEE Trans. Antennas Propag.* **68** 7711–6

[46] Shirman Y D *et al* 2002 *Computer Simulation of Aerial Target Radar Scattering Recognition, Detection and Tracking* (Norwood, MA: Artech House)

[47] Ufimtsev P Y 2003 *Theory of Edge Diffraction in Electromagnetics* (Encino, CA: Tech Science)

[48] Knott E F, Shaeffer J F and Tuley M T 2004 *Radar Cross Section* 2nd edn (Raleigh, NC: SciTech)

[49] Sukharevsky O I, Vasilets V A and Zalevsky G S 2015 Electromagnetic wave scattering by aerial and ground radar objects *2015 IEEE Int. Radar Conf. (Arlington, VA, 10–15 May)* pp 162–7

[50] Sukharevsky O, Vasilets V, Orlenko V and Ryapolov I 2020 Radar scattering characteristics of a UAV model in X-band *IET Radar Sonar Navig* **14** 532–7

[51] Zalevsky G S, Muzychenko A V and Sukharevsky O I 2012 Method of radar detection and identification of metal and dielectric objects with resonant sizes located in dielectric medium *Radioelectron. Commun. Syst.* **55** 393–404
[52] Zalevsky G S, Sukharevsky O I, Vasilets V A and Nechitaylo S V 2014 Secondary radiation of resonance perfectly conducting objects *J. Commun. Technol. Electron.* **59** 1321–32
[53] Zalevsky G S, Sukharevsky O I, Vasilets V A and Surgai M V 2019 Estimation of radar scattering characteristics of artillery shells in meter, decimeter and centimeter wavelength ranges *Radioelectron. Commun. Syst.* **62** 356–67
[54] Zalevsky G *et al* 2020 High-resolution range profiles of artillery shells in UHF and VHF bands *IEEE Ukrainian Microwave Week, IEEE 10th Int. Conf. on Ultrawideband and Ultrashort Impulse Signals* vol 4 *(Kharkiv, 21–25 September)* pp 1136–9
[55] Zalevsky G S, Sukharevsky O I and Vasilets V A 2021 Integral equation modelling of unmanned aerial vehicle radar scattering characteristics in VHF to S frequency bands *IET Microw. Antennas Propag.* **15** 1299–309
[56] Smirnov V I 1964 *A Course of Higher Mathematics* vol 4 (Oxford: Pergamon)
[57] Colton D and Kress R 1983 *Integral Equation Methods in Scattering Theory* (New York: Wiley-Interscience)
[58] Sirenko Y K, Sukharevsky I V, Sukharevsky O I and Yashina NP 2000 *Fundamental and Applied Problems of the Electromagnetic Wave Scattering Theory* (Kharkov: Krok) (in Russian)
[59] Sukharevsky O and Zalevsky G 1997 Electromagnetic wave scattering by resonance-size buried objects *Telecommun. Radio Eng.* **51** 80–7
[60] Harrington R F 1989 Boundary integral formulations for homogeneous material bodies *J. Electromagn. Waves Appl.* **3** 1–15
[61] Ylä-Oijala P and Taskinen M 2005 Well-conditioned Müller formulation for electromagnetic scattering by dielectric objects *IEEE Trans. Antennas Propag.* **53** 3316–23
[62] Altair Altair Feko: Solving Connectivity, Compatibility, and Radar Challenges
[63] Sukharevsky I, Nosich A and Altintas A 2015 Manipulation of backscattering from a dielectric cylinder of triangular cross-section using the interplay of GO-like ray effects and resonances *IEEE Trans. Antennas Propag.* **63** 2162–9
[64] Sewell P *et al* 2015 Nystrom method for the Muller boundary integral equations on a dielectric body of revolution: axially symmetric problem *IET Microw. Antennas Propag.* **9** 1186–92
[65] Markarian G and Staniforth A 2021 *Countermeasures for Aerial Drones* (Norwood, MA: Artech House)
[66] Ritchie M *et al* 2015 Micro-drone RCS analysis *Proc. IEEE Radar Conf. (Johannesburg, October)* pp 452–56
[67] Patel J, Fioranelli F and Anderson D 2018 Review of radar classification and RCS characterisation techniques for small UAVs or drones *IET Radar Sonar Navig.* **12** 911–9
[68] Cui Z *et al* 2019 Low-altitude UAV air-ground propagation channel measurement and analysis in a suburban environment at 3.9 GHz *IET Microw. Antennas Propag.* **13** 1503–8
[69] Gong J *et al* 2019 Interference of radar detection of drones by birds *Prog. Electromagn. Res.* **81** 1–11
[70] Laučys A *et al* 2019 Investigation of detection possibility of UAVs using low cost marine radar *Aviation* **23** 48–53
[71] Patel J *et al* 2019 Multi-time frequency analysis and classification of a micro-drone carrying payloads using multistatic radar *J. Eng.* **2019** 7047–51

[72] Khristenko A *et al* 2018 Magnitude and spectrum of electromagnetic wave scattered by small quadcopter in X-band *IEEE Trans. Antennas Propag.* **66** 1977–84

[73] Yang Y *et al* 2020 Experimental analysis of fully polarimetric radar returns of a fixed-wing UAV *IET Radar Sonar Navig.* **14** 525–31

[74] Riabukha V 2020 Radar surveillance of unmanned aerial vehicles (review) *Radioelectron. Commun. Syst.* **63** 561–73

[75] Sukharevsky O I, Zalevsky G S and Vasilets V A 2017 Iterative algorithm for simulation of EM scattering by objects, contained constructive elements of different electric sizes *22nd Int. Seminar/Workshop on Direct and Inverse Problems of Electromagnetic and Acoustic Wave Theory (Dnipro, 25–28 September)* pp 190–3

[76] Zalevsky G *et al* 2020 Combined calculation method for the scattering characteristics of complex shaped objects and its application to model helicopter rotor modulation spectra *IEEE Ukrainian Microwave Week, IEEE 6th Int. Symp. on Microwaves, Radar and Remote Sensing* vol 2 *(Kharkiv, 21–25 September)* pp 473–7

[77] Kenyon C and Dogaru T 2015 Numerical computation of the radar cross section of rockets and artillery rounds *Report* No. ARL-TR-7468 Army Research Laboratory, Adelphi, MD https://apps.dtic.mil/dtic/tr/fulltext/u2/a622300.pdf

[78] Barret T W 2016 Ultrawideband (UWB) time-frequency signal processing *Advanced Ultrawideband Radar: Signals, Targets, and Applications* ed J D Taylor (Boca Raton/ London: CRC Press/Taylor and Francis) ch 4 pp 105–94

[79] Gaunaurd G C, Uberall H and Nagl A 1983 Complex-frequency poles and creeping-wave transients in electromagnetic-wave scattering *Proc. IEEE* **31** 172–4

[80] Baum C, Rothwell E, Chen K and Nyquist D 1991 The singularity expansion method and its application to target identification *Proc. IEEE* **79** 1481–92

[81] Mooney J E, Ding Z and Riggs L S 1998 Robust target identification in white Gaussian noise for ultra wide-band radar systems *IEEE Trans. Antennas Propag.* **46** 1817–23

[82] Mooney J E, Ding Z and Riggs L S 2001 Performance analysis of a CLRT automated target discrimination scheme *IEEE Trans. Antennas Propag.* **49** 1827–35

[83] Choi I-S, Lee J-H, Kim H-T and Rothwell E J 2003 Natural frequency extraction using late-time evolutionary programming-based CLEAN *IEEE Trans. Antennas Propag.* **51** 3285–92

[84] Chen W C and Shuley N V Z 2014 Robust target identification using a modified generalized likelihood ratio test *IEEE Trans. Antennas Propag.* **62** 264–73

[85] Daniels D J 2004 Ground penetrating radar *IEE Radar, Sonar and Navigation Series 15* 2nd edn (London: The Institution of Electrical Engineers) p 752

[86] Robinson L A, Weir W B and Young L 1974 Location and recognition of discontinuities in dielectric media using synthetic RF pulses *Proc. IEEE* **62** 36–44

[87] Chan L C, Moffatt D L and Peters L 1979 A characterization of subsurface radar targets *Proc. IEEE* **67** 991–1000

[88] Borysenko A, Borysenko E, Ivashchuk V and Prokhorenko V 1920 Transmitter/receiver pulse-driven antenna array with near-field beam-forming for UWB subsurface imaging radar *Proc. of 2000 Antenna Applications Symp., Montichello (Italy)* pp 55–82

[89] Sugak V, Bukin A, Ovchinkin O, Pedenko Y and Silaev Y S 2005 Use of the georadar for groundwater table determination and mappings of the territories polluted with oil *Telecommun. Radio Eng.* **64** 959–72

[90] Taylor J D 2012 *Ultrawideband Radar: Application and Design* ed J D Taylor (Boca Raton/ London: CRC Press/Taylor and Francis) p 520

[91] Sugak V 2003 Reconstruction of electrical constants of soil and depth of subsurface objects using data of subsurface sounding *Telecommun. Radio Eng.* **59** 54–63

[92] Vitebsky S and Carin L 1996 Resonances of perfectly conducting wires and bodies of revolution buried in a lossy dispersive half-space *IEEE Trans. Antennas Propag.* **44** 1575–83

[93] Gaunaurd G C and Uberall H 1981 Solution of the inverse electromagnetic scattering problem in the resonance case *IEEE Trans. Antennas Propag.* **29** 293–7

[94] Moser P J *et al* 1984 Complex eigenfrequencies of axisymmetric objects: physical interpretation in terms of resonances *Proc. IEEE* **72** 1652–3

[95] Moser P J and Uberall H 1983 Complex eigenfrequencies of axisymmetric perfectly conducting bodies: radar spectroscopy *Proc. IEEE* **71** 171–2

[96] Varganov M E *et al* 1985 *Radar Characteristics of Air Vehicles* ed L T Tuchkov (Moscow: Radio i svyaz) (in Russian)

[97] Marple S L 1987 *Digital Spectral Analysis: With Applications* (Englewood Cliffs, NJ: Prentice-Hall)

[98] Astanin L Y and Kostylev A A 1992 Ultra-wideband signals - a new step in radar development *IEEE Aerosp. Electron. Syst. Mag.* **7** 12–5

[99] Hipp J E 1974 Soil electromagnetic parameters as functions of frequency, soil density, and soil moisture *Proc. IEEE* **62** 98–103

[100] Atlas D (ed) 1990 *Radar in Meteorology: Battan Memorial and 40th Anniversary Radar Meteorology Conference* (Boston, MA: American Meteorological Society)

[101] Oguchi T 1975 Rain depolarization studies at centimeter and millimeter wavelengths: theory and measurement *J. Radio Res. Lab.* **22** 165–211

[102] Rachkov D S, Lekhovytskiy D I, Semeniaka A V, Atamanskiy D V, Laurukevich U U and Pushkov A A 2013 Estimation of meteorological objects energy spectra in pulse Doppler weather radar *2013 14th Int. Radar Symp. (IRS)* pp 811–7

[103] Laurukevich U, Pushkov A, Vylegzhanin I, Vovshin B, Lekhovytskiy D and Rachkov D 2011 Estimation of energy, spectral and polarimetric characteristics of meteorological echoes in DMRL-C *2011 12th Int. Radar Symp. (IRS)* pp 267–72

[104] Khlopov G I and Veselovska G B 2013 Double-frequency sounding of rain with ellipsoidal water drops *2013 Signal Processing Symp. (SPS)* pp 1–4

[105] Khlopov G I and Veselovska G B 2014 On scattering cross-section calculation accuracy of nonspherical particles of rain precipitation using the dipole scattering approximation *Radioelectron. Commun. Syst.* **57** 189–96

[106] Magono C and Lee C 1966 Meteorological classification of natural snow crystals *J. Fac. Sci. Hokkaido Univ. Japan, Ser. 7* **2** 321–35

[107] Rozenberg V I 1974 *Scattering and Attenuation of Electromagnetic Radiation by Atmospheric Particles* (Washington, DC: NASA) 339

[108] Libbrecht K G 2015 *The Snowflake: Winter's Frozen Artistry*vol 144 (Minneapolis, MN: Voyageur Press)

[109] Aragones J L, MacDowell L G and Vega C 2011 Dielectric constant of ices and water: a lesson about water interactions *J. Phys. Chem A* **115** 5745–58

[110] Kozakoff D J 2010 *Analysis of Radome-Enclosed Antennas* 2nd edn (Norwood, MA: Artech House Publishers)

[111] Mrdakovic B L and Kolundzija B M 2016 A method for full wave analysis of electrically large transparent radomes *Proc. 2016 IEEE Int. Symp. Antennas Propagat. (APS-URSI)* pp 1331–2

[112] Sukharevsky O I, Vasylets V A and Nechitaylo S V 2017 Scattering and radiation characteristics of antenna systems under nose dielectric radomes *Prog. Electromagn. Res. B* **76** 141–57
[113] Xu W, Duan B Y, Li P and Qiu Y 2017 Study on the electromagnetic performance of inhomogeneous radomes for airborne applications—part I: characteristics of phase distortion and boresight error *IEEE Trans. Antennas Propag.* **65** 3162–74
[114] Xu W, Duan B Y, Li P and Qiu Y 2017 Study on the electromagnetic performance of inhomogeneous radomes for airborne applications—part II: the overall comparison with variable thickness radomes *IEEE Trans. Antennas Propag.* **65** 3175–83
[115] Rokhlin V 1983 Solution of acoustic scattering problems by means of second kind integral equations *Wave Motion* **5** 257–72
[116] Boriskin A V *et al* 2008 Small hemielliptic dielectric lens antenna analysis in 2-D: boundary integral equations versus geometrical and physical optics *IEEE Trans. Antennas Propag.* **56** 485–92
[117] Sukharevsky I O, Shapoval O V, Altintas A and Nosich A I 2014 Validity and limitations of the median-line integral equation technique in the scattering by material strips of subwavelength thickness *IEEE Trans. Antennas Propag.* **62** 3623–31
[118] Smotrova E I *et al* 2013 Spectra, thresholds and modal fields of a kite-shaped microcavity laser *J. Opt. Soc. Am.* B **30** 1732–42
[119] Spiridonov A O *et al* 2017 Symmetry accounting in the integral-equation analysis of lasing eigenvalue problems for two-dimensional optical microcavities *J. Opt. Soc. Am.* B **34** 1435–43
[120] Burghignoli P, Di Nallo C, Frezza F and Galli A 2003 A simple Nystrom approach for the analysis of 3-D arbitrarily shaped conducting and dielectric bodies *Int. J. Numer. Modelling* **16** 179–94
[121] Vico F, Greengard L and Ferrando M 2018 Decoupled field integral equations for electromagnetic scattering from homogeneous penetrable obstacles *Commun. PDE* **43** 159–84
[122] van de Hulst H C 1957 *Light Scattering by Small Particles* (New York: Wiley)
[123] Bohren C F and Huffman D R 2004 *Absorption and Scattering of Light by Small Particles* (Weinheim: Wiley)
[124] Johnson P B and Christy R W 1972 Optical constants of the noble metals *Phys. Rev.* B **6** 4370–8
[125] Sukharevsky O I *et al* 2018 3-D electromagnetic scattering by ellipsoidal silver nanoparticles in optical band *IEEE 17th Int. Conf. Mathematical Methods in Electromagnetic Theory (Kyiv, 2–5 July)* pp 151–5

IOP Publishing

Applied Problems in the Theory of Electromagnetic Wave Scattering

Oleg I Sukharevsky

Chapter 2

Asymptotic methods for solving some applied problems

Sergey Nechitaylo, Valery Orlenko, Oleg Sukharevsky, Vitaly Vasilets and Gennady Zalevsky

2.1 Electromagnetic wave scattering by a cylindrical object immersed in a dielectric half-space

The problem of plane monochromatic electromagnetic wave (EMW) scattering at a well-conducting electrically large cylindrical object, which is immersed in a dielectric half-space (the EM parameters of the said half-space coinciding with those of soil in particular), appear in a number of practical applications. In particular, one can come across such a problem while looking for underground pipelines using ground penetrating radar (GPR).

In this situation, the limit case scenario is when the object's conductivity tends to infinity and the object can be considered as perfectly conducting. In this chapter we obtain asymptotical expressions regarding the field scattered by the perfectly conducting object given that the diameter of its cross-section is large. We also point out the fact that one can employ these results to figure out the scatterer's parameters (depth in the soil, curvature radius of the specular reflection surface, and the object's orientation in space). Next, we derive similar asymptotic expressions for the field scattered by a well-conducting (or impedance) object, as well as analyzing the physical sense of summands pertaining to the E- and H-polarizations of the incident wave that appear in the final expression [1, 2].

2.1.1 Calculation method for a perfectly conducting object

Let us consider an infinite perfectly conducting cylindrical scatterer (figure 2.1) with surface L, which is buried at the depth d in homogeneous half-space V_2 characterized by permittivity ε_2, permeability μ_2, and the wavenumber k_2 (for the sake of simplicity and transparency of results, we assume that k_2 will always be real valued) [1]. We put

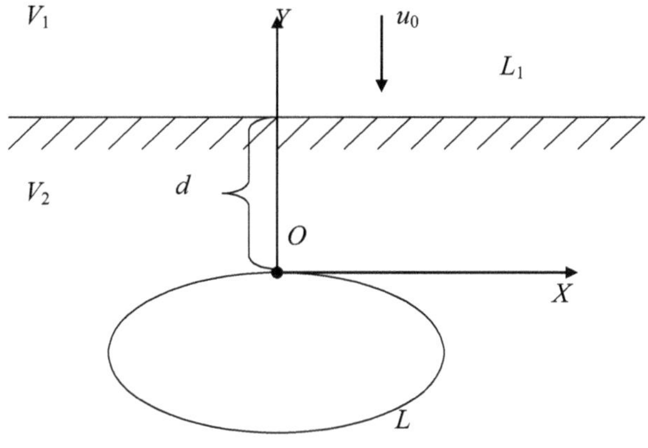

Figure 2.1. Geometry of the problem.

the origin of the coordinate system $OXYZ$ in such manner that it would be sitting on the object's surface right above its center. We directed axis OZ along the cylinder generatrix. Another half-space V_1, its parameters coinciding with those of free space, is adjacent to the half-space V_2, the plane L_1 representing the interface between the two half-spaces. The plane monochromatic EMW of unit intensity propagating from the half-space V_1 is incident onto the interface at a right angle (the direction of incidence coinciding with the normal to the interface). In addition, its wavelength is much shorter than the characteristic curvature radius of the scatterer. The objective is to find the field scattered by the cylinder at an observation point in half-space V_1, which is moved far away from the interface along the vertical line passing through the center of the cylinder's cross-section (figure 2.1).

To solve the problem stated above, we use the integral representations [3] for the total EM field (EMF) given the presence of a cylindrical object immersed in the half-space V_2,

$$u(\vec{r}) = u_0(\vec{r}) + \int_L \left[u(\vec{\xi}) \frac{\partial G(\vec{r}, \vec{\xi})}{\partial n} - G(\vec{r}, \vec{\xi}) \frac{\partial u(\vec{\xi})}{\partial n} \right] \mathrm{d}l, \tag{2.1}$$

where functions u and u_0 are the z-components of the electrical field intensity vector (given the case of E-polarization) or of magnetic field intensity vector (given the case of H-polarization) for the total field and incident field, respectively, function $G(\vec{r}, \vec{\xi})$ is the Green's function pertaining to this particular problem, $\vec{r}$ is the radius-vector of the observation point with coordinates x_0 and y_0, $\vec{\xi}$ is the radius-vector of the point at contour L with coordinates x and y, and dl is the element of contour's L arc length. By E (or H)-polarization we denote the case when the electric (or magnetic) field intensity vector of the incident EMW is parallel to the cylinder's generatrix.

Below, we present the solution to the problem for the case of only E-polarization.

Taking into account the fact that for the problem at hand $u(\vec{\xi})|_{\vec{\xi} \in L} = 0$, we can present the total EMF as follows:

$$u(\vec{r}) = u_0(\vec{r}) - \int_L G(\vec{r}, \vec{\xi}) \frac{\partial u(\vec{\xi})}{\partial n} \mathrm{d}l. \tag{2.2}$$

Here, the function $G(\vec{r}, \vec{\xi})$ is the field of the auxiliary source shaped as a thread of electric current passing through the point $\vec{r}$ inside the half-space V_1 given the presence of the half-space V_2.

Having differentiated expression (2.2) along the normal $\vec{n}$ to the cylinder given the fact that the observation point tends towards the scatterer's surface along the normal, and having taken into account the properties of the normal derivative of the single-layer potential, we obtain the integral equation (IE) with respect to surface current density at contour L. The equation thus obtained is the Fredholm's IE of the second kind. We used iterations to solve it. In the book [4], it was shown that the series obtained as a solution of the IE generates the so-called short-wave asymptotic terms. When we limit ourselves by only two iterations, we can write down the following expression

$$v(\vec{\tau}) \approx 2v_0(\vec{\tau}) - 4\int_L \frac{\partial G(\vec{\tau}, \vec{\xi})}{\partial n_\tau} v_0(\vec{\xi})\,\mathrm{d}l, \tag{2.3}$$

where $\vec{\tau}$ is the radius-vector of the observation point at the cylinder's surface (its coordinates being x_1, y_1), $v_0(\vec{\tau}) = \dfrac{\partial u_0(\vec{\tau})}{\partial n}$, $u_0(\vec{\xi}) = pe^{-jk_2(y-d)}$, and p is the coefficient indicating the share of EMW passing through into half-space V_2. In our case of normal wave incidence upon the plane interface between the two half-spaces, it is determined by the following expression [5]:

$$p = \frac{2W_2}{W_1 + W_2},$$

where $W_1 = 120\pi$ ohm and $W_2 = \sqrt{\dfrac{\mu_2}{\varepsilon_2}}$ ohm are the wave impedances of the corresponding half-spaces.

Thus, in order to solve the problem at hand, all we need is to compute the electric current density at the scatterer's surface using expression (2.3) and substitute it into the expression for the scattered field, which has the following form:

$$u_{\text{scat}}(\vec{r}) = u(\vec{r}) - u_0(\vec{r}) = \int_L G(\vec{r}, \vec{\xi})v(\vec{\xi})\,\mathrm{d}l. \tag{2.4}$$

In this case (the observation point $\vec{r}$ being in the free half-space, and the integration point $\vec{\xi}$ being in the medium filled half-space V_2), function $G(\vec{r}, \vec{\xi})$ has the following form [3]:

$$G(\vec{r}, \vec{\xi}) = -\frac{1}{\pi}\int_0^\infty \mathrm{e}^{-\eta_1(y_0+d)}\mathrm{e}^{\eta_2(y-d)}\frac{\cos\lambda(x - x_0)}{\eta_1 + \eta_2}\,\mathrm{d}\lambda, \tag{2.5}$$

where $\eta_i = \sqrt{\lambda^2 - k_i^2}$, $i = 1,2$.

Since the observation point is considered to be quite far from the interface (in other words, $k_1y_0 > >1$), then the integral in expression (2.5) can be evaluated asymptotically with a satisfactory accuracy as a first-order estimate by means of the stationary phase method. We also took into account the fact that the scatterer was electrically large, which means we can neglect the currents induced at the shadowed part of its surface. The expression (2.4) can be presented in the following form

$$u_{\text{scat}}(\vec{r}) \approx \Omega(y_0) \int_{L_{\text{ill}}} \mathrm{e}^{-jk_2 y} v(x, y)\ \mathrm{d}l,$$

where

$$\Omega(y_0) = \frac{\mathrm{e}^{jk_1(y_0+d)}}{\sqrt{y_0 + d}} \sqrt{\frac{k_1}{2\pi}} \frac{\mathrm{e}^{jk_2 d}}{k_1 + k_2} \mathrm{e}^{j\frac{\pi}{4}},$$

and L_{ill} is the illuminated part of the scatterer's surface. The value of this integral depends on the scatterer's shape.

Later in this section we are going to consider scattering from a circular cylinder with the base radius $R\left(y = \sqrt{R^2 - x^2} - R,\ \mathrm{d}l = \sqrt{1 + (y')^2}\ \mathrm{d}x = \dfrac{R}{\sqrt{R^2 - x^2}}\ \mathrm{d}x\right)$. In this particular case, the expression for a scattered field can be presented as follows:

$$u_{\text{scat}}(\vec{r}) \approx \Omega(y_0) \int_{-R}^{R} \mathrm{e}^{-jk_2 R\left(\sqrt{1-\frac{x^2}{R^2}}-1\right)} v(x, y) \frac{R}{\sqrt{R^2 - x^2}}\ \mathrm{d}x. \tag{2.6}$$

Evaluating the integral (2.6) by means of the stationary phase method, we notice that the scattered field came mostly from a relatively small area $(-\delta, \delta)$ around the point $x = 0$. We approximate the scatterer's surface around this particular point by an osculating paraboloid, i.e. $y \approx -ax^2$ $\left(\text{for the case of circular cylinder, } a = \dfrac{1}{2R}\right)$. Given these approximations, we can rewrite the scattered field expression as follows

$$u_{\text{scat}}(\vec{r}) \approx \Omega(y_0) \int_{-\infty}^{\infty} \mathrm{e}^{j\frac{k_2 x^2}{2R}} v(x, -ax^2)\ \mathrm{d}x. \tag{2.7}$$

Thus, it follows from the above considerations that to evaluate the scattered field we need to compute the values of function $v(x, -ax^2)$ in the vicinity of $x = 0$.

For that purpose we use expression (2.3). Now, both the observation and the integration points belong to half-space V_2, and the Green's function takes the following form [3]:

$$G(\vec{\tau}, \vec{\xi}) = G_0(\vec{\tau}, \vec{\xi}) + G_1(\vec{\tau}, \vec{\xi}), \tag{2.8}$$

where

$$G_0(\vec{\tau}, \vec{\xi}) = \frac{1}{4j} H_0^{(1)}(k_2 r), G_1(\vec{\tau}, \vec{\xi}) = -\frac{1}{2\pi} \int_0^{\infty} \mathrm{e}^{\eta_2(y+y_1-2d)} \cos \lambda(x - x_1) \frac{\eta_1 - \eta_2}{\eta_1 + \eta_2} \frac{\mathrm{d}\lambda}{\eta_2},$$

r is the distance between the points $\vec{\tau}$ and $\vec{\xi}$, $H_0^{(1)}(k_2 r)$ is the Hankel function of the first kind and zero order, x_1 and y_1 are the coordinates of the observation point $\vec{\tau}$, and x and y are the coordinates of integration point $\vec{\xi}$.

Now we differentiate function $G(\vec{\tau}, \vec{\xi})$ term by term along the normal to the scatterer's surface at the point of observation. We also account for the fact that in the vicinity of point $x = 0$ the distance between the observation and integration points can be written approximately as $r = |\vec{\xi} - \vec{\tau}| \approx |x - x_1|$. Then the first term of the sum takes the form

$$\frac{\partial G_0(\vec{\tau}, \vec{\xi})}{\partial n_\tau} \approx \frac{1}{4j} \frac{k_2|x - x_1|}{2R} H_1^{(1)}(k_2|x - x_1|). \tag{2.9}$$

Computation of the second term leads us to the following expression

$$\frac{\partial G_1(\vec{\tau}, \vec{\xi})}{\partial n_\tau} \approx -\frac{\partial G_1(\vec{\tau}, \vec{\xi})}{\partial y} = \frac{1}{2\pi} \int_0^\infty \mathrm{e}^{\eta_2(y+y_1-2d)} \frac{\eta_1 - \eta_2}{\eta_1 + \eta_2} \cos \lambda(x - x_1)\, \mathrm{d}\lambda. \tag{2.10}$$

Thus the surface current density can be expressed as follows:

$$v(\vec{\tau}) = 2v_0(\vec{\tau}) - 4(I_0(\vec{\tau}) + I_1(\vec{\tau})), \tag{2.11}$$

where

$$I_0(\vec{\tau}) = \int_L \frac{\partial G_0(\vec{\tau}, \vec{\xi})}{\partial n_\tau} v_0(\vec{\xi})\, \mathrm{d}l_\xi,\ I_1(\vec{\tau}) = \int_L \frac{\partial G_1(\vec{\tau}, \vec{\xi})}{\partial n_\tau} v_0(\vec{\xi})\, \mathrm{d}l_\xi.$$

Let us consider these summands that appear in the brackets of (2.11) individually while taking into account formulas (2.9) and (2.10). The first integral is

$$I_0(\vec{\tau}) = \frac{k_2^2 \mathrm{e}^{jk_2 d}}{8R} p \int_L |x - x_1| H_1^{(1)}(k_2|x - x_1|) \mathrm{e}^{-jk_2 y}\, \mathrm{d}l_\xi. \tag{2.12}$$

In expression (2.12) we replace integration over the entire scatterer's surface by integration over small vicinity of a 'specular' point and we take into account that the integrand is the even function. Hence we obtain the following:

$$I_0(\vec{\tau}) = \frac{k_2^2 \mathrm{e}^{jk_2 d}}{8R} p I_{01},$$

where

$$I_{01} = \int_{x_1}^{x_1+\delta} (x - x_1) H_1^{(1)}(k_2(x - x_1)) \mathrm{e}^{j\frac{k_2 x^2}{2R}}\, \mathrm{d}x.$$

After some mathematical transformation the latter integral can be presented as

$$I_{01} = \frac{2R}{k_2} \mathrm{e}^{j\frac{k_2 x_1^2}{2R}} \int_0^{\delta_1} \gamma H_1^{(1)}(\varepsilon\gamma) \mathrm{e}^{j(\gamma^2 + 2\gamma\beta_0)}\, \mathrm{d}\gamma$$

or

$$I_{01} = -\frac{2R}{k_2} \mathrm{e}^{j\frac{k_2 x_1^2}{2R}} \frac{\partial}{\partial \varepsilon} \int_0^{\delta_1} H_0^{(1)}(\varepsilon\gamma) \mathrm{e}^{j(\gamma^2 + 2\gamma\beta_0)}\, \mathrm{d}\gamma, \tag{2.13}$$

where $\varepsilon = \sqrt{2k_2R}$, $\delta_1 = \delta\sqrt{\frac{k_2}{2R}}$, $\gamma = (x - x_1)\sqrt{\frac{k_2}{2R}}$, and $\beta_0 = \frac{k_2}{2R}x_1^2$. The second multiplier of the integrand at the integration interval is a relatively slowly changing function compared to the first multiplier. Therefore, the whole integral can be evaluated by means of the method described in [6]. According to this method, we consider the auxiliary integral

$$I(\zeta, \gamma) = \int_0^\infty \mathrm{e}^{j(\gamma^2+2\gamma\beta_0)} H_0^{(1)}(\varepsilon\gamma)\mathrm{e}^{-\zeta\gamma}\,\mathrm{d}\gamma.$$

Having replaced the function $\mathrm{e}^{j(\gamma^2+2\gamma\beta_0)}$ by its Taylor series representation in the vicinity of point $\gamma = 0$ and having retained only the major term of the series, we obtain

$$I(\zeta, \gamma) \approx \int_0^\infty H_0^{(1)}(\varepsilon\gamma)\mathrm{e}^{-\zeta\gamma}\,\mathrm{d}\gamma.$$

The value of the latter integral can be evaluated using the following formula [7]:

$$I(\zeta, \gamma) = \frac{1}{\sqrt{\zeta^2 + \varepsilon^2}}\left(1 - \frac{2j}{\pi}\ln\left[\frac{\zeta}{\varepsilon} + \sqrt{1 + \left(\frac{\zeta}{\varepsilon}\right)^2}\right]\right).$$

Now we pass to the limit for $\zeta \to 0$ to obtain

$$I(0, \gamma) = \frac{1}{\varepsilon}.$$

The latter expression is the asymptotical value of the integral in formula (2.13). Taking into account this asymptote, we can rewrite the expression for $I_0(\vec{\tau})$ in the following form:

$$I_0(\vec{\tau}) = \frac{\mathrm{e}^{-jk_2(y_1-d)}}{4R}p.$$

Now we proceed to computing the integral $I_1(\vec{\tau})$

$$I_1(\vec{\tau}) = \frac{1}{2\pi}jk_2p\mathrm{e}^{jk_2d}\int_0^\infty \frac{\eta_1 - \eta_2}{\eta_1 + \eta_2}\mathrm{e}^{\eta_2(y_1-2d)}\int_L \mathrm{e}^{\eta_2 y - ik_2 y}\cos(\lambda(x - x_1))\,\mathrm{d}l\,\mathrm{d}\lambda.$$

Changing the variable of the inner integral, we reduce it to the standard integral. Then the expression for $I_1(\vec{\tau})$ takes the following form:

$$I_1(\vec{\tau}) = jk_2p\sqrt{\frac{R}{2\pi}}\mathrm{e}^{jk_2d}\int_0^\infty \frac{\eta_1 - \eta_2}{\eta_1 + \eta_2}\frac{\mathrm{e}^{\eta_2(y_1-2d)}\mathrm{e}^{-\frac{\lambda^2R}{2(\eta_2-jk_2)}}}{\sqrt{\eta_2 - jk_2}}\cos(\lambda x_1)\,\mathrm{d}\lambda.$$

After some simplifications, this integral can be evaluated using the saddle-point method [8]

$$I_1(\vec{\tau}) \sim j\frac{k_2}{2}p\frac{k_1 - k_2}{k_1 + k_2}\mathrm{e}^{-jk_2y_1}\mathrm{e}^{jk_23d}\sqrt{\frac{R}{4d + R - 2y_1}}.$$

Having substituted the latter expression into formula (2.7) and having evaluated the resulting function by means of a stationary phase method, we obtain the following expression for the field scattered by the cylinder in the far-field zone

$$u_{\text{scat}}(\vec{r}) = -\frac{e^{jk_1(y_0+d)}}{\sqrt{y_0+d}}\frac{e^{j2k_2d}}{k_1+k_2}\sqrt{2k_1k_2Rp}\left(1+\frac{j}{2k_2R}-\frac{k_1-k_2}{k_1+k_2}\sqrt{\frac{R}{R+4d}}e^{j2k_2d}\right).$$

The latter expression can be improved by taking into account the next expansion term of the phase function in the integral (2.5) $\left(\text{i. e. } y \approx -\frac{x^2}{2R}-\frac{x^4}{8R^3}\right)$

$$\begin{aligned} u_{\text{scat}}(\vec{r}) = &-\frac{e^{jk_1(y_0+d)}}{\sqrt{y_0+d}}\frac{e^{j2k_2d}}{k_1+k_2}\sqrt{2k_1k_2Rp} \\ &\times\left(1+\frac{5j}{16k_2R}+\frac{k_1-k_2}{k_1+k_2}\sqrt{\frac{R}{R+4d}}e^{j2k_2d}\left(\frac{3j}{16k_2R}-1\right)\right). \end{aligned} \tag{2.14}$$

In the same manner one obtains the expression for the field scattered by a cylinder immersed in a dielectric half-space for H-polarization of the incident wave:

$$\begin{aligned} u_{\text{scat}}(\vec{r}) = &\frac{e^{jk_1(y_0+d)}}{\sqrt{y_0+d}}\frac{e^{j2k_2d}}{k_1+k_2}\sqrt{2k_1k_2Rp} \\ &\times\left(1+\frac{5j}{16k_2R}-\frac{k_1-k_2}{k_1+k_2}\sqrt{\frac{R}{R+4d}}e^{j2k_2d}\left(\frac{3j}{16k_2R}-1\right)\right). \end{aligned} \tag{2.15}$$

In order to check the veracity of the proposed method, we compared our computation results with those obtained using the method of IEs [9]. The computation error involved in the above expressions decreases as the carrier frequency of the incident wave increases, and for the value $|k_2R| = 8$ this error amounted to just 7.8%. This makes the method developed above applicable to provisional engineering computations.

2.1.2 Calculation features for an impedance object

In the case of well-conducting (or impedance) object [2], we start our consideration from the results obtained above for the perfectly conducting cylinder [1]. As before, we employ the integral representation (2.1) presented in subsection 2.1.1 and consider the problem for the E-polarization of the incident wave, which means that its electric field intensity vector is parallel to the cylinder's generatrix.

The corresponding boundary condition has the form [10],

$$\frac{\partial u(\vec{\xi})}{\partial n} - jk_2\frac{W_2}{W}u(\vec{\xi}) = 0 \text{ given that } \vec{\xi} \in L,$$

where $W_2 = 120\pi\sqrt{\frac{\mu_2}{\varepsilon_2}}$ is the wave impedance of the half-space V_2, $W = \frac{1-j}{\sqrt{2}}\sqrt{\frac{\omega\mu}{4\pi\sigma}}$ is the scatterer's impedance.

The above boundary condition can be presented as follows:

$$u(\vec{\xi}) = \frac{1}{jk_2}\alpha\frac{\partial u(\vec{\xi})}{\partial n},$$

where $\alpha = \frac{W}{W_2} << 1$.

Accounting for the above, we obtain the following integral representation of the total field:

$$u(\vec{r}) = u_0(\vec{r}) + \int_L \left[\frac{\alpha}{jk_2}\frac{\partial G(\vec{r}, \vec{\xi})}{\partial n} - G(\vec{r}, \vec{\xi})\right]\frac{\partial u(\vec{\xi})}{\partial n}\, \mathrm{d}l_\xi. \quad (2.16)$$

Now, we see that to solve the problem we need to figure out the values of function $\frac{\partial u(\vec{\xi})}{\partial n}$ at points belonging to contour L. In the case of E-polarization, the function $\frac{\partial u(\vec{\xi})}{\partial n}$ is proportional to the surface current density. To find the latter, we differentiated expression (2.16) along the normal $\vec{n}$ to the cylinder's surface and then we assume that the observation point tends (also along the normal) to the scatterer's surface. Additionally, we accounted for the properties of the normal derivative of the simple layer potential and the result was the sought for surface current density:

$$v(\vec{\tau}) = 2v_0(\vec{\tau}) + 2\frac{\partial}{\partial n_\tau}\int_L \frac{\alpha}{jk_2}\frac{\partial G(\vec{\tau}, \vec{\xi})}{\partial n_\xi}v(\vec{\xi})\, \mathrm{d}l_\xi - 2\int_L \frac{\partial G(\vec{\tau}, \vec{\xi})}{\partial n_\tau}v(\vec{\xi})\, \mathrm{d}l_\xi, \quad (2.17)$$

where $\vec{\tau} = (x_1, y_1)$ is the radius-vector of the observation point at the cylinder's surface, $\vec{n}_\tau$ is the normal to its surface at this point, $v_0(\vec{\tau}) = \frac{\partial u_0(\vec{\tau})}{\partial n}$, and $v(\vec{\tau}) = \frac{\partial u(\vec{\tau})}{\partial n}$.

Given a perfectly conducting scatterer ($\alpha = 0$) we can refer to [4], which proves the asymptotic property of the formulas we obtained as a result. Now, it is reasonable to assume that given small enough values of α, the formulas obtained while iterating the equation (2.17) would also produce a correct asymptotic expansion of the solution. Having limited ourselves to only two iterations, we obtained the following expression:

$$v(\vec{\tau}) \approx 2v_0(\vec{\tau}) + 4\frac{\partial}{\partial n_\tau}\int_L \frac{\alpha}{jk_2}\frac{\partial G(\vec{\tau}, \vec{\xi})}{\partial n_\xi}v_0(\vec{\xi})\, \mathrm{d}l_\xi - 4\int_L \frac{\partial G(\vec{\tau}, \vec{\xi})}{\partial n_\tau}v_0(\vec{\xi})\, \mathrm{d}l_\xi. \quad (2.18)$$

Formula (2.18) allows us to compute the values of the surface current density at points belonging to the scatterer's surface, these particular values can be substituted into expression (2.16) in order to find the sought for scattered field using the following expression:

$$u_{\text{scat}}(\vec{r}) = u(\vec{r}) - u_0(\vec{r}) = \int_L \left[\frac{\alpha}{jk_2}\frac{\partial G(\vec{r}, \vec{\xi})}{\partial n} - G(\vec{r}, \vec{\xi})\right]\frac{\partial u(\vec{\xi})}{\partial n}\, \mathrm{d}l_\xi. \quad (2.19)$$

The function $G(\vec{r}, \vec{\xi})$ is determined, as before, by formula (2.5).

According to the assumptions made above, the observation point was located at a great distance from the interface between the media. Evidently (due to the real nature of k_2) the major input into the integral (as was the case for a perfectly conducting object) is going to be from the vicinity of point $\lambda = 0$ [8], and this particular integral can be asymptotically computed by the stationary phase method. In addition, the formula (2.19) then takes the following form:

$$u_{\text{scat}}(\vec{r}) = \Omega_1(y_0) \int_L \mathrm{e}^{-jk_2 y} v(\vec{\xi})\, \mathrm{d}l_\xi, \tag{2.20}$$

where $\Omega_1(y_0) = (1 - \alpha)\dfrac{\mathrm{e}^{jk_1(y_0+d)}}{\sqrt{y_0 + d}}\sqrt{\dfrac{k_1}{2\pi}}\dfrac{\mathrm{e}^{jk_2 d}}{k_1 + k_2}\mathrm{e}^{j\frac{\pi}{4}}$.

In reference [1] it was shown that the integral in formula (2.20) can be evaluated using the stationary phase method. In addition, the major input into the resulting integral comes from the small vicinity of point $x = 0$ or the so-called 'specular' line at the cylinder's surface. Approximation of the scatterer's surface in the vicinity of such a line by an osculating paraboloid $y = -ax^2$ gives us the following formula:

$$u_{\text{scat}}(\vec{r}) = \Omega_1(y_0) \int_L \mathrm{e}^{jk_2 a x^2} v(\vec{\xi})\, \mathrm{d}l_\xi$$

Next, similar to reference [1], we consider a circular cylinder of radius R, for which the coefficient for the osculating paraboloid equals to $a = \dfrac{1}{2R}$. In this case, the expression for the sought for field is

$$u_{\text{scat}}(\vec{r}) \approx \Omega_1(y_0) \int_{-\varepsilon}^{\varepsilon} \mathrm{e}^{j\frac{k_2 x^2}{2R}} v(x, -ax^2)\, \mathrm{d}x.$$

Thus, to compute $u_{\text{scat}}(\vec{r})$ we need to evaluate function $v(x, -ax^2)$ in the vicinity of point $x = 0$. For this purpose, we used expression (2.18).

Having performed transformations similar to those we in the case of a perfectly conducting object, we obtained the following expression for the surface current density in the vicinity of 'specular' line:

$$v(\vec{\tau}) \approx 2jk_2 \mathrm{e}^{-ik_2(y_1-d)} p\Bigg(1 + \frac{\alpha - 1}{2jk_2 R} + (\alpha - 1)\frac{k_1 - k_2}{k_1 + k_2}\mathrm{e}^{jk_2 2d}\sqrt{\frac{R}{R+4d}}\left(1 + \frac{y_1}{R+4d}\right)$$
$$- \frac{\alpha}{jk_2(R+4d)}\frac{k_1 - k_2}{k_1 + k_2}\mathrm{e}^{jk_2 2d}\sqrt{\frac{R}{R+4d}}\left(1 + \frac{3y_1}{R+4d}\right)\Bigg).$$

Having substituted the latter expression into (2.20) and having asymptotically evaluated the resulting integral, we obtained the following expression for the field scattered by an impedance circular cylinder immersed into dielectric half-space given the E-polarization of the incident wave:

$$u_{\text{scat}}(\vec{r}) = -\frac{\mathrm{e}^{jk_1(y_0+d)}}{\sqrt{y_0 + d}}\frac{\mathrm{e}^{j2k_2 d}}{k_1 + k_2}\sqrt{2k_1 k_2 R}\, p\Bigg(1 - \alpha + j\frac{(1-2\alpha)}{2k_2 R}$$
$$- \frac{k_1 - k_2}{k_1 + k_2}\sqrt{\frac{R}{R + 4d}}\mathrm{e}^{j2k_2 d}\left(1 - j\frac{1}{4k_2(R + 4d)} - \alpha\frac{3 - 4k_2(R + 4d)}{2k_2(R + 4d)}\right)\Bigg) + O(\alpha^2). \tag{2.21}$$

In addition, given $\alpha = 0$ the expression (2.21) coincides with the formula for the perfectly conducting scatterer [1].

Let us analyze the above expression. We single out a few terms from its structure and consider them individually.

The first term before brackets just tells us that the scattered field is characterized by the cylindrical wave originated at the 'specular' line at the surface of our cylindrical scatterer. Multiplier e^{j2k_2d} before the brackets describes the phase and attenuation of the EMW propagating from the interface between the media to the scatterer and back.

The first summand inside the brackets corresponds to the field scattered given the physical optics (PO) approximation that also accounts for the impedance nature of the scatterer.

The second summand is the correction to the first one and it accounts for the curvature of the scatterer's surface. In addition, it includes the multiplier $(1-2\alpha)$, which, in this particular case, corresponds to the EMW reflection coefficient [5].

The third summand also accounts for the curvature of the scatterer's surface but its main function here is to take into account the 'secondary' scattered field component, which was generated by the 'primary' scattered field bouncing back towards the scatterer from the interface and being scattered by the object's surface for the second time. The evidence of the latter is the presence of multiplier $\frac{k_1 - k_2}{k_1 + k_2}$, which is basically the interface reflection coefficient for the EMW propagating towards the interface from the half-space V_2, the latter being filled with dielectric (i.e. $\mu \approx 1$). Then the multiplier e^{i2k_2d} describes an additional phase difference and attenuation of that 'secondary' scattered wave with respect to the 'primary' one. Coefficient $\sqrt{\frac{R}{R + 4d}}$ takes into account the influence of the object's immersion depth and the object's radius on the intensity of the 'secondary' scattered field.

In order to clarify physical sense of this 'secondary' scattered field component, let us trace the corresponding ray for situations involving various values of this square root.

Figure 2.2 shows that as the immersion depth of the scatterer increases, the angular spread of the cone of these 'secondary' rays widens (figure 2.2(a)), thus reducing the power density of the EMW scattered towards the receiving antenna of a monostatic radar.

Figure 2.3 illustrates the case where the cylinder's radius changes but its immersion depth d stays unchanged. It can be seen that the decrease in the cylinder's radius (figure 2.2(b)) results in a wider cone of 'secondary' rays, which, as in the previous situation, provides for a weaker radar echo.

Let us proceed now to the case of H-polarization of the incident wave (the magnetic field intensity vector is parallel to the cylinder's generatrix). Now, the impedance boundary condition can be presented as follows:

$$\frac{\partial u(\vec{\xi})}{\partial n} = jk_2\alpha u(\vec{\xi}).$$

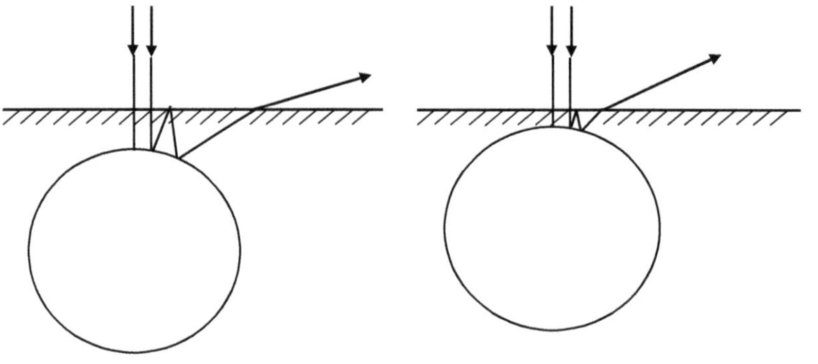

a) b)

Figure 2.2. Ray tracing for the field scattered by a cylindrical object immersed into a dielectric medium given a varying depth of the object's immersion.

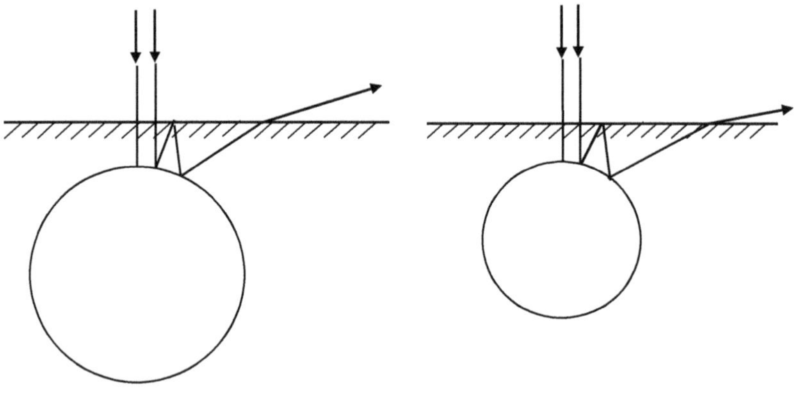

a) b)

Figure 2.3. Ray tracing for the field scattered by a cylindrical object immersed into a dielectric medium given varying cylinder's radius.

Substituting it in the integral representation (2.1), we obtain the following expression for the scattered field:

$$u_{\text{scat}}(\vec{r}) = \int_L \left[\frac{\partial G(\vec{r}, \vec{\xi})}{\partial n} - \mathrm{i}k_2 \alpha G(\vec{r}, \vec{\xi}) \right] u(\vec{\xi})\, \mathrm{d}l_\xi. \tag{2.22}$$

The expression for the field at points belonging to the scatterer's surface $u(\vec{\xi})$ is obtained similar to the case of E-polarization and has the form

$$u(\vec{\tau}) \approx 2u_0(\vec{\tau}) + 4\int_L \left[\frac{\partial G(\vec{\tau}, \vec{\xi})}{\partial n} - jk_2 \alpha G(\vec{\tau}, \vec{\xi}) \right] u_0(\vec{\xi})\, \mathrm{d}l_\xi. \tag{2.23}$$

The asymptotic evaluation of the integrals appearing in the right-hand side of formula (2.23) is carried out in a similar manner as before in the case of

E-polarization. Once the result is substituted into expression (2.22), we obtain the following expression for the scattered field given the H-polarization of the incident wave:

$$u_{\text{scat}}(\vec{r}) = \frac{e^{jk_1(y_0+d)}}{\sqrt{y_0+d}}\frac{e^{j2k_2d}}{k_1+k_2}\sqrt{2k_1k_2Rp}\left(1+j\frac{(1-\alpha)}{2k_2R}\right.$$
$$\left.+\frac{k_1-k_2}{k_1+k_2}\sqrt{\frac{R}{R+4d}}e^{j2k_2d}\left(1+\frac{1}{4k_2(R+4d)}\right)\right)+O(\alpha^2). \tag{2.24}$$

The physical sense of the constituents is the same as in the case of E-polarization.

Comparative analysis of formulas (2.21) and (2.24) allows us to conclude that the conductivity of the material of which the scatterer is made affects the intensity of the scattered field less given the H-polarization of the incident wave. In that case, the PO term in the expression and the term responsible for the 'secondary' scattered field and the relation of the latter to the scatterer's surface curvature are exactly the same as the corresponding terms of the scattered field expression for the perfectly conducting scatterer [1]. However, given E-polarization of the incident wave the conductivity of the scatterer's material affects all three spatial components of the scattered field (unlike in the case of H-polarization).

2.1.3 The possibility of figuring out the parameters of the scattering cylinder

As seen from formulas (2.14), (2.15), (2.21), and (2.24), the intensity of the scattered field explicitly depends on the geometrical parameter of the scatterer (namely, its radius R) and the depth d of its immersion in the soil. This justifies the creation of an algorithm for solving the inverse problem, namely, figuring out the parameters of an unknown cylindrical scatterer using the far-field zone measurements of its scattered field characteristics [11].

Let us suppose that the antenna system of a monostatic radar is placed at point $\vec{x}$. Considering the assumptions made above we understand that this antenna system is placed at a great distance from the ground (for instance, this can be an airborne antenna). Let us further assume that somehow we have known the location of the underground object and it is also known to us that this object has a cylindrical form. Our objective is to figure out its radius, immersion depth, and spatial orientation. To achieve this objective, it is sufficient to probe the object on two orthogonal polarizations characterized by unit vectors $\vec{u}_1^0$ and $\vec{u}_2^0$. The radar receiver at the reception point measures the scattered field at these two polarizations and provides us with two complex amplitudes of field intensities $\vec{u}_1$ and $\vec{u}_2$. Next, for the sake of definiteness, let us assume that the cylinder's axis makes up angle φ with respect to the direction of the polarization vector $\vec{u}_1$. Let us introduce the notation of $\vec{u}_{\|}$ and $\vec{u}_{\perp}$, which are the scattered field intensity vectors given that the polarization vector is parallel and perpendicular to the cylinder's axis, respectively. Now we can write

$$|\vec{u}_1|^2 = |\vec{u}_{\|}|^2\cos^2\varphi + |\vec{u}_{\perp}|^2\sin^2\varphi,$$
$$|\vec{u}_2|^2 = |\vec{u}_{\|}|^2\sin^2\varphi + |\vec{u}_{\perp}|^2\cos^2\varphi,$$

whence

$$|\vec{u}_1|^2 + |\vec{u}_2|^2 = \left|\vec{u}_{||}\right|^2 + |\vec{u}_{\perp}|^2, \tag{2.25}$$

$$|\vec{u}_1|^2 - |\vec{u}_2|^2 = \left(\left|\vec{u}_{||}\right|^2 - |\vec{u}_{\perp}|^2\right)\cos 2\varphi. \tag{2.26}$$

The left-hand parts of both expressions above are the known parameters (by measurement). The right-hand parts are functions of the sought for parameters, the analytical form of which is also known (formulas (2.14), (2.15) or (2.21), (2.24)). In addition, equation (2.25) does not include the angle. A further approach depends upon the presence and composition of *a priori* data.

If the cylindrical scatterer's immersion depth is known (for instance, it was measured before by other means) and if we also obtained a clue regarding its impedance, one can use equation (2.25) to compute the radius of the cylinder. Having achieved that, one can use expression (2.26) to compute the spatial orientation of the cylinder's axis.

Accounting for the fact that all the calculations above were obtained under the assumption that a planar monochromatic EMW was used for radar probing, the best way to implement the above algorithm in a practical application is to use continuous wave signals.

It should be noted that if the measurements were conducted at more than one probing frequency, one would receive enough data to be able to solve the problem of finding the cylindrical scatterer parameters without any need for *a priori* data regarding its depth of immersion and impedance.

Another point to be made here is that for the above algorithm to function properly, one needs to know the electric properties of the soil the scatterer is immersed in. Such data can be extracted from digital terrain maps or it can be obtained by remote radar measurements.

Thus in this section we derived asymptotic expressions regarding an EMF scattered by an impedance cylinder, its radius being much greater than the wavelength of the incident EMW. In addition, the source of the incident wave was in free space and at a great distance (great enough for the incident EMW to be considered planar at the point where it reaches the interface between the two media). The scatterer itself was immersed in a dielectric half-space (lossy in general) and the receiving antenna was also in free space and at a great distance from the interface.

As a result, we have obtained quite simple expressions that have an obvious physical sense, which described the scattered field given the *E*- and *H*-polarizations of the incident (primary) field.

Moreover, we pointed out the possibility of solving (based on the expressions derived) the so-called inverse problem of figuring out particular features of cylindrical perfectly conducting scatterers immersed in a dielectric half-space.

2.2 Reconstruction of the 'illuminated' surface part of a convex perfectly conducting scatterer

In quite a few radar and electrodynamics problems, one finds it necessary to determine the geometric shape of the scatterer using only its backscattering data [12].

A number of approaches have been proposed to date to solve such problems. These include, for example, methods that work in the time domain and that are based on the relationship between the impulse response of an object and the second derivative of the profile function describing its surface [13, 14]. Also in references [15, 16] an approach has been proposed to that end that utilized the backscattering data obtained at various polarizations. Reference [17] describes the finite difference time domain method and the corresponding iterative method for reconstructing the surface of a perfectly conducting convex body of revolution, the dimensions of which fall into the resonant frequency range. These methods are based on solving a non-stationary IE for the surface current density. However, the applicability of the above methods is limited to simple geometric shapes (bodies of revolution) and the objects of resonant dimensions.

In this section, we propose a technique for reconstructing the 'illuminated' surface part of a convex perfectly conducting object, the geometrical dimensions and radii of curvature of which are much greater than the length of the incident EMW [12]. This technique is based on the relationship between the geometry of the scatterer's surface and the high-frequency approximation of its unit step response (USR) [18]. Using the data obtained from the full polarization radar reception, one can find the principal curvatures of the object's surface at the point of specular reflection, as well as their orientation in space (principal directions). The latter information is obtained by iterating the Fock equation with respect to the current density induced on the surface of a perfectly conducting scatterer and by finding the two terms of the ray asymptotics of the scattered field given the general case of bistatic radar layout (reception diversity) [19]. This makes it possible to construct an osculating paraboloid for the surface being reconstructed, its vertex being at the point of the stationary phase. This osculating paraboloid is taken as an initial approximation of the object's surface being reconstructed in the vicinity of a stationary phase point. Next, several time steps (increments) are taken. At each step, the initial shape of the sought surface is approximated by a ruled surface, and later it is refined using a correction determined by the difference between the actual USR of the object and the USR calculated given the approximate surface thus reconstructed. The actual USR can be found easily since the probing and echo signals are known. This, for example, can be achieved by solving the convolution equation

$$\Omega(t) = \int_{\infty}^{t} \Omega'_0(\tau)H(t - \tau)\, \mathrm{d}\tau,$$

where $\Omega'_0(t)$ is the time derivative of the probing signal, $\Omega(t)$ is the echo signal, and $H(t)$ is the actual USR of the scatterer.

One needs to carry out the above procedure until the boundary of the 'illuminated' part of the object's surface is reached.

2.2.1 Finding the principle curvatures at the point of specular reflection

The algorithm for obtaining the principal curvatures of the scatterer's surface as well as the orientation angle of its principal directions at the point of specular reflection is based on iterating the Fock equation, which can be obtained given that the object's surface was assumed to be convex [19]. The method developed in [19] gives us the two terms of the ray expansion for the general case of bistatic radar layout. As a result, we obtain the following expression for the magnetic intensity vector of the scattered field:

$$\vec{H}(t) = \vec{A}_0\,\Omega_0(t - t_0) + \vec{A}_1\Omega_1(t - t_0),$$

where function $\Omega_0(t)$ describes the structure of the probing signal in the time domain, t_0 is the lag time of the echo signal, $\Omega_1(t) = \int_0^t \Omega_0(\tau)\, d\tau$, and $\vec{A}_0$ and $\vec{A}_1$ are the vector coefficients of the ray expansion, the directions of which depend on the orientation of the principal directions at the point of specular reflection $\vec{\tau}_1$, $\vec{\tau}_2$ and on the polarization of the magnetic intensity vector of the scattered wave $\vec{p}^{\,0}_{\text{scat}}$:

$$\vec{A}_0 = \frac{1}{2d\sqrt{\kappa_1\kappa_2}}\vec{p}^{\,0}_{\text{scat}},$$

$$\vec{A}_1 = \frac{1}{4d\sqrt{\kappa_1\kappa_2}\cos\theta}\Big[-\vec{\tau}_1 V_{12}(\theta)\cos\theta + \vec{\tau}_2 V_{11}(\theta)\cos\theta - \vec{n}_0(V_{11}(\theta)\sin\alpha$$

$$- V_{12}(\theta)\cos\alpha)\sin\theta - -\frac{\vec{p}^{\,0}_{\text{scat}}}{2}(\kappa_1 + \kappa_2)\Big].$$

Here θ is the half bistatic angle, κ_1 and κ_2 are the principal curvatures of the scatterer's surface at the point of specular reflection, α is the angle between the projection of the probing direction unit vector $\vec{R}_0$ onto the plane, which includes both unit vectors of principal curvatures at the stationary phase point $\vec{\tau}_1$, $\vec{\tau}_2$ and the unit vector $\vec{\tau}_1$, $\vec{n}_0$ is the unit vector of the external normal to the scatterer's surface at the point of specular reflection, and d is the distance from the point of reception to the object itself,

$$V_{11}(\theta) = \Phi_0(\theta)\left[\frac{p_1^0\kappa_2}{2}\sin 2\alpha - \frac{p_2^0(\kappa_1 + \kappa_2)}{4}\cos 2\alpha\right] + \frac{p_2^0}{4}(\kappa_2 - \kappa_1)\Phi_1(\theta),$$

$$V_{12}(\theta) = \Phi_0(\theta)\left[-\frac{p_2^0\kappa_1}{2}\sin 2\alpha - \frac{p_1^0(\kappa_1 + \kappa_2)}{4}\cos 2\alpha\right] + \frac{p_1^0}{4}(\kappa_2 - \kappa_1)\Phi_1(\theta),$$

$$\Phi_0(\theta) = 2\left[\frac{tg^2\frac{\theta}{2}}{\cos\theta} + \frac{1}{\sin^2\theta}\cdot\left(2 + \frac{3\sin^2\theta - 2}{\cos^3\theta}\right)\right], \Phi_0(0) = 0, \Phi_1(\theta) = 2\frac{1 + \cos^2\theta}{\cos^3\theta}.$$

Let us consider particular case of monostatic reception, which means that bistatic angle $\theta = 0°$. Accounting for the polarization of scattered wave $\vec{p}^{\,0}_{\text{scat}} = \vec{p}^{\,0}$, where

$\vec{p}^{\,0}$ is the unit polarization vector that shows direction of the incident wave's magnetic intensity vector, we obtain the following:

$$\vec{A}_0 = \frac{1}{2d\sqrt{\kappa_1\kappa_2}}\vec{p}^{\,0},$$

$$\vec{A}_1 = \frac{1}{4d\sqrt{\kappa_1\kappa_2}}\left[-\vec{\tau}_1(\kappa_2 - \kappa_1)\cos\beta + \vec{\tau}_2(\kappa_2 - \kappa_1)\sin\beta - \frac{\vec{p}^{\,0}}{2}(\kappa_2 + \kappa_1)\right],$$

where β is the angle between the unit vectors $\vec{p}^{\,0}$ and $\vec{\tau}_1$, $(0 \leqslant \beta \leqslant \pi)$.

Thus we have expressions for the vector coefficients $\vec{A}_0$ and $\vec{A}_1$, which represent a solution to the problem of EMW scattering by a convex perfectly conducting body of large electric sizes. These expressions include the sought values κ_1, κ_2, and β.

Now we proceed to solving the inverse problem—finding the geometrical parameters of the scatterer's surface at the stationary phase point, namely κ_1, κ_2, and β.

Taking into account the vector nature of coefficients $\vec{A}_0$ and $\vec{A}_1$, it is advisable to receive the echo signals using independent antennas matched with two orthogonal polarizations: $\vec{p}^{\,0}$ and $\vec{p}^{\,0}_{\perp}$. As a result, one would obtain the corresponding values of projections as the receiver outputs:

$$\left(\vec{H}\cdot\vec{p}^{\,0}\right) = \frac{1}{2d\sqrt{\kappa_1\kappa_2}}\Omega_0(t - t_0) + \frac{(\kappa_1 - \kappa_2)\cos 2\beta - \frac{1}{2}(\kappa_1 + \kappa_2)}{4d\sqrt{\kappa_1\kappa_2}}\Omega_1(t - t_0), \tag{2.27}$$

$$\left(\vec{H}\cdot\vec{p}^{\,0}_{\perp}\right) = \frac{\sin 2\beta}{4d\sqrt{\kappa_1\kappa_2}}(\kappa_2 - \kappa_1)\Omega_1(t - t_0). \tag{2.28}$$

Let us introduce the following notation:

$$B = \frac{1}{2d\sqrt{\kappa_1\kappa_2}},\ C = \frac{(\kappa_1 - \kappa_2)\cos 2\beta - \frac{1}{2}(\kappa_1 + \kappa_2)}{4d\sqrt{\kappa_1\kappa_2}},\ D = \frac{\sin 2\beta}{4d\sqrt{\kappa_1\kappa_2}}(\kappa_2 - \kappa_1).$$

These three constants depend on the three parameters which are our objective—principal curvatures κ_1, κ_2, and the angle β.

Taking into account the above notation, we rewrite the expressions (2.27) and (2.28) as follows:

$$\Omega_{\text{rec}}(t - t_0) = B\Omega_0(t - t_0) + C\Omega_1(t - t_0), \tag{2.29}$$

$$\Omega_{\text{rec}1}(t - t_0) = D\Omega_1(t - t_0), \tag{2.30}$$

where $\Omega_{\text{rec}}(t)$ describes the signal structure in the time domain that was received at polarization $\vec{p}^{\,0}$, and $\Omega_{\text{rec}1}(t)$ describes a similar structure of the signal received at polarization $\vec{p}^{\,0}_{\perp}$.

Having sampled equations (2.29) and (2.30), we obtained two overdetermined sets of linear algebraic equations (OSLE): the first allows us to find B and C, the second allows us to find D. Now, we present the ith equation of the first system as

$$B\Omega_0(\tau_i) + C\Omega_1(\tau_i) = \Omega_{rec}(\tau_i),$$

and the ith equation of the second system as

$$D\Omega_1(\tau_i) = \Omega_{rec1}(\tau_i),$$

where $\tau_i = t_i - t_0$ $(1 \leqslant i \leqslant n)$ are the corresponding time samples of $\Omega_{rec}(\tau_i)$, $\Omega_{rec1}(\tau_i)$, $\Omega_1(\tau_i)$, and $\Omega_0(\tau_i)$.

The above systems of equations are solved under the condition of minimizing the root-mean-square error. The solution has the form

$$D = \frac{m}{k},\; C = \frac{af - bl}{(ak - b^2)},\; B = \frac{l - bC}{a},$$

where $a = \sum_{i=1}^{n} \Omega_0^2(\tau_i)$, $b = \sum_{i=1}^{n} \Omega_0(\tau_i)\Omega_1(\tau_i)$, $l = \sum_{i=1}^{n} \Omega_{rec}(\tau_i)\Omega_0(\tau_i)$, $f = \sum_{i=1}^{n} \Omega_{rec}(\tau_i)\Omega_1(\tau_i)$, $m = \sum_{i=1}^{n} \Omega_{rec1}(\tau_i)\Omega_1(\tau_i)$, and $k = \sum_{i=1}^{n} \Omega_1^2(\tau_i)$.

As soon as the approximate values of B, C, and D are found, one can try to find the values of κ_1, κ_2, and β using the following expressions:

$$\frac{1}{2d\sqrt{\kappa_1\kappa_2}} = B,$$

$$\frac{(\kappa_1 - \kappa_2)\cos 2\beta - \frac{1}{2}(\kappa_1 + \kappa_2)}{4d\sqrt{\kappa_1\kappa_2}} = C, \tag{2.31}$$

$$\frac{(\kappa_2 - \kappa_1)\sin 2\beta}{4d\sqrt{\kappa_1\kappa_2}} = D,$$

the combination of which is the system of three equations with respect to three unknowns. We introduce the notation of $w = \kappa_1 + \kappa_2$. Having solved the above system, we obtain the following:

$$\kappa_1 = \frac{w}{2} + \frac{1}{dB}\sqrt{w^2d^2B^2 - 1},\; \kappa_2 = \frac{w}{2} - \frac{1}{dB}\sqrt{w^2d^2B^2 - 1},$$

$$\cos 2\beta = \frac{4dC\sqrt{\kappa_1\kappa_2} + \frac{1}{2}(\kappa_1 + \kappa_2)}{\kappa_1 - \kappa_2}, \text{ if } \kappa_1 \neq \kappa_2,$$

$$\beta = \begin{cases} \frac{1}{2}\arcsin\frac{4dD\sqrt{\kappa_1\kappa_2}}{\kappa_2 - \kappa_1}, & \text{if } \cos 2\beta \geqslant 0, \\ \pi - \frac{1}{2}\arcsin\frac{4dD\sqrt{\kappa_1\kappa_2}}{\kappa_2 - \kappa_1}, & \text{if } \cos 2\beta < 0, \end{cases}$$

where $w = \frac{4}{3B}\left[C + \sqrt{4C^2 + 3\left(D^2 + \frac{1}{4d^2}\right)}\right]$.

It is obvious from the expressions thus obtained that the necessary condition for finding the principle curvatures κ_1, κ_2 is the non-negativity of the expression appearing under the sign of the radical. The latter condition holds better as the wavelength of the probing signal decreases. In that latter case, the measured values of B, C, and D will be predicted much better by the above formulas, since the formulas themselves are obtained from the short-wave asymptote of the scattered field. Given the measurement errors, usually unavoidable, which lead to a negative value of the expression under the sign of the radical, it is advisable to find a solution to the system (2.5) approximately. One can find such an approximate solution by minimizing the sum of squared errors that would appear in the left- and right-hand parts of the equations constituting the whole system, given that values κ_1, κ_2 must be non-negative.

Thus, for a combined reception of the echo signals at two orthogonal polarizations, it is possible to obtain the values of the principal curvatures and the orientation of the principal directions at the scatterer surface's point of specular reflection.

2.2.2 The algorithm for reconstructing the 'illuminated' part of the object's surface

The consecutive steps of the proposed algorithm are shown in figure 2.4. The illuminated part of the object is reconstructed in a local coordinate system tied to the object itself. Its origin is placed at the stationary phase point, axis z coinciding with

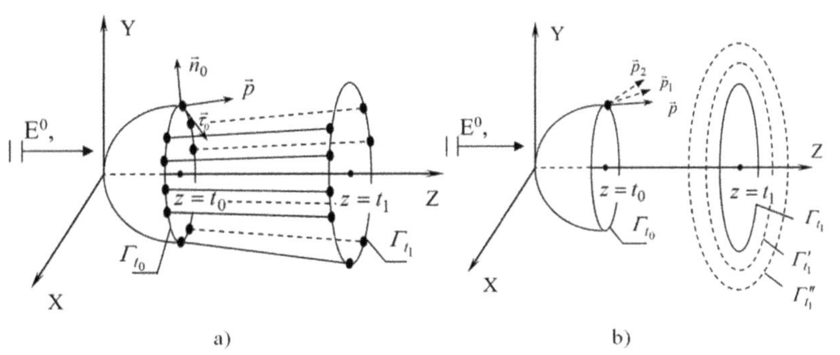

Figure 2.4. Consecutive steps to be taken in order to reconstruct the 'illuminated' part of the scatterer's surface.

the direction of radar probing (incidence direction). Along this axis, we read out the current time of EMW propagation as well as its longitudinal coordinate in lightmeters.

Axes x and y are directed in such a manner that they coincide with the principal axes of the scatterer's surface at the specular point and their unit vectors form the right three with the unit vector of axis z. The surface to be reconstructed is going to be defined in every time section $z = t$ by a set of polar coordinates ρ, φ:

$$\rho = \rho(\varphi, t) \quad (0 \leqslant \varphi < 2\pi).$$

We assume that in the vicinity of the stationary phase point the sought surface is sufficiently well approximated by an osculating paraboloid defined by the following equation:

$$z = ax^2 + by^2,$$

where a and b are the principal curvatures of the surface at the point of specular reflection, the algorithm for finding which was presented above. Now, we choose such a value of t_0 that given $z \leqslant t_0$ the sought surface would be approximated quite well by this osculating paraboloid. In every particular case, a specific value of t_0 must be chosen by balancing out the absolute values of principal curvatures. Our analysis showed that if the principle curvatures are close to each other (the difference does not exceed a few percent), then t_0 can be set to be small in order to improve the accuracy. In this case, the actual values of t_0 and Δt are determined by the available computer power. However, if the principle curvatures differ significantly, then the value of t_0 cannot be very low. This is due to the fact that given low values of z one deals mostly with the tip of the osculating paraboloid, which is almost flat and its horizontal cross-sections are very close to circles. This flat part of the osculating paraboloid works well when the principal curvatures are close to each other, yet otherwise the lack of depth in the osculating paraboloid results in a poor initial approximation. That, in its turn, necessitates the higher values of $z = t_0$ in order to add some depth and thus improve the accuracy of the whole surface reconstruction procedure. Our analysis showed that one achieves a decent approximation accuracy given the values of $t_0 = \dfrac{0.1 - 0.3}{\kappa_{\max}}$, where $\kappa_{\max}$ is the greatest value of the principal curvature at the 'specular' point.

Let us denote the contour limiting the paraboloid at level $z = t_0$ as Γ_{t0}. For every point of this contour we define the unit vector that is normal to the paraboloid's surface $\vec{n}_0(\varphi)$ and the unit vector $\vec{\tau}_0(\varphi)$ that is tangent to the contour Γ_{t0}. Finding the vector product $\vec{p} = \vec{n}_0 \times \vec{\tau}_0$, we obtain the vector which is perpendicular to contour Γ_{t0} and which, at the same time, belongs to the plane tangent to the object's surface. Next, at every point of contour Γ_{t0}, we draw a straight line with the directing vector $\vec{p}$. Now we let all these lines intersect with plane $z = t_1 = t_0 + \Delta t$. It is obvious that every one of these lines belongs to the plane tangent to the paraboloid. A set of intersection points between these straight lines and the plane $z = t_0 + \Delta t$ generates the contour, which we denote as $\Gamma_{t0+\Delta t}$. The segments of these lines that are confined between Γ_{t0} and $\Gamma_{t0+\Delta t}$, in their turn, generate the ruled surface. Coordinates of the contour $\Gamma_{t_0+\Delta t}$ points can be found from the coordinates of the contour Γ_{t_0} points as follows:,

$$x = x_0 + \frac{p_x}{p_z}\Delta t, y = y_0 + \frac{p_y}{p_z}\Delta tz = t_0 + \Delta t,$$

where x_0, y_0, z_0 are the coordinates of the points belonging to contour Γ_{t_0}, and p_x, p_y, p_z are the corresponding components of vector $\vec{p}$.

In addition, equation of the contour $\Gamma_{t_0+\Delta t}$ has the form

$$\rho = \rho(\varphi, t_0 + \Delta t) = \sqrt{x^2 + y^2}. \tag{2.32}$$

The way the contour $\Gamma_{t_0+\Delta t}$ is constructed is explained in figure 2.4(a). However, the actual values of function $\rho(\varphi, t_0 + \Delta t)$ will differ from those computed using formula (2.32). We can account for the corresponding corrections by introducing the function $\alpha(\varphi)$ $(\alpha(\varphi) < <1)$, which corrects the vector $\vec{p}$ as follows:

$$\tilde{\vec{p}} = \vec{p} + \alpha(\varphi)(\vec{p} \times \vec{\tau}_0).$$

It is shown in [20] that projection of a unit step response of a perfectly conducting scatterer onto the direction perpendicular to that of wave propagation can be presented, in the approximation of PO, as the following contour integral:

$$\Psi(t) = \vec{R}^0 \int_{\Gamma_t} \vec{f}(\vec{n})\, dl,$$

where $\vec{R}^0$ is the unit vector of the incident wave propagation direction, $\vec{f}(\vec{n}) = \frac{\vec{n}}{\sqrt{1 - \left(\vec{R}^0 \cdot \vec{n}\right)^2}}$, and $\vec{n}$ is the unit vector of the external normal to the scatterer's surface at points belonging to the contour Γ_t.

Accounting for the correction $\alpha(\varphi)$, one can present the scatterer's unit step response (USR) at time $t_0 + \Delta t$ in the following form:

$$\Psi_i(t_0 + \Delta t) = \int_0^{2\pi} \alpha(\varphi)(\Phi_1(\varphi) + \Phi_2(\varphi))\, d\varphi + \Psi^0(t_0 + \Delta t), \tag{2.33}$$

where $\Psi_i(t_0 + \Delta t)$ is the value of the USR projection onto the direction of ith polarization of the receiving antenna, here $i = (1,2)$;

$$\Phi_1(\varphi) = \frac{\left(\vec{R}^0 \cdot \vec{p}\right)\left(1 + \frac{\left(\vec{R}^0 \cdot \vec{n}_0\right)^2}{1 - \left(\vec{R}^0 \cdot \vec{n}_0\right)^2}\right)\sqrt{\rho^2 + \rho_\varphi^2} + \left(\vec{R}^0 \cdot \vec{n}_0\right)\Delta t(Q(\varphi)\cos\chi + Q'(\varphi)\sin\chi)}{\sqrt{1 - \left(\vec{R}^0 \cdot \vec{n}_0\right)^2}},$$

$$\Phi_2(\varphi) = -\Delta t\frac{\partial}{\partial\varphi}\left(\frac{\left(\vec{R}^0 \cdot \vec{n}_0\right)}{\sqrt{1 - \left(\vec{R}^0 \cdot \vec{n}_0\right)^2}}Q(\varphi)\sin\chi\right),$$

$$Q(\varphi) = q_x \cos\varphi + q_y \sin\varphi Q'(\varphi) = \frac{\partial Q(\varphi)}{\partial\varphi}\cos\chi = \frac{\rho}{\sqrt{\rho^2 + \rho_\varphi{}^2}}, \quad \sin\chi = \frac{\rho_\varphi}{\sqrt{\rho^2 + \rho_\varphi{}^2}},$$

$$q_x(\varphi) = \frac{(\vec{p}(\varphi) \times \vec{\tau}_0(\varphi))_x}{p_z(\varphi)} - \frac{p_x(\varphi)(\vec{p}(\varphi) \times \vec{\tau}_0(\varphi))_z}{p_z^2(\varphi)},$$

$$q_y(\varphi) = \frac{(\vec{p}(\varphi) \times \vec{\tau}_0(\varphi))_y}{p_z(\varphi)} - \frac{p_y(\varphi)(\vec{p}(\varphi) \times \vec{\tau}_0(\varphi))_z}{p_z^2(\varphi)}, \text{ and}$$

$$\Psi^0(t_0+\Delta t) = \vec{R}^0 \int_0^{2\pi} \frac{\vec{n}_0(\varphi, t_0 + \Delta t)}{\sqrt{1-(\vec{R}^0 \cdot \vec{n}_0(\varphi, t_0+\Delta t))^2}} \sqrt{\rho^2(\varphi, t_0+\Delta t) + \rho_\varphi^2(\varphi, t_0+\Delta t)}\ d\varphi$$

is the value of the USR function that was computed using the part of the scatterer's surface reconstructed by that time moment, and ρ_φ is the derivative of the radius-vector ρ by the angle φ. Since the unit vector of the normal to the ruled surface that moves along that surface's generatrix do not change, then $\vec{n}_0(\varphi, t_0 + \Delta t) = \vec{n}_0(\varphi, t_0) = \vec{n}_0(\varphi)$.

It is worth noting that significant errors can occur while calculating the derivatives. The latter is due to the error in determining the actual USR of the scatterer. Therefore, numerical differentiation was carried out using the formula with smoothing:

$$\overline{f}'(x_0) = \frac{2(f(x_0 + 2\Delta x) - f(x_0 - 2\Delta x)) + (f(x_0 + \Delta x) - f(x_0 - \Delta x))}{10\Delta x}.$$

Expression (2.33) was obtained under the assumption that contour $\Gamma_{t_0+\Delta t}$ belongs to the surface of the sought scatterer. Having found function $\alpha(\varphi)$ from expression (2.33), we can define more accurately contour $\Gamma_{t_0+\Delta t}$ itself, this function representing the line of the intersection of the sought surface with the plane $z = t_0 + \Delta t$.

Thus we have the following equation for computing the value of correction $\alpha(\varphi)$:

$$\int_0^{2\pi} \alpha(\varphi)\Phi(\varphi)\ d\varphi = \Delta\Psi_i, \tag{2.34}$$

where $\Delta\Psi_i = \Psi_i - \Psi^0, \Phi(\varphi) = \Phi_1(\varphi) + \Phi_2(\varphi)$.

Considering that values of $\alpha(\varphi)$ are small enough and they vary pretty smoothly with the change of angle φ, we can limit ourselves with only those solutions (out of an infinite number of possible solutions to equation (2.34)) that possess the minimal norm. This approach is the same as the one used in [21]. This minimal norm is to be found in the metric C or the metric L_2. Then, the solution to equation (2.34) can be presented in explicit form: for the metric C the solution is $\alpha_0(\varphi) = \dfrac{\Delta\Psi_i \operatorname{sgn}\Phi(\varphi)}{\int_0^{2\pi} |\Phi(\varphi)| d\varphi}$, for the metric L_2 the solution is $\alpha_0(\varphi) = \dfrac{\Delta\Psi_i\Phi(\varphi)}{\int_0^{2\pi} \Phi^2(\varphi)\ d\varphi}$.

In addition, of the two values of $\alpha_0(\varphi)$, we choose the one possessing the lowest value of $|\Delta\Psi_i|$.

As soon as the elaborated value of vector $\vec{p}$ is found, we draw the new ruled surface with the generatrix being parallel to this vector $\vec{p}$, and we find the intersection points of the new surface with plane $z = t_0 + \Delta t$. In this manner, the contour $\Gamma_{t_0+\Delta t}$ is refined. Having done that, we proceed to the next iteration of refining the contour $\Gamma_{t_0+\Delta t}$. In addition, to find the value of refined error $\Delta\Psi(t_0 + \Delta t)$, we use the refined values of vector $\vec{p}$, the refined radius-vector ρ, and the normal $\vec{n}_0$ to the refined ruled surface. The iterative procedure to refine the contour $\Gamma_{t_0+\Delta t}$ is carried on until the value of error $\Delta\Psi(t_0 + \Delta t)$ goes below some predefined small threshold ε. The way the described operations are carried out is shown in figure 2.4(b). Once this threshold is achieved, we take another step in time. For this purpose, we find the unit vector of normal to the final ruled surface $\vec{n}_0$ and the unit vector of tangent to the final contour $\vec{\tau}_0$, using which we compute the vector $\vec{p}$. Next, we find the contour $\Gamma_{t_0+2\Delta t}$, which is refined in the same manner as described above.

The whole procedure of reconstructing the surface is carried on until the 'light-shadow' boundary at the object's surface is reached.

2.2.3 Results of numerical simulation

To test the performance of the algorithm described above, we tried to reconstruct the 'illuminated' surface parts of the following shapes:

(I) sphere of 3 m radius (figure 2.5);

(II) ellipsoid of revolution with semi-axes of 3 m, 3 m, and 1 m (figure 2.6) (the incidence direction is parallel to the smallest semi-axis);

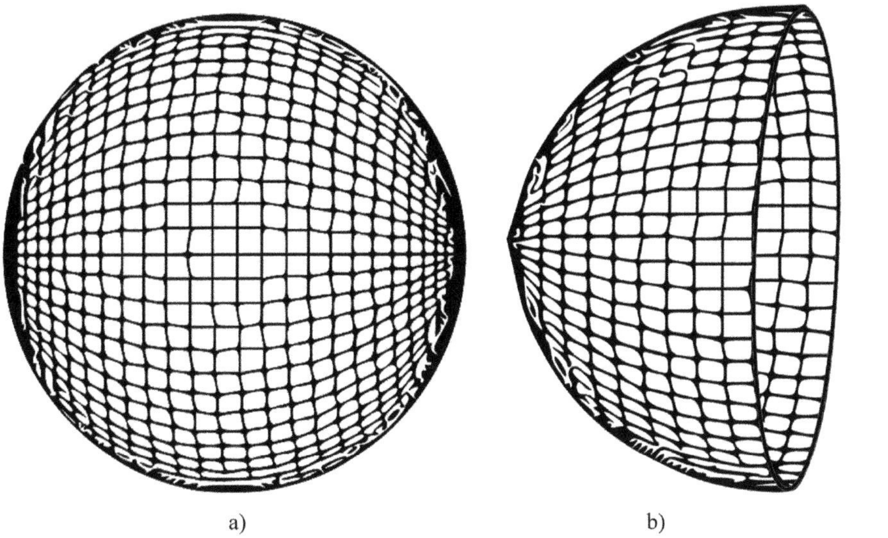

a) b)

Figure 2.5. The actual shape of the sphere with radius 3 m (a) and the result of the reconstruction of its 'illuminated' part (b).

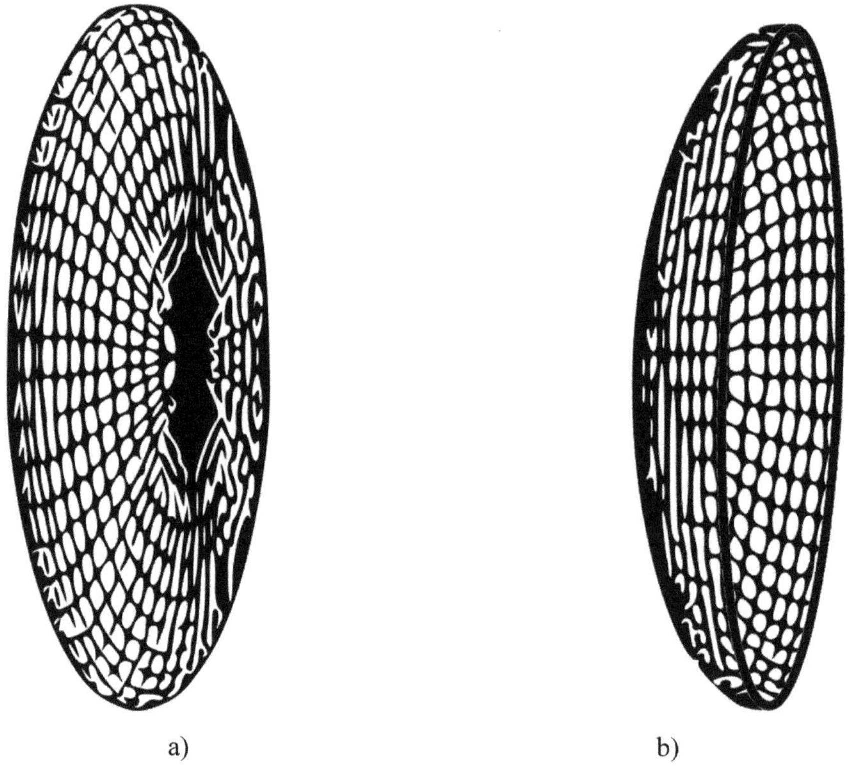

Figure 2.6. The actual shape of the ellipsoid with semi-axes of 3 m, 2 m, and 1 m (a) and the result of the reconstruction of its 'illuminated' part (b).

(III) three-axial ellipsoid with semi-axes of 3 m, 2 m, and 1 m (figure 2.7) (the incidence direction is also parallel to the smallest semi-axis);
(IV) three-axial ellipsoid with semi-axes of 1 m, 2 m, and 3 m (figure 2.8) (the incidence direction is parallel to the greatest semi-axis).

In the above cases, figure 2.6(a) shows the actual surface of the object, and figure 2.6(b) shows the result of the object's surface reconstruction using the algorithm described above.

As an actual USR function of the object, we used its high-frequency approximation that was computed under the assumption of PO given the corresponding sizes of the object and the corresponding direction of incidence.

Our results showed that the surface reconstruction accuracy depended on the procedure parameters mentioned above (the depth of the osculating paraboloid t_0 and the time increment Δt) and also depended on the stretch of the 'illuminated' part of the object's surface l along the direction of the wave incidence. As a performance characteristic of the procedure at every particular point, we used the relative reconstruction error δ of the following form:

$$\delta(\varphi, t) = \frac{\rho(\varphi, t) - \widehat{\rho}(\varphi, t)}{\rho(\varphi, t)} \cdot 100\%,$$

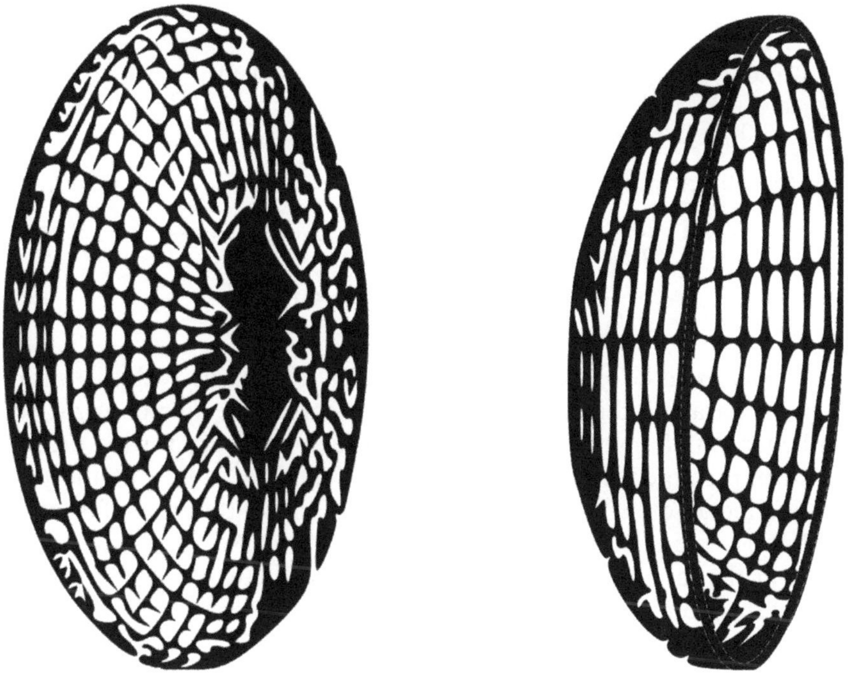

a) b)

Figure 2.7. The actual shape of the ellipsoid with semi-axes of 3 m, 2 m, and 1 m (a) and the result of the reconstruction of its 'illuminated' part (b).

a) b)

Figure 2.8. The actual shape of the ellipsoid with semi-axes of 1 m, 2 m, and 3 m (a) and the result of the reconstruction of its 'illuminated' part (b).

where $\rho(\varphi, t)$ are the actual values of the radius-vector at point $M(\varphi, t)$, and $\widehat{\rho}(\varphi, t)$ are the reconstructed values of the corresponding radius-vector at point $M(\varphi, t)$.

Finally, the average relative error of the surface reconstruction was as follows:

- for the objects of clauses (I) and (II) above:

given $\dfrac{l}{t_0} = \dfrac{l}{\Delta t} = 10$ it was $\delta = 13.4\%$, given $\dfrac{l}{t_0} = \dfrac{l}{\Delta t} = 20$ it was $\delta = 9.1\%$, given $\dfrac{l}{t_0} = \dfrac{l}{\Delta t} = 100$ it was $\delta = 2.7\%$.

- for the object of clause (III) above:

  given $\frac{l}{t_0} = \frac{l}{\Delta t} = 10$ it was $\delta = 17.0\%$, given $\frac{l}{t_0} = \frac{l}{\Delta t} = 20$ it was $\delta = 13.0\%$, given $\frac{l}{t_0} = \frac{l}{\Delta t} = 100$ it was $\delta = 8.6\%$.
- for the object of clause (IV) above:

  given $\frac{l}{t_0} = \frac{l}{\Delta t} = 10$ it was $\delta = 25.8\%$, given $\frac{l}{t_0} = \frac{l}{\Delta t} = 20$ it was $\delta = 23.4\%$, given $\frac{l}{t_0} = \frac{l}{\Delta t} = 100$ it was $\delta = 14.6\%$.

These results could be explained by the fact that for smaller values of its depth t_0, the osculating paraboloid is getting closer to the surface being reconstructed, which means a better initial approximation to the surface necessary for the procedure.

The values presented here also demonstrate another feature of the procedure, namely it is a reduction in reconstruction quality if the scatterer's surface in the form of an ellipsoid stretched along the direction of EMW incidence (figure 2.8). Moreover, in the latter case one also observes an increased error of finding the boundary of the illuminated part of the scatterer's surface. Provisional performance analysis of the reconstruction procedure of the illuminated surface part showed that such a reduced accuracy was due to the fact that the error in determining the calculated USR becomes comparable to that of the calculation error. The latter leads to the situation where the error $\Delta\Psi$ and the corresponding correction $\alpha_0(\varphi)$ (which participate in the refinement of the surface being reconstructed) cannot be determined correctly and even become negative at some points. This can lead to a distortion of the surface instead of its refinement. Obviously, in a situation like this, we can talk about the applicability boundary for the method proposed above.

Figure 2.9 presents typical qualitative plots of error averaged over all the time cross-sections versus coordinate z along the direction of incidence. Figure 2.9(a) corresponds to the objects presented in clauses (I), (II), and (III); figure 2.9(b) corresponds to the object from clause (IV). The plots show that the error attained greater values at first. This was due to a poor initial approximation of the scatterer's

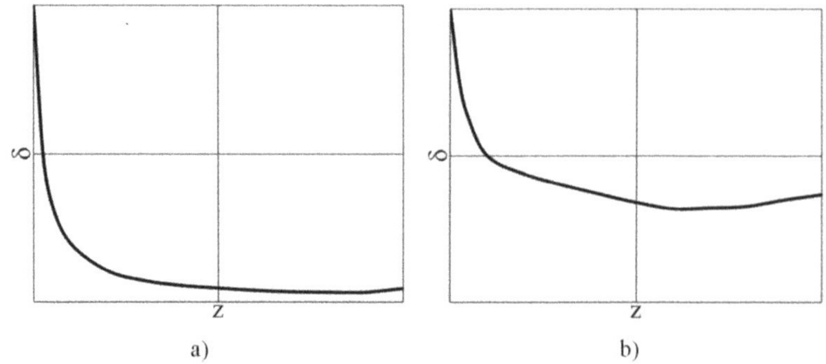

Figure 2.9. Qualitative dependences of error δ averaged over all the time cross-sections versus coordinate z along the direction of incidence.

surface by the osculating paraboloid. Obviously, it can be expected that by improving the initial approximation (for example, using information about closely spaced 'specular' points collected during several radar probings) one can significantly improve the quality of the object's surface reconstruction. It is also seen that as we move along the direction of wave incidence, the reconstruction error decreases rapidly, which indicates that the reconstruction algorithm was designed correctly.

When reconstructing the surface of an ellipsoid stretched along the direction of incidence, one obtains the average error that decreases until a certain moment, after which it increases again. The analysis showed that this was due to the incorrect determination of the boundary of the 'illuminated' part of the scatterer's surface. The reasons for that were explained above.

Thus in the case of symmetric bodies probed along one of their axes of symmetry, it is sufficient to carry out a single radar probing given that the echoes would be received at two orthogonal polarizations to obtain such an approximation of the scatterer's surface. The echoes thus received would be used to find the values of the main curvatures of the reconstructed surface at the point of 'specular' reflection. Such curvatures would be used to construct a paraboloid, which could be used as an initial approximation of the object's surface. However, this approach turns out to be unacceptable when reconstructing the 'illuminated' surface parts of objects that are not symmetrical with respect to the direction of observation, since the paraboloid carries no information about the asymmetry of the surface to be reconstructed. Let us make a remark on how we can overcome the above issue.

In order to deal with non-symmetrical bodies, we can use the radar information obtained by probing the object from several directions. Taking into account the specifics of the problem, namely the large electrical dimensions and the convexity of the object surface, we can argue that every new probing from a slightly different perspective provides us with information about the coordinates of the corresponding 'specular' point as well as the direction of the normal to the scatterer's surface at that particular point. This would make it possible to construct, with a sufficient degree of accuracy, an approximation of the surface in the vicinity of that point. Further steps could be carried out as before in the case of symmetrical scatterer.

Thus, in this section, we introduced a genuine algorithm for reconstructing the 'illuminated' surface part of a smooth convex perfectly conducting scatterer located in free space, its radar echoes being received at two orthogonal polarizations (full polarization reception). Examples have been presented of the 'illuminated' surface part reconstruction for a spherical scatterer as well as scatterers shaped as triaxial ellipsoids given that the direction of the incident wave coincided with one of their axes.

2.3 Scattering and radiation characteristics of antenna systems under nose dielectric radomes

Some content in this section has been reproduced with permission from [136].

In this section we consider the method for computing an EMF scattered by an antenna placed inside the nose dielectric radome. We also present the radiation characteristics of such an antenna using the original method for computing the field

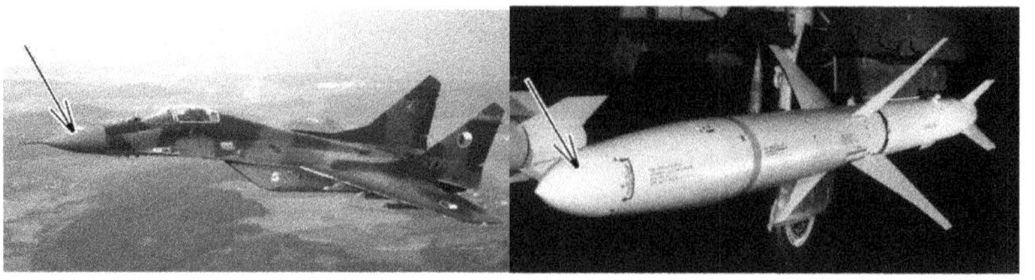

a b

Figure 2.10. ASs under dielectric radomes borne by an aircraft (a) and a missile (b).

radiated by an active aperture given that an arbitrary system of scatterers (the radome in particular), which can exist in its vicinity, scatter the near-field zone fields thus taking part in shaping the directivity pattern (DP) of such an aperture. Using our method, we analyze the influence of radomes on the radiation performance of the antenna and find the ways to bring that influence to a minimum. We also show that radomes can be used to reduce the radar observability of antenna systems (ASs) for radar frequency ranges that lie outside the antenna operating frequency range.

The majority of airborne objects (for example, aircraft, and missiles with homing heads) have nose radomes that cover the AS (figure 2.10). Existing antenna systems can increase the radar cross section (RCS) significantly, even of an otherwise low radar visibility object [22].

Thus the development of the calculation method for evaluating the scattering characteristics (particularly the RCS) of ASs under radomes is a very important problem to be solved while designing detection algorithms for airborne targets. The internal wave interactions appearing once the 'antenna–radome' system is illuminated lead to highly complicated calculation algorithms. The geometrical-optics methods usually used for calculating the scattering characteristics of such systems [23, 24] do not account for these interactions (multiple reflections between the antenna and radome). At the same time, the electrodynamic interaction between the antenna reflector and the radome can lead to a significant increase (often with resonant behavior) in the AS RCS [25], and, consequently, in the RCS of the entire radar object. The iterative method proposed in [26] for an 'antenna–radome' system was developed under the following limitations: first, all considerations were given for two-dimensional models and, second, the radome shape had to have very small curvatures. In [25] a bounded dielectric plate was substituted for the radome. Thus it is necessary to develop an asymptotic method for calculating the RCS of the reflector antenna system under the nose cone (a dielectric radome) in a three-dimensional configuration. Such a method would need to account for at least first-order re-reflections between the antenna and radome.

Some computational models were described in recent papers that were used to assess the impact of dielectric radomes on the performance of antennas in terms of their radiation patterns [27–34]. However, in the literature known to us there is no discussion concerning the RCS of reflector antennas placed under nose radomes (which is our primary concern in this section).

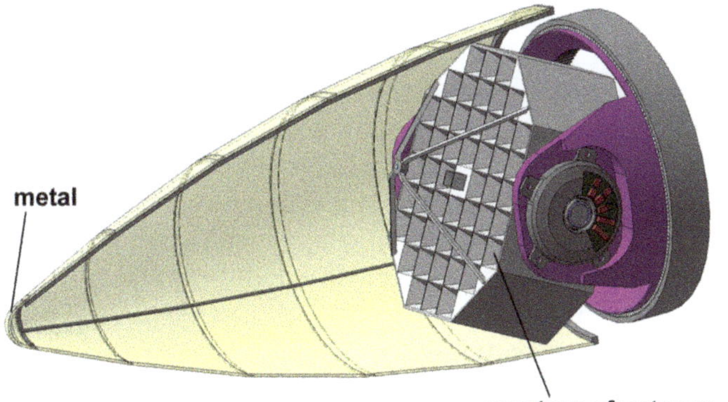

Figure 2.11. The general appearance of the antenna under a compound dielectric radome.

In addition to calculating the scattering characteristics (the RCS) (problem #1), it is also necessary to calculate the DP of such an AS (problem #2). From the viewpoint of a manufacturer, the manufacturability is another very important issue in radome production. Therefore, three kinds of radome shapes are considered in this paper: ogive, compound cone (consisting of several truncated cones), and pyramid (having a regular octagon as its base). Both the scattering characteristics (the RCS) and DPs are obtained for these kinds of radomes. The methods for solving these two problems (#1 and #2) are based on using the same 'standard' problems. In addition, the general approach to calculations regarding the 'antenna–radome' system is the same for different kinds of radomes. Some computation details will be described below along with the results of these computations.

In the beginning, we consider an AS under the nose radome, which is the conjugation (not smooth) of truncated cones (figure 2.11). The relative permittivity of the cones' material is $\varepsilon' = 3.8 + j0.0076$. Inside the radome, there is a phased-array antenna with an octagonal aperture (figure 2.10). As a first approximation, we assume the antenna to be represented by an octagonal plate and the field scattered by the antenna itself (as is usually done in engineering) will be multiplied by a factor of ≈ 0.9. Such a representation provides a satisfactory approximation to the field scattered by the antenna surface (if the wavelength of the incident wave is much greater than that the antenna operation wavelength). In the nose part of the radome, there is a metallic spherical cap conjugated smoothly with the top truncated cone (figure 2.11). The presence of this cap is accounted for in the EM calculations by the beam tracing method. The latter means that a wave passing through the radome (in a geometrical-optics approximation) is altered as follows: the incident beam hitting the metallic cap surface does not pass to the radome inside. The field scattered by the entire radome includes the wave scattered by this metallic half spherical cap.

Considering all of the above, the 'standard' problem here is the calculation of the wave scattered by the side surface of a conical radome given that the octagonal plate, which could be rotated around some point outside that plate, was placed inside it (the scattering of which was considered to be simple reflection with the coefficient equal to 0.9).

Moreover, the method proposed here accounts for the influence of beams that pass through the radome and do not hit the antenna aperture directly, but rather are rescattered by the radome's inner surface. It should be noted that the same beam can pass through the radome at a surface point belonging to one truncated cone but it can rescatter from the inner surface belonging to another one. Below, we consider a solution procedure for the 'key' problem stated above [19].

2.3.1 Electromagnetic wave scattering by an antenna system under the nose dielectric radome

In this subsection we consider the scattering from the antenna under different types of radomes.

Electromagnetic wave scattering by an antenna system under the conical dielectric radome

We consider a plane EMW

$$\vec{E}^{0}(\vec{x}) = \vec{p}^{\,0} \exp\left(jk_0\left(\vec{R}^{0} \cdot \vec{x}\right)\right),$$
$$\vec{H}^{0}(\vec{x}) = \sqrt{\frac{\varepsilon_0}{\mu_0}}\left(\vec{R}^{0} \times \vec{p}^{\,0}\right) \exp\left(jk_0\left(\vec{R}^{0} \cdot \vec{x}\right)\right), \quad (2.35)$$

which is incident from without on the model of the reflector AS under the cone radome (figure 2.12).

Application of the Lorentz reciprocity theorem to the sought total field $(\vec{E}, \vec{H})$ and to auxiliary field $\left(\hat{\vec{E}}, \hat{\vec{H}}(\vec{x} \mid \vec{x}_0, \vec{p}\,)\right)$, the latter corresponding to that of the

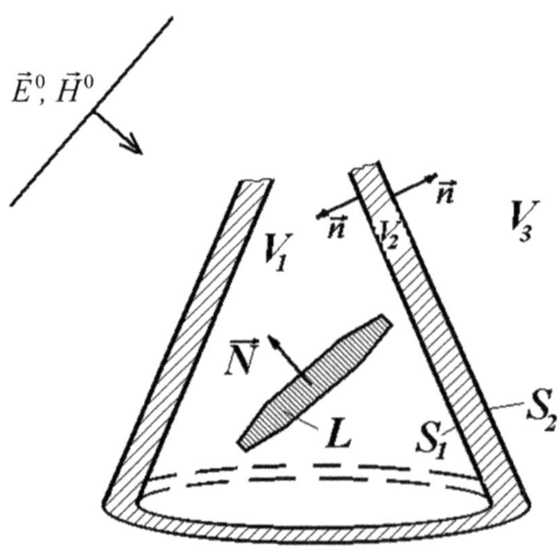

Figure 2.12. The 'antenna–radome' system.

electric dipole placed at point $\vec{x}_0$ that has the vector-moment $\vec{x}_0 = -r\vec{R}^0$ given that only the radome is present, allows us to obtain the integral representation of the sought for field [136]:

$$j\omega\vec{p} \cdot \vec{E}(\vec{x}_0) = j\omega\vec{p} \cdot \vec{E}^{\text{scat.}}_{\text{radome}}(\vec{x}_0) + \int_L \left(\vec{K}(\vec{x}) \cdot \vec{\hat{E}}^T(\vec{x} \mid \vec{x}_0, \vec{p})\right) \mathrm{d}S, \quad (2.36)$$

where $\vec{E}^{\text{scat.}}_{\text{radome}}(\vec{x}_0)$ is the field scattered by the radome only, and $\vec{K}(x)$ is the surface current density over the antenna reflector's surface. The integral summand of expression (2.36) represents the response of the antenna reflector onto the incident wave that takes into account the EMW interaction with the radome. Having assumed $\vec{x}_0 = -r\vec{R}^0$ and having further assumed that $r \to \infty$, we receive the expression for the resulting field scattered from the 'antenna–radome' system in the far-field zone:

$$\vec{p} \cdot \vec{E}\left(\vec{R}^0\right)\vec{p} \cdot \vec{E}^{\text{scat.}}_{\text{radome}}\left(\vec{R}^0\right) - jk_0\frac{\mathrm{e}^{jk_0r}}{4\pi r}\sqrt{\frac{\mu_0}{\varepsilon_0}} \int_L \left(\vec{\hat{E}}(\vec{x}) \cdot \vec{K}(\vec{x})\right) \mathrm{d}S. \quad (2.37)$$

Here $\vec{\hat{E}}(\vec{x})$ is the field excited by the primary incident plane wave (2.35) at points of the reflector's surface L given that only the radome is present. We are going to compute this field by the method of geometrical optics.

Under the approximation assumed above, we represent $\left(\vec{\hat{E}}(\vec{x}), \vec{\hat{H}}(\vec{x})\right)$ as a sum of two fields, one that passes to the reflector directly through the illuminated part of the radome (propagation path 1 in figure 2.13) and the other that passes to the reflector via a single reflection from the inner wall of the radome (propagation path 2 in figure 2.13).

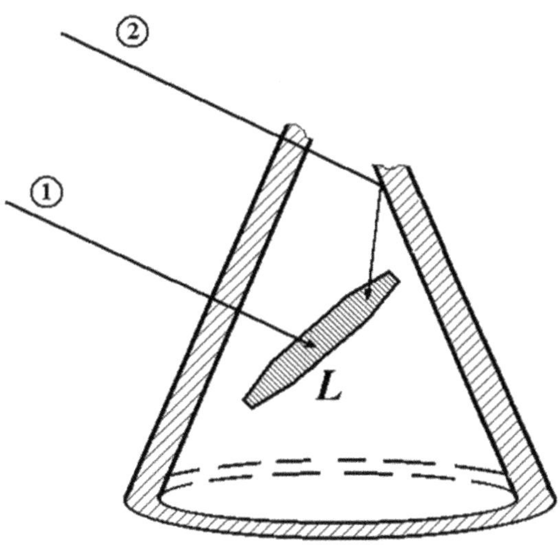

Figure 2.13. The propagation paths of the incident wave to reach the antenna beneath the radome.

The field corresponding to propagation path 1 in figure 2.13 can be presented as:

$$\vec{E}_1(\vec{x}) = \left[\tau_\perp p_\perp \vec{e}_\perp + \tau_\parallel p_\parallel \left(\vec{R}^0 \times \vec{e}_\perp\right)\right] \exp\left(jk_0\left(\vec{R}^0 \cdot \vec{x}\right)\right), \tag{2.38}$$

$$\vec{H}_1(\vec{x}) = \sqrt{\frac{\varepsilon_0}{\mu_0}}\left[\tau_\perp p_\perp\left(\vec{R}^0 \times \vec{e}_\perp\right) - \tau_\parallel p_\parallel \vec{e}_\perp\right] \exp\left(jk_0\left(\vec{R}^0 \cdot \vec{x}\right)\right), \tag{2.39}$$

where $\vec{e}_\perp = \dfrac{\vec{R}^0 \times \vec{n}}{\left|\vec{R}^0 \times \vec{n}\right|}$, $\vec{e}_\parallel = \left(\vec{R}^0 \times \vec{e}_\perp\right)$, $p_\perp = (\vec{p} \cdot \vec{e}_\perp)$, $p_\parallel = (\vec{p} \cdot \vec{e}_\parallel)$, and $\vec{n}$ is the vector normal to the radome's surface.

The complex valued coefficients $\tau_\perp$, $\tau_\parallel$ describe the propagation of the plane EMW through a flat uniform layer with the properties of the radome at two mutually orthogonal polarizations. For parallel polarization marked by subscript $\parallel$, we consider the situation where the electric field intensity vector of the incident wave belongs to the plane passing through vector $\vec{R}^0$ and normal $\vec{n}$ at every particular point of the radome surface. Accordingly, the perpendicular polarization, marked by subscript $\perp$, corresponds to the situation where the electric field intensity vector is orthogonal to the plane specified above. The general expression for the propagation coefficient can be presented as follows:

$$\tau = \left(\left(\cos \kappa\delta + \frac{j}{c}\sin \kappa\delta\right) + \left(\cos \kappa\delta - \frac{j}{c}\sin \kappa\delta\right)\rho\right)\exp(-jk_0\delta \cos \theta), \tag{2.40}$$

where ρ is the complex valued coefficient of reflection from the plane uniform layer of material that has the properties of the radome. This coefficient can be expressed as

$$\rho = \frac{j(c^2 - 1)\sin \kappa\delta}{2c \cos \kappa\delta - j(c^2 + 1)\sin \kappa\delta}. \tag{2.41}$$

Here $c = \dfrac{\sqrt{\varepsilon' - \sin^2 \theta}}{\beta \cos \theta}$, $\kappa = k_0\sqrt{\varepsilon' - \sin^2 \theta}$, $\cos \theta = \left|\left(\vec{R}^0 \cdot \vec{n}\right)\right|$, $\sin^2 \theta = 1 - \left(\vec{R}^0 \cdot \vec{n}\right)^2$, ε' is the relative permittivity of the radome material, and δ is the radome thickness:

$$\beta = \begin{cases} 1 & \text{in case} \quad \perp \quad \text{polarization,} \\ \varepsilon' & \text{in case} \quad \parallel \quad \text{polarization.} \end{cases}$$

In the case of $\perp$ polarization, we have $\rho = \rho_\perp$ and $\tau = \tau_\perp$, and in the case of $\parallel$ polarization we have $\rho = \rho_\parallel$ and $\tau = \tau_\parallel$.

If the ray passing through radome at some point $\vec{x}_0$ does not hit the reflector, then it must pass through the radome once again at some point $\vec{x}_1$ at the 'shadow' part of the radome surface. Taking into account that the equation of the conical surface has the following form:

$$x = z \tan \theta \cos s,$$

$$y = z \tan\theta \sin s, \qquad (2.42)$$

$$(0 \leqslant z \leqslant h,\ 0 \leqslant s \leqslant 2\pi),$$

and using the equation of a beam in parametric form

$$\begin{aligned} x &= x_0 + R_1^0 t, \\ y &= y_0 + R_2^0 t, \\ z &= z_0 + R_3^0 t, \end{aligned} \qquad (2.43)$$

where (x_0, y_0, z_0) is a point at the 'illuminated' part of the cone's surface, we obtain the value t_1, which corresponds to point $\vec{x}_1$ at the 'shadow' part of the radome's surface:

$$t_1 = 2\frac{R_3^0 z_0 \tan^2\theta - R_1^0 x_0 - R_2^0 y_0}{1 - R_3^{0^2}\big/\cos^2\theta} R_1^0 x_0 + R_1^0 t. \qquad (2.44)$$

If we find, in this case, the values of $\tau_\perp$, $\tau_\|$, $\rho_\perp$, $\rho_\|$, $\vec{e}_\perp$, and $\vec{e}_\|$ at point $\vec{x}_0$, then we can compute the intensity vector of the electric field that has propagated through the radome at point $\vec{x}_0$ and that is incident onto its inner surface at point $\vec{x}_1$:

$$\vec{p}_1 \exp\left(jk_0\left(\vec{R}^0 \cdot \vec{x}_1\right)\right), \quad \vec{p}_1 = \tau_\perp p_\perp \vec{e}_\perp + \tau_\| p_\| \vec{e}_\|. \qquad (2.45)$$

Vector $\vec{p}_1$, wave incidence direction $\vec{R}_0$, and the normal $\vec{n}(\vec{x}_1)$ to the inner surface of the radome S_1 at point $\vec{x}_1$ can be used for finding $\tau_{1\perp}$, $\tau_{1\|}$, $\rho_{1\perp}$, $\rho_{1\|}$, $\vec{e}_{1\perp}$, and $\vec{e}_{1\|}$ by means of formulas (2.40) and (2.41). The field that has been reflected from the inner surface of radome at point $\vec{x}_1$ and that is incident onto the antenna's reflector (path 2 in figure 2.13) can be represented as follows:

$$\vec{\hat{E}}_2(\vec{x}) = \left[\rho_{1\perp} p_{1\perp} \vec{e}_{1\perp} + \rho_{1\|} p_{1\|}\left(\vec{R}^1 \times \vec{e}_{1\perp}\right)\right] \exp\left(jk_0\left[\left(\vec{R}^0 \cdot \vec{x}_1\right) + \left(\vec{R}^1 \cdot \vec{x}\right)\right]\right), \qquad (2.46)$$

$$\begin{aligned} \vec{\hat{H}}_2(\vec{x}) = \sqrt{\frac{\varepsilon_0}{\mu_0}}&\left[-\rho_{1\|} p_{1\|} \vec{e}_{1\perp} + \rho_{1\perp} p_{1\perp}\left(\vec{R}^1 \times \vec{e}_{1\perp}\right)\right] \\ &\exp\left(jk_0[(\vec{R}^0 \cdot \vec{x}_1) + (\vec{R}^1 \cdot \vec{x})]\right), \end{aligned} \qquad (2.47)$$

where $\vec{R}^1 = \vec{R}^0 - 2\vec{n}(\vec{x}_1)\left(\vec{R}^0 \cdot \vec{n}(\vec{x}_1)\right)$.

It should be noted that once the planar EMW bounces off the inner surface of the radome, then a caustic surface may form. The computation of such a caustic surface that appears in the case of oblique incidence of the plane wave onto the conical radome has been carried out in [19]. The ray passing through such a caustic surface changes its phase by $\pi/2$ [35, 36], which has to be taken into account for the wave hitting the antenna after reflection from the radome's inner surface.

The surface current density $\vec{K}(\vec{x})$ over the antenna reflector in expression (2.37) is computed in the form of the sum of currents excited on the reflector's surface by 'direct' and 'reflected' waves (path 1 and path 2 in figure 2.13). In the PO approximation, the current surface density can be expressed as

$$\vec{K}(\vec{x}) = 2(\vec{N} \times \vec{\hat{H}}), \tag{2.48}$$

where $\vec{N}$ is the vector of the normal to the reflector's surface at point $\vec{x}$, and $\vec{\hat{H}}$ can be computed as a sum of the magnetic field intensities for the first and second paths of incident wave propagation according to expressions (2.39) and (2.47), respectively.

In our case, the vector of normal $\vec{N}$ is constant, and the direction of incidence for the direct and reflected beam to the octagonal antenna aperture can be calculated as follows.

Let the plane containing the antenna aperture (figure 2.14) be described by equation

$$-x \sin\alpha - z \cos\alpha = d, \tag{2.49}$$

and the incident wave direction be described by unit vector $\vec{R}^0 = (R_1^0, R_2^0, R_3^0)$.

The direct (reflected) beam can be expressed in parametric form (2.43).

We find the point where those beams cross the plane (2.49):

$$x^* = x_0 + R_1^0 t_0,$$
$$y^* = y_0 + R_2^0 t_0,$$
$$z^* = z_0 + R_3^0 t_0,$$

where

$$t_0 = -\frac{x_0 \sin\alpha + z_0 \cos\alpha + d}{R_1^0 \sin\alpha + R_3^0 \cos\alpha}.$$

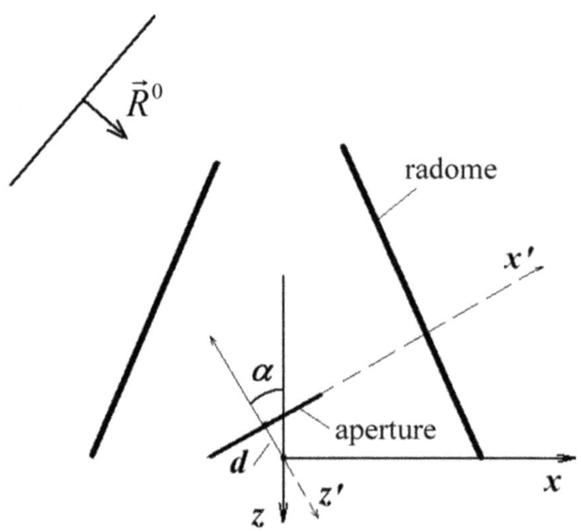

Figure 2.14. Geometry of the relative antenna aperture position inside the radome.

Next, we switch to local coordinate system (x', y', z'), which results from the rotation of coordinate system (x, y, z) by angle α in plane (xOz):

$$\begin{cases} x' = x\cos\alpha - z\sin\alpha \\ y' = y \\ z' = x\sin\alpha + z\cos\alpha + d \end{cases}.$$

Figure 2.15 shows the appearance of the octagonal aperture L.

Now we find the point where the beam crosses the aperture in plane $(x'Oy')$:

$$x' = x^*\cos\alpha - z^*\sin\alpha, \qquad y' = y^*. \tag{2.50}$$

Then the region L can be described by the system of inequalities

$$L: |x'| \leqslant D_1; \quad |y'| \leqslant D_2; \quad |\gamma x' + y'| \leqslant \beta; \quad |-\gamma x' + y'| \leqslant \beta. \tag{2.51}$$

Thus if conditions (2.50) and (2.51) are met, then we can conclude that the beam actually crosses the octagonal antenna aperture.

The field $\vec{p}\cdot\vec{E}^{\,\mathrm{scat}}_{\mathrm{radome}}\left(\vec{R}^0\right)$ scattered by the radome can be computed using Kirchhoff's approximation [19]:

$$\vec{p}\cdot\vec{E}^{\,\mathrm{scat}}_{\mathrm{radome}}\left(\vec{R}^0\right) \approx -jk_0\frac{e^{jk_0r}}{4\pi r} \times \iint_{S_{\mathrm{illum.}}}\left[\left(\vec{p}\cdot\left(\vec{n}\times\vec{H}'(\vec{x})\right)\right)\sqrt{\frac{\mu_0}{\varepsilon_0}} + \vec{E}'(\vec{x})\cdot\left(\vec{n}\times\left(\vec{p}\times\vec{R}^0\right)\right)\right]\exp\left(jk_0\left(\vec{R}^0\cdot\vec{x}\right)\right)\,\mathrm{d}S. \tag{2.52}$$

Here $(\vec{E}', \vec{H}')$ is the field at (or close to) the illuminated surface of a radome, which, in the Kirchhoff approximation, can be expressed as

$$\vec{E}'(\vec{x}) \approx \left[\rho_\perp(\vec{x})p_\perp(\vec{x})\frac{\left(\vec{R}^1\times\vec{n}\right)}{\left|\vec{R}^1\times\vec{n}\right|} + \rho_\parallel(\vec{x})p_\parallel(\vec{x})\frac{\vec{R}^1\times\left(\vec{R}^1\times\vec{n}\right)}{\left|\vec{R}^1\times\left(\vec{R}^1\times\vec{n}\right)\right|}\right]\exp\left(jk_0\left(\vec{R}^1\cdot\vec{x}\right)\right), \tag{2.53}$$

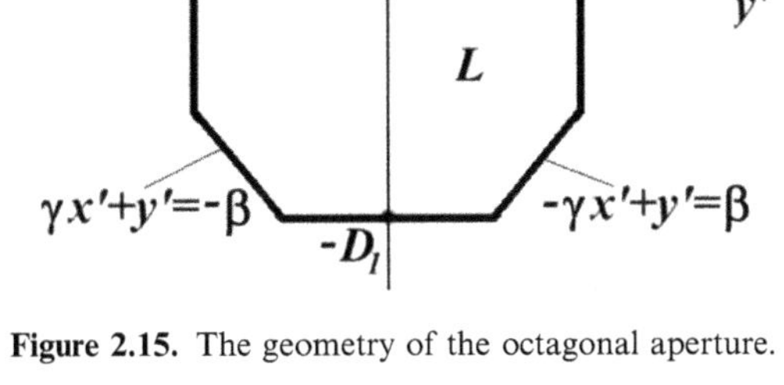

Figure 2.15. The geometry of the octagonal aperture.

$$\vec{H}'(\vec{x}) = \frac{1}{j\omega\mu_0}\vec{\nabla} \times \vec{E}'(\vec{x}), \tag{2.54}$$

where $\vec{R}^1 = \vec{R}^0 - 2\vec{n}\left(\vec{R}^0 \cdot \vec{n}\right)$, and $\vec{n} = \vec{n}(x)$ is the normal to outer radome's surface S_2. For a dielectric radome of conical shape, representation (2.52) can be simplified and reduced to a single integral by the angular coordinate α tied to the illuminated part of the radome:

$$\vec{p} \cdot \vec{E}^{\text{scat.}}_{\text{radome}}\left(\vec{R}^0\right) \approx -jk_0 \frac{e^{jk_0 r}}{4\pi r}\frac{\sin\theta}{\cos^2\theta}\int_{\alpha_0}^{\alpha_1}\Psi(\alpha)\,\mathrm{d}\alpha, \tag{2.55}$$

where

$$\Psi(\alpha) = F(\alpha)\left[\frac{h\exp\left(j2k_0 h\varphi(\alpha)\right)}{2jk_0\varphi(\alpha)} + \frac{\exp\left(j2k_0 h\varphi(\alpha)\right) - 1}{4k_0^2\varphi^2(\alpha)}\right],$$

$$F(\alpha) = \left(\rho_\perp(\alpha)p_\perp^2(\alpha) - \rho_{||}(\alpha)p_{||}^2(\alpha)\right)\left(\vec{R}^0 \cdot \vec{n}(\alpha)\right),$$

$$\varphi(\alpha) = tg\,\theta(R_1^0\cos\alpha + R_2^0\sin\alpha) + R_3^0,$$

$$\alpha_0 = \operatorname{arcctg}\frac{\eta}{\sqrt{1 - \eta^2}}\alpha_1 = 2\pi - \alpha_0\eta = \frac{tg\,\theta}{tg\,\gamma},$$

h is the radome height, θ is the half angle of the cone, γ is the angle between the cone axis and the vector $\vec{R}^0$, and $\vec{n}(\alpha)$ is the vector of the normal to the outer surface of the radome S_2.

Some calculation peculiarities for radomes shaped as an ogive and as a pyramid
To perform the calculations for an ogival shaped radome, it is necessary to know the exact shape of the radome surface and the unit vector perpendicular to it (the normal) at every surface point. In other respects, the calculations for the ogival radome are similar to those for the conical radome.

The ogive can be described by the function

$$z - c = a\left(d - \sqrt{x^2 + y^2}\right)^{\alpha}, \tag{2.56}$$

where parameters a, c, α, and d are set in such manner that the surface of ogival radome would follow as close as possible the surface of the compound cone radome (passing through the conjunction lines of the compound cone). In our case, these parameters were as follows: $a = -1.579$ m, $c = 0.054$ m, $\alpha = 0.5593$, and $d = 0.12$ m. The normal unit vector to the side surface of the ogival radome has the form

$$\overline{N} = \left(F_x, F_y, F_z\right) \Big/ \sqrt{F_x^2 + F_y^2 + F_z^2}, \tag{2.57}$$

$$F_x = a\alpha\left(d - \sqrt{x^2 + y^2}\right)^{\alpha-1} \cdot \frac{x}{\sqrt{x^2 + y^2}},$$

$$F_y = a\alpha\left(d - \sqrt{x^2 + y^2}\right)^{\alpha-1} \cdot \frac{y}{\sqrt{x^2 + y^2}},$$

$$F_z = 1.$$

The point where the incident wave beam crosses the inner ('shadow') part of the ogive surface is found similarly to that of the conical surface, but instead of obtaining a quadratic equation for finding parameter t_1 (such as (2.44)) we obtain, generally speaking, a non-algebraic equation, which can be solved using the bisection (Bolzano) method [37].

When performing calculations for the system with a pyramidal radome, the radome model was designed as follows. An N-gon is used as the base of the pyramid. In our calculation we used $N = 8$. The origin of the coordinate system coincided with the polygon center. The radius-vector of the pyramid vertex is $\vec{r}_0 = (0,0, H)$. The radius-vectors of the polygon apices are $\vec{r}_i = (x_i, y_i, z_i)$, $(i = 1,..., N)$. Here, $x_i = \rho_0 \cos\varphi_i$, $y_i = \rho_0 \sin\varphi_i$, $z_i = 0$, $\varphi_i = \alpha + \frac{2\pi}{N}(i - 1)$, $\varphi_{N+1} \equiv \varphi_1$, and ρ_0 is the radius of the circle circumscribing the N-gon. In this case, the unit vector that is normal to the ith side has the form

$$\vec{n}_i = \frac{((\vec{r}_0 - \vec{r}_i) \times (\vec{r}_0 - \vec{r}_{i+1}))}{|(\vec{r}_0 - \vec{r}_i) \times (\vec{r}_0 - \vec{r}_{i+1})|}, \tag{2.58}$$

and the equation for the ith side plane is

$$\vec{n}_i \cdot (\vec{r}_0 - \vec{r}_i) = 0. \tag{2.59}$$

If the incident plane wave propagates in the direction $\overrightarrow{R}_0$, then the side is checked for being 'illuminated' or not by the sign of the dot product $(\overrightarrow{R}_0 \cdot \vec{n})$.

A simple calculation allows the surface integral for the illuminated side of the pyramid $\Sigma(\varphi_i < \varphi < \varphi_{i+1})$ to be written:

$$\iint_\Sigma f(x, y, z)\, \mathrm{d}s = \frac{1}{\cos\beta} \int_{\varphi_i}^{\varphi_{i+1}} \mathrm{d}\varphi \int_0^{\rho_1(\varphi)} f(x, y, z)\rho\, \mathrm{d}\rho. \tag{2.60}$$

Here $\rho_1(\varphi) = \rho_0 \cos(\pi/N)/\cos(\pi/N - (\varphi - \varphi_i))$.

The integral of (2.60) type is used for obtaining the field scattered by the pyramid side. The parameters of the function f depend on ρ, φ as follows:

$$x = \rho\cos\varphi\,; y = \rho\sin\varphi\,; z = \frac{\rho_1(\varphi) - \rho}{\rho_1(\varphi)} \cdot H; \text{ and } \beta = \mathrm{arctg}\left(H/\left(\rho_0 \cos(\pi/N)\right)\right).$$

Also, we can find the point where the beam propagating in direction $\vec{R}_0$ and passing through the 'illuminated' side crosses the 'shadow' side. Let us assume that we check the jth 'shadow' side. Then, the radius-vector of the crossing point at the jth side is

$$\vec{r}^* = \vec{r}_0 + \vec{R}^0 \frac{\vec{n}_j \cdot (\vec{r}_j - \vec{r}_0)}{(\vec{R}^0 \cdot \vec{n}_j)}, \tag{2.61}$$

where $\vec{r}_0$ is the radius-vector of the crossing point at the 'illuminated' side where that beam passed through.

The results of the scattering characteristics calculation for an antenna system under a nose dielectric radome

We computed the scattering characteristics of the antenna under several radomes shaped as a compound cone, ogive, and pyramid in order to find the impact of the radome type on those fields. When evaluating the field radiated by the antenna, we computed the DPs of the antenna under the same radome types.

Our computation model included an AS with the following parameters. The total radome height was h= 454 mm, the radius of the metallic spherical cap at the radome's apex was 18 mm, the radius of the radome base (the radius of the circle circumscribing the base of the pyramidal radome) was 120 mm. The dielectric part of the compound cone radome consisted of five truncated cones (figure 2.11), each having the height of 75–90 mm and their half-aperture angles (the angle between the cone's axis and generatrix) being equal to 22.1°, 19°, 13.4°, 8.8°, and 4°. The side surface of the ogival radome was set in such a manner that it would circumscribe the compound cone radome (passing through every conjunction line of the compound cone). The equation describing the side surface of the ogival radome was presented in (2.56). The octahedral pyramidal radome was inscribed into the circular cone of the same height.

The relative permittivity of the radome material was ε'= 3.8 + j0.0076. The distance from the radome base to the antenna aperture's center of rotation was 54 mm (figure 2.11), the distance from the aperture's center of rotation to the aperture plane was 40 mm (figure 2.14), the length of the octagonal antenna aperture along axis x' (figure 2.15) was 188 mm, the width of the aperture along axis y' was 145 mm, and the parameters γ, β in expression (2.51) were equal to 0.96511 and 0.12172, respectively. The wavelength of the illumination signal was 30 mm (10 GHz carrier frequency). The antenna was being rotated in plane Oxz by an angle of 15° with respect to the radome axis.

In the next subsection, the optimal (with regard to the transmission coefficient) thicknesses of the radome walls are calculated for an antenna operating frequency of 36.6 GHz, that thickness varies from 2.4 mm through 4.75 mm. For such thicknesses, we evaluated the RCS figures versus the elevation angle of illumination direction that are presented in figures 2.16–2.19. The plots represent three kinds of radomes and two polarizations of the incident wave. A pyramidal radome is represented by two curves rather than one: the pyramid 1 curve corresponds to

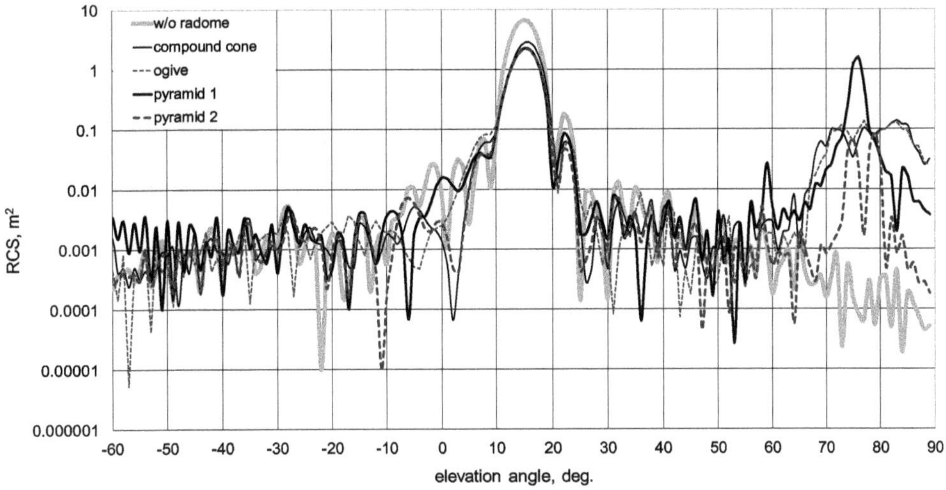

Figure 2.16. The RCS of the AS versus elevation angle for the radomes with a wall of 2.4 mm thickness given perpendicular polarization of the incident wave.

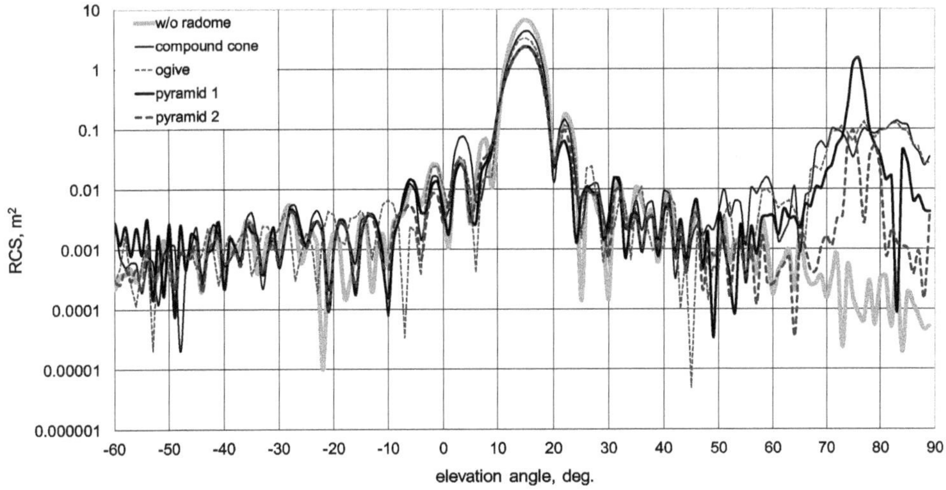

Figure 2.17. The RCS of the AS versus elevation angle for the radomes with a wall of 2.4 mm thickness, given parallel polarization of the incident wave.

the vector of the wave incidence direction $\vec{R}_0$ belonging to the plane perpendicular to one of the pyramid side faces, and the pyramid 2 curve corresponds to the wave incidence direction vector $\vec{R}_0$ belonging to the plane bisecting the angle between two radome side faces.

The plots in figures 2.16–2.19 show that if there is no radome, the RCS of the antenna aperture attains its maximum of 6.60 m^2 for the normal wave incidence upon its surface. The use of radomes that are 2.4 mm and 4.75 mm thick reduces this maximum to 2–3 m^2 for the parallel polarization of the incident wave and to 2–4 m^2

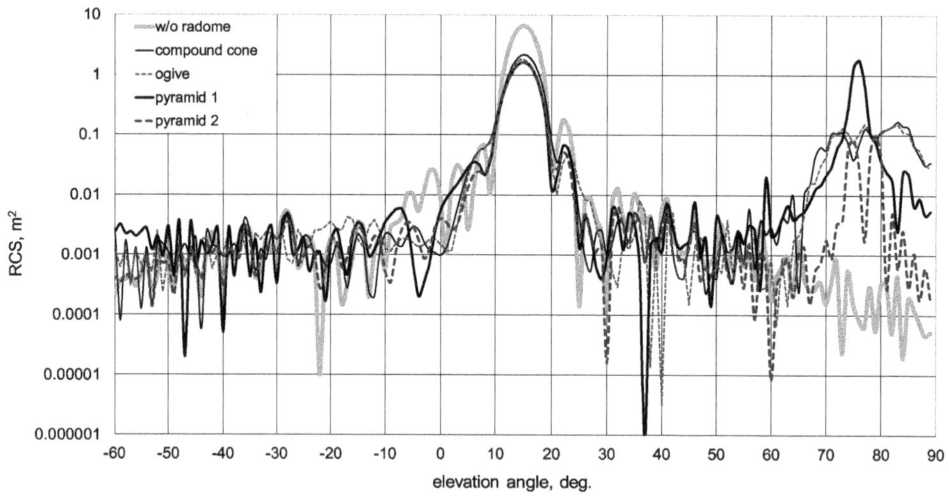

Figure 2.18. The RCS of the AS versus elevation angle for the radomes with a wall of 4.75 mm thickness given perpendicular polarization of the incident wave.

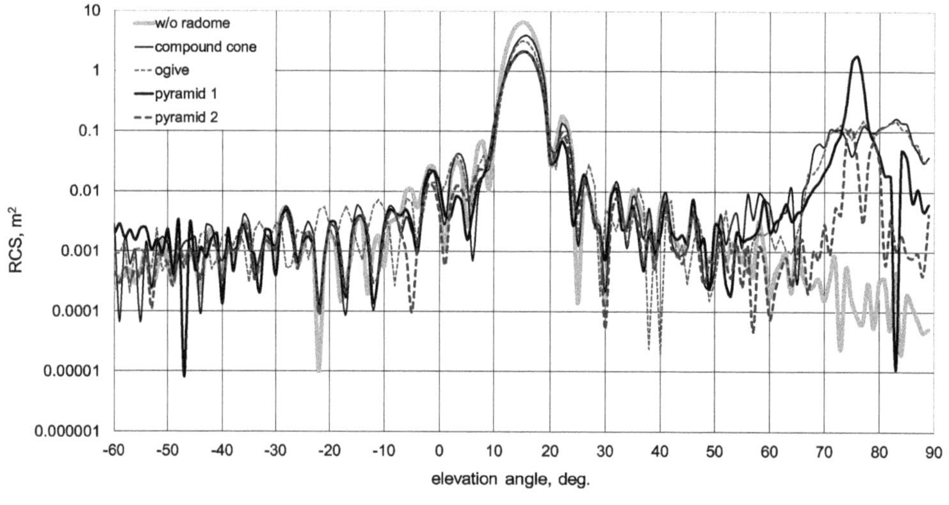

Figure 2.19. The RCS of the AS versus elevation angle for the radomes with a wall of 4.75 mm thickness given parallel polarization of the incident wave.

for the perpendicular polarization of the incident wave. The maximum RCS of the pyramidal radome is slightly lower than that of the compound cone and ogival radomes. The contribution of a metallic cap into the total AS RCS was very small (~0.001 m^2), and its impact on the general scattering level was negligible.

Reflections from the radome contribute significantly to the RCS for small values of elevation angles. A similar situation is observed for an elevation angle of 70°, which corresponds to normal wave incidence on the cone's generatrix. The side surfaces of the radomes shaped as a compound cone and ogive are very close to a pure conical surface. At the same time, the scattering from such radomes becomes quite pronounced for illumination directions close to normal wave incidence on that conical surface.

In a similar way the RCS of the pyramidal radomes surges when the incidence direction of the illumination wave approaches the normal vector to its side faces or edges. Such a surge in the RCS due to wave reflection from the side face may create another peak, which would be as high as the RCS of the entire antenna–radome system (1.8 m^2 at the elevation of 76° in this particular case). However, that surge only occurs if the incidence direction vector $\vec{R}^0$ becomes perpendicular to one of the pyramid faces. In addition, the lobe width corresponding to that RCS surge is less than 3°.

Contribution of reflections from the 'shadow' part to the total scattered field is quite small for the majority of illumination directions and it goes as high as hundredth of square meter for elevation angles of 58°–60° only in two cases: parallel polarization for a compound cone radome and perpendicular polarization for a pyramid when the illumination direction vector $\vec{R}^0$ belongs to the plane perpendicular to this 'shadow' pyramid face.

The width of the main lobe of the antenna scattering diagram (SD) is less than a few degrees. In addition, the strong scattering is concentrated in the vicinity of the direction that is perpendicular to the antenna aperture. For the majority of incident aspects, the RCS of the AS in question is less than 0.1 m^2 regardless of the radome type.

2.3.2 Radiation of an antenna system under a nose dielectric radome

In this subsection we consider the radiation properties of an antenna under a radome, the design of which was described in the previous subsection.

A method for calculating the wave radiated by an antenna under a radome

Let the antenna be an octagonal aperture S_0 (figure 2.20) cut in a perfectly absorbing plane. According to the generalized mirror reflection principle [19] one needs to consider a 'symmetrized' radome (figure 2.20) and the field scattered by it at every point of aperture S_0.

The field scattered by the 'symmetrized' radome at the aperture points includes two components. The first component is the field scattered by the upper (really existing) part of the radome, while the second component is the field scattered by its lower part, which is the notional radome's mirror image symmetrical with respect to the aperture plane (figure 2.20).

Then, as was shown in [19], the field radiated by that AS in direction $\vec{R}^0$ can be represented by formula

$$\vec{p} \cdot \vec{E}\left(\vec{R}^0\right) = \int_{S_0} \left(\left(\vec{E}^T(x) \times \vec{H}_0\left(x, \vec{R}^0, \vec{p}\right)\right) - \left(\vec{H}^T(x) \times \vec{E}_0\left(x, \vec{R}^0, \vec{p}\right)\right)\right) \mathrm{d}\vec{S}. \quad (2.62)$$

Here $\vec{E}_0\left(x, \vec{R}^0, \vec{p}\right)$, $\vec{H}_0\left(x, \vec{R}^0, \vec{p}\right)$ is the field scattered by the 'symmetrized' radome (figure 2.20) at the aperture points x (from which the perfect absorber screen

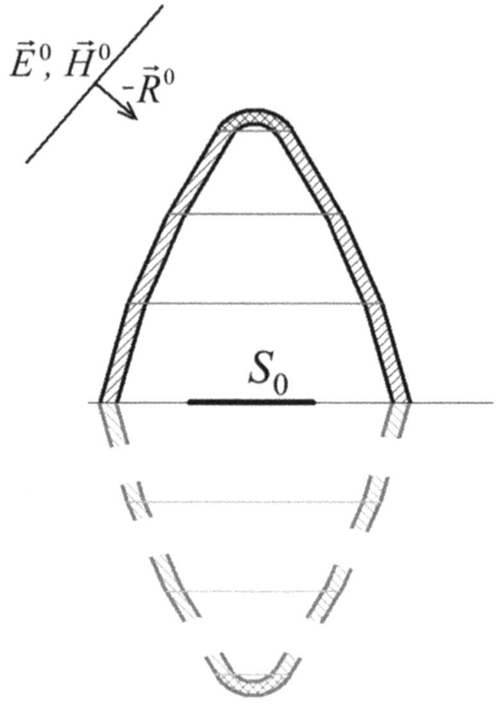

Figure 2.20. 'Symmetrized' radome.

was removed). This field is generated by an incident plane wave with the direction unit vector $\left(-\vec{R}^0\right)$, or

$$\vec{E}_0 = \left(\vec{R}^0 \times \left(\vec{p} \times \vec{R}^0\right)\right) \exp\left(-jk_0\left(\vec{R}^0 \cdot \vec{x}\right)\right),$$

$$\vec{H}_0 = \left(\vec{p} \times \vec{R}^0\right)\sqrt{\varepsilon_0/\mu_0} \exp\left(-jk_0\left(\vec{R}^0 \cdot \vec{x}\right)\right). \tag{2.63}$$

The functions $\vec{E}^T$, $\vec{H}^T$ are the distributions of the tangential components of the field intensity vectors along the aperture for Kirchhoff's approximation.

As long as the conical radome shape is accounted for, it turns out that the field scattered by the radome's 'mirror reflection' (the bottom part of the 'symmetrized' radome in figure 2.20) at every point of the aperture becomes significantly less intense than that passed directly through the real radome. Thus, as a first approximation, we can assume that $\vec{E}_0\left(x, \vec{R}^0, \vec{p}\right)$, $\vec{H}_0\left(x, \vec{R}^0, \vec{p}\right)$ is the field at the aperture point x, which is excited by waves propagating along two paths (see figure 2.13): the wave propagated through the radome's wall to hit the aperture directly (path 1) and the wave that propagated through the radome's wall to reach its inner surface and bounced back from it to hit the aperture indirectly (path 2). The integrals in (2.62) were calculated by the method of mean rectangles.

Thus, to calculate the field $\vec{E}_0$, $\vec{H}_0$, we can use the results obtained when modeling the scattering characteristics of the antenna system (subsection 2.3.1).

We note that antenna polarization (the direction of vector $\vec{E}^T(x)$) was chosen to be aligned with axis Ox' (see figure 2.15), i.e. to be pointing in the direction for which the aperture size is the greatest. We computed the cross-section of the radiation pattern (RP) by the plane orthogonal to the aperture and passing through axis Ox' (the unit vector $\vec{p}$ belongs to that plane and it is orthogonal to the direction $\vec{R}^0$).

Additionally, we also note that the plane containing the perfectly absorbing screen rotates simultaneously with the antenna if the latter is turned by some angle. In the case of small angles of antenna rotation ($\leqslant 25°$), all assumptions made with regard to the structure of the field $\vec{E}_0\left(x, \vec{R}^0, \vec{p}\right)$, $\vec{H}_0\left(x, \vec{R}^0, \vec{p}\right)$ remain valid.

Results of numerical calculation of field radiated by an antenna system under the nose dielectric radome

As the computation model, we used an AS with the same parameters and the same radome types as in previous sections. The operating frequency of the antenna was assumed to be 36.6 GHz (the 8.2 mm wavelength). The antenna was rotated by the angles from 0° to 25°.

An example of the computation result pertaining to the field radiated by an AS covered by various types of radomes is shown in figure 2.21. This figure also visualizes the radomes along with their 'illuminated' (marked as gray) and 'shadow' (marked as black) zones. These images correspond to a radiation direction of 30° with respect to the radome's axis. Further, the appearance of an octagonal antenna aperture under various radomes is shown in figure 2.22. Here, the parts of the antenna aperture that were illuminated by the field reflected from the radome's 'shadow' zone are blackened. It should be noted that even the presence of wide swaths of the antenna surface that were illuminated by the reflected wave (as for the pyramidal radome) does not lead to significant distortion of the antenna RP.

Figure 2.23 shows the normalized antenna RP for different thicknesses of the compound cone radome. The antenna aperture was shifted by the rotation angle of 15° with respect to the radome axis. All the RPs obtained for various radomes were normalized by the 'global' maximum of the antenna's radiated power attained for the case of no radome. We can see that the main lobe of the radiation pattern stays

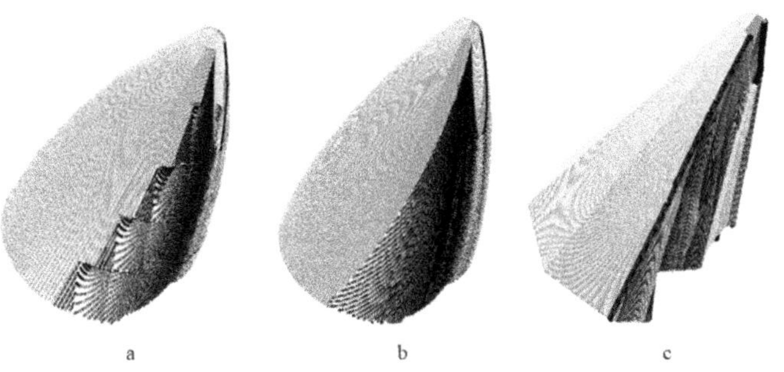

Figure 2.21. Appearance of 'illuminated' and 'shadow' zones on radomes of different types : (a) compound cone; (b) ogival radome; and (c) pyramidal radome.

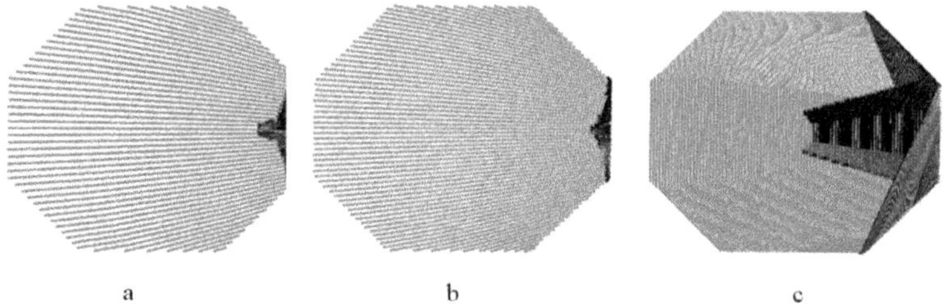

Figure 2.22. Appearance of zones at the antenna aperture that are illuminated indirectly by the wave reflected from the radome's 'shadow' part for an antenna under: (a) a compound cone radome; (b) an ogival radome; and (c) a pyramidal radome.

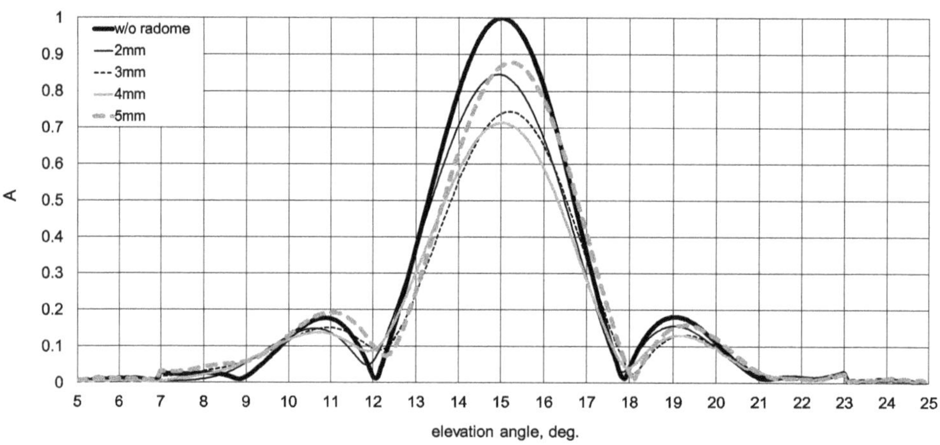

Figure 2.23. Antenna RP versus elevation angle for various values of the wall thickness, the case of a compound cone radome.

the same regardless of the radome's wall thickness. The impact of the radome manifests itself as a decrease in the peak level of radiation attained by the main lobe of the antenna pattern. To reduce the negative impact of the radome on the power radiated by the antenna, we need to know what thickness of the radome's wall would be optimal as far as the radiation power is concerned. Thus, below we present antenna RPs computed for various values of the radome's wall thickness.

Figure 2.24 shows the RPs of the antenna normalized to the 'global' maximum versus the radome's wall thickness for various types of radomes. The antenna was rotated by an angle of 15°. The pyramidal radome is represented by two curves that differed by the orientation of plane the radiation direction vector belonged to: pyramid 1—radiation direction vector $-\vec{R}^0$ belonged to the plane perpendicular to one of the pyramid faces; pyramid 2—radiation direction vector $-\vec{R}^0$ belonged to a bisector plane between two radome faces.

The results obtained here show that the transmission coefficient has two maxima at a radome thickness of 2.4 mm and of 4.75 mm regardless of the radome type. It should also be noted that these maxima do not remain static. As the antenna rotates

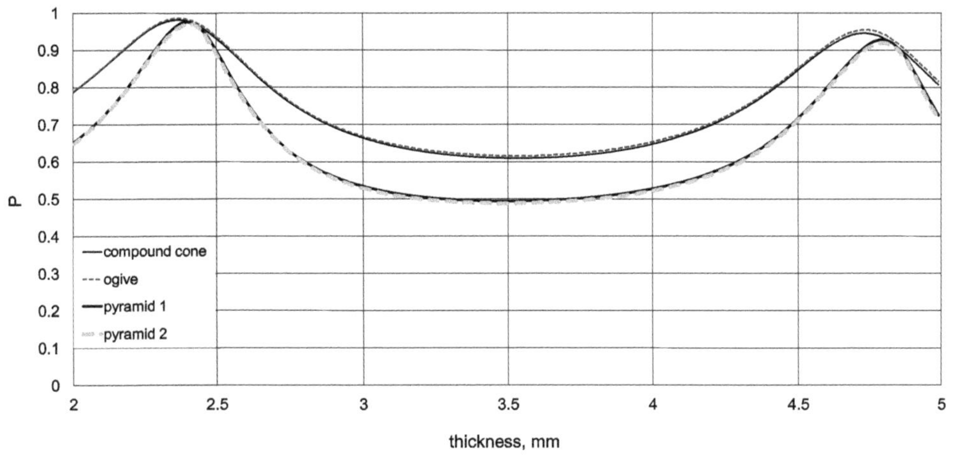

Figure 2.24 The maxima of the antenna RP normalized to the 'global' power maximum versus the radome wall thickness for various radome types, the case of an antenna rotated by an angle of 15°.

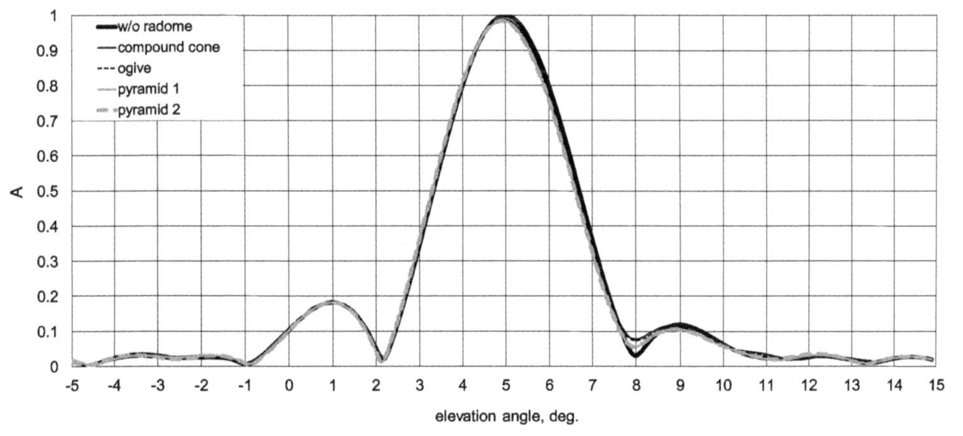

Figure 2.25. Antenna RPs versus elevation angle for an antenna rotated by 5° and for a radome wall thickness of 2.4 mm.

inside the radome, the maxima shift slightly to the left or right depending on the radome type. Thus once the radome type to house the AS is selected, it would be necessary to carry out additional research in order to accurately evaluate the optimal thickness of the radome.

Figures 2.25–2.30 show the antenna RPs for various angular positions of the antenna and for the optimal thicknesses of the radome walls.

Analysis of the plots in figures 2.25–2.30 shows that the maximum power radiated by the antenna under the radome is very close to that radiated by a standalone antenna, in particular for radomes with a thin (2.4 mm) wall. In the case of a thicker wall (4.75 mm), the impact of the radome became more pronounced but the antenna still retained more than 91% of its maximum radiated power. This fact is to be taken into account if there is a need to increase the radome thickness.

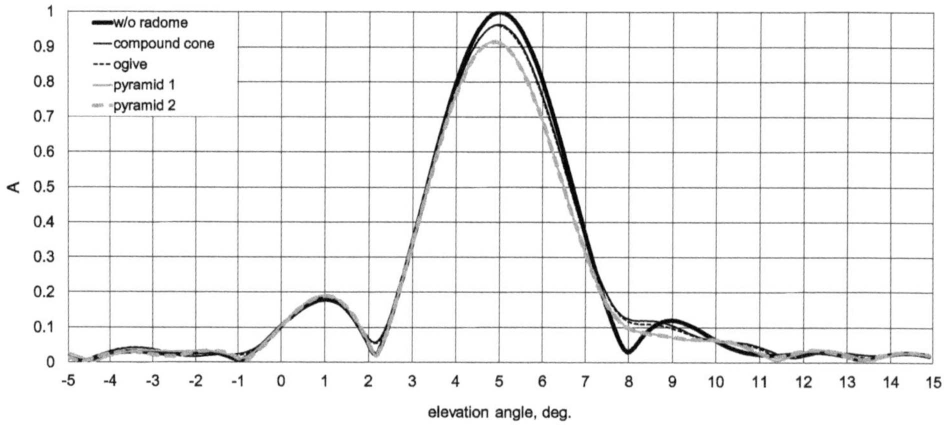

Figure 2.26. Antenna RPs versus elevation angle for an antenna rotated by 5° and for a radome wall thickness of 4.75 mm.

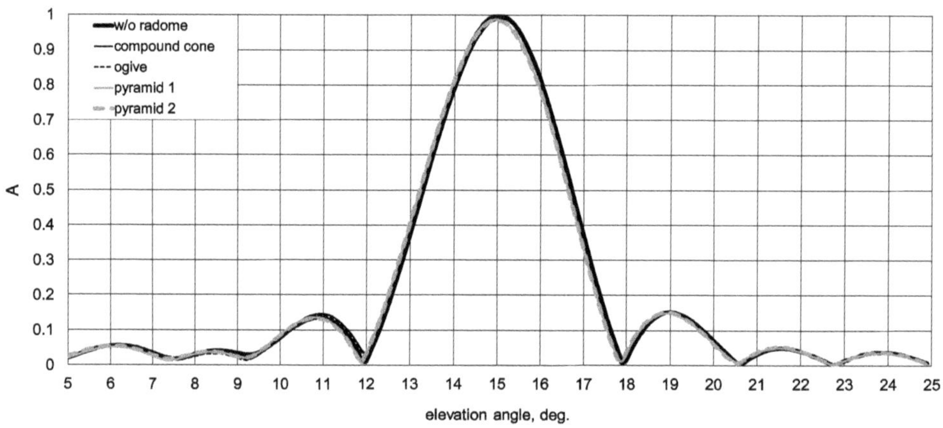

Figure 2.27. Antenna RPs versus elevation angle for an antenna rotated by 15° and for a radome wall thickness of 2.4 mm.

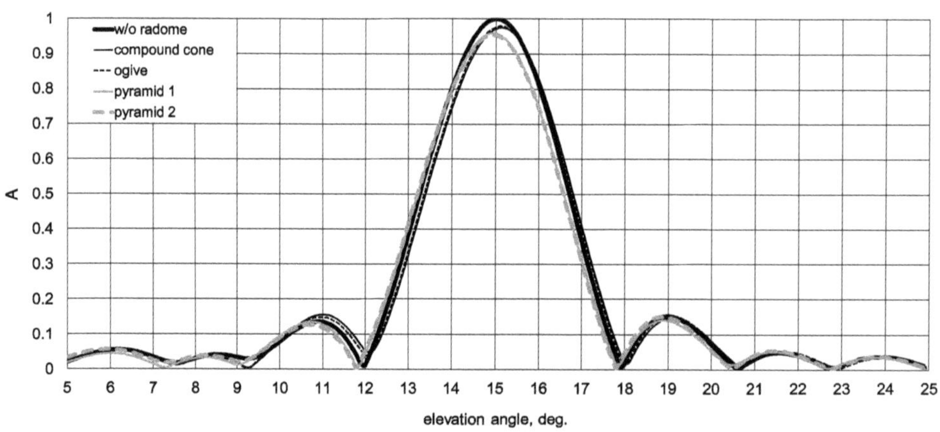

Figure 2.28. Antenna RPs versus elevation angle for an antenna rotated by 15° and for a radome wall thickness of 4.75 mm.

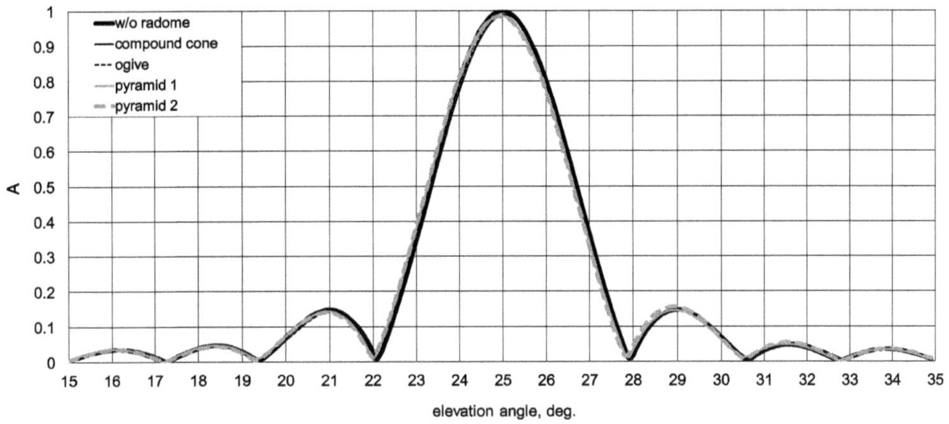

Figure 2.29. Antenna RPs versus elevation angle for an antenna rotated by 25° and for a radome wall thickness of 2.4 mm.

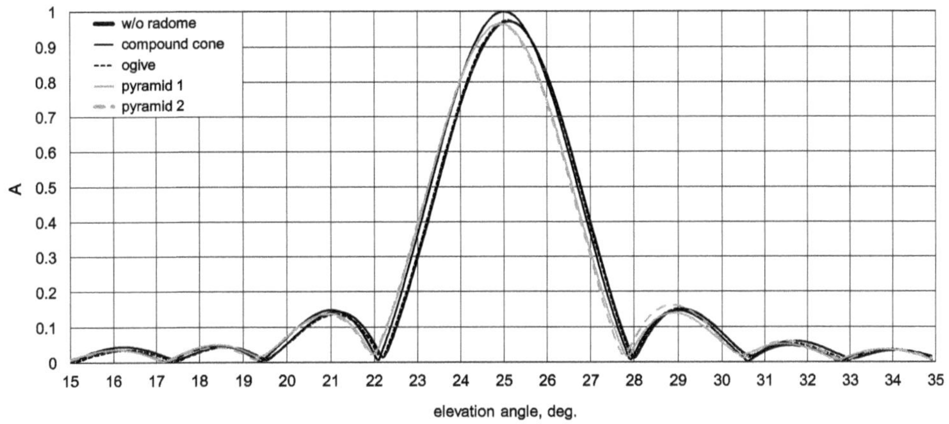

Figure 2.30. Antenna RPs versus elevation angle for an antenna rotated by 25° and for a radome wall thickness of 4.75 mm.

The angular position of the antenna under the radome affects its RP in two basic ways. First, the notches of the pattern may become very shallow or may disappear. Second, the antenna boresight radiation direction may become slightly shifted with respect to the antenna's axis of symmetry (the normal to the aperture in our example). Given small angles of antenna rotation, 5° in our example, it is the notches of the antenna RP that become less pronounced (this effect is aggravated by the thicker radome wall of 4.75 mm). Given greater angles of antenna rotation of 15°–25°, it is the antenna boresight radiation direction that slightly shifts with respect to the antenna normal, the shift not exceeding a value of about 0.1° (as before, this effect is aggravated by the thicker radome wall of 4.75 mm).

This subsection presents the original method for calculating an EMF scattered by an antenna placed beneath the nose dielectric radome. The radiation performance of such an antenna was computed using the method for finding the field excited by the radiation aperture given the presence of an arbitrary system of scatterers in its vicinity (the radome in particular).

The field scattered by such an AS was evaluated for the three types of nose dielectric radomes, the types differing by their shapes. We considered three shapes: compound cone, ogive, and pyramid. The RCS of a standalone antenna (without a radome) for normal incidence of a radio wave upon its aperture was equal to 6.60 m^2. That RCS figure represented the maximum, which was used as a reference. When the antenna was put under various radomes with walls 2.4 mm and 4.75 mm thick, this maximum decreased to 2–3 m^2 for parallel polarization of the illumination signal and to 2–4 m^2 for perpendicular polarization of the illumination signal. In addition, the pyramidal radome provided a slightly better reduction in the maximum RCS figure compared to the radomes shaped as a compound cone and ogive.

The contribution of the metal cap to the RCS of the entire AS is very small and its influence on the overall scattering level can be neglected. The main lobe of the AS scattering pattern is just a few degrees wide and the antenna's main radiation is concentrated in a narrow sector around a direction that is perpendicular to the antenna aperture. For the majority of sounding aspect angles, the RCS of the AS considered here is less than 0.1 m^2 regardless of the radome type.

The radiation characteristics of the antenna system under the radome were estimated for different parameters of the system. The application of the radome can affect the RP of the AS. The impact of the radome manifests itself in reduced power radiated by the antenna's main lobe, in a widened antenna pattern lobe, in shallow notches (dips) of the antenna RP, and in a slightly shifted maximum of the antenna RP with respect to its symmetry axis (of about 0.1°).

Using the results of calculations, we determined that the 'optimal' radome thicknesses are equal to 2.4 mm and to 4.75 mm given an antenna operating frequency of 36.6 GHz. The RP maxima are almost the same as those for an antenna without a radome given that radomes with walls of such thicknesses are to be employed. For an antenna turn angle of 5°, the notches (dips) of the antenna RP start to disappear. For antenna turn angles of 15°–25°, there is slight shift in the antenna RP maximum amounting to 0.1°.

2.4 The influence of precipitation on the gain of the reflector antennas

Accurate meteorological forecast is of great importance in many domains of human activity. Meteorological support plays a core role in aviation [38]. In civil aviation, meteorological support is aimed at providing safe regular flights of aircraft by feeding accurate meteorological information to air traffic controllers, aircrews, and other authorities responsible for flight planning and support. Such information is vital for all these players in order to perform their functions well. To serve its purpose, meteorological information intended for aviation users must be timely, concise, and easily interpretable.

In addition, immediate meteorological support of flights (overflights) at airports is usually provided by special meteorological authorities (units). According to [38], the actual weather monitoring must be close, accurate, and continuous, which means it needs to be conducted for as long as aircraft are being operated in the vicinity of an airport. Meteorological data that are obtained on a short notice basis include not only on-site weather measurements, but also include measurements provided by ground and airborne weather radar (WR).

Multifunctional WR are used widely in many countries (USA, UK, France, etc) for generating warnings about dangerous weather phenomena [39–44], which include heavy precipitation in the form of hail, snow, and rain. Reflector antennas are not uncommon features of such WR (figure 2.31).

Sometimes, WR are protected from harsh weather conditions by radomes of mostly spherical shape (figure 2.32). However, such radomes make ASs bulky, which is not suitable for mobile WR.

If the antennas of WR are used without any protective shelter (figure 2.31), the surfaces of these antennas are fully exposed and, therefore, they are frequently covered with a layer of precipitation. However, the prime objective of WR is to measure the absolute reflectivity of meteorological formations. The latter means that the precipitation layer can alter the power calibration of the radar. Thus a method is

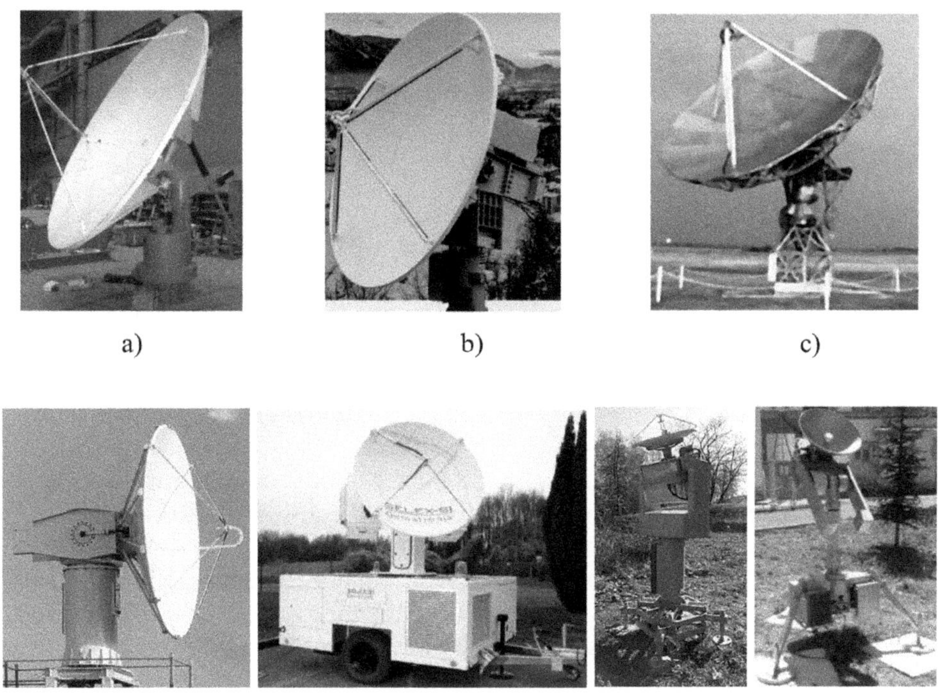

Figure 2.31. Examples of ASs employed in WR: (a) DMRL-S (Russia); (b) DWSR 2501C (USA); (c) WSR-88D (USA); (d) Meteor 500C (Germany); (e) Meteor 50DX (Germany); (f) MRIAN-Ka-S-1.2 (Ukraine); and (g) MRIAN-Ka-C-S-0.5 (Ukraine).

Figure 2.32. Examples of antenna radomes housing WRs: (a) Meteor 50DX; (b) Meteor 635C; (c) DMRL-S; (d) MRL-5.

needed for assessing the impact of the precipitation layers that can settle on the antenna reflector on its radiation performance in order to calibrate the radar's power potential correctly.

Figure 2.33 shows an example of radar images featuring a fall of snow. These images were obtained at a radar signal wavelength of 3 cm [45] (the water equivalent of the precipitation rate was $W \cong 2$ mm h^{-1}). Each image has two dimensions: height and range. The first image (figure 2.33(a)) was obtained right after the beginning of snowfall so there was practically no snow deposited on the reflector, the diameter of which was 3 m. The second image (figure 2.33(b)) was taken 40 min from the beginning of snowfall and the antenna reflector had a snow deposit of $\cong$ 3–4 cm.

It is obvious that a layer of snow deposit on the antenna reflector's surface significantly distorts the information regarding the radar reflectivity of precipitation and its special distribution measured by the radar. The end result of this distortion is the incorrect assessment of the type and intensity of precipitation obtained by the radar operator. Comprehensive data regarding the effect of the snow structure, density, moisture content, and other parameters on its electrical properties are presented in many sources, see [46–49] for instance. Some results of experimental studies of how snow and rain influence the performance of reflector antennas are

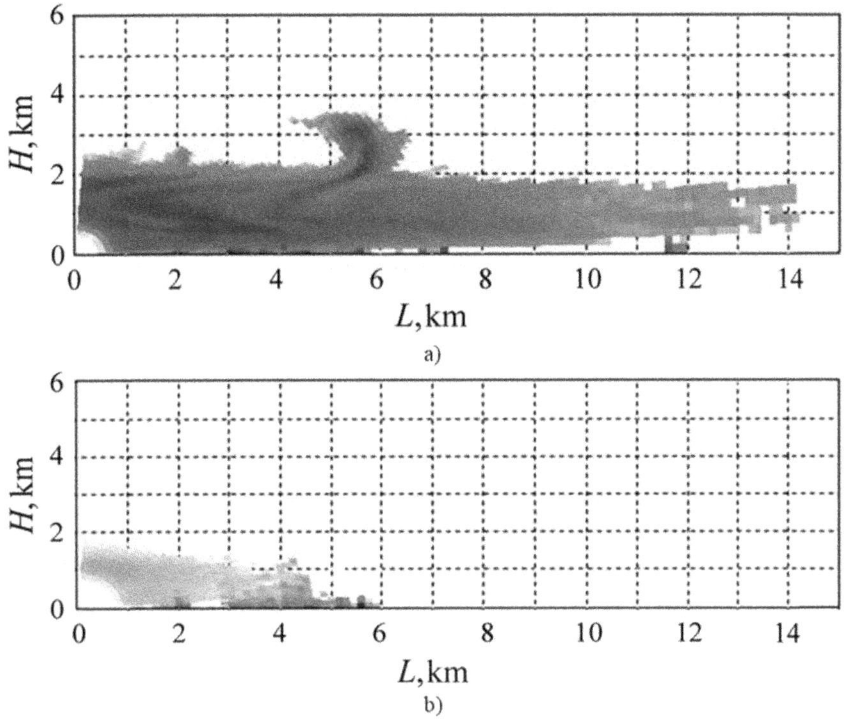

Figure 2.33. Radar images of the fall of snow: (a) the reflector was free from snow and (b) the reflector had a snow deposit on its surface.

provided in [50]. These results demonstrate that precipitation can significantly alter the radiation performance of reflector antennas.

2.4.1 A method for computing the fields radiated by reflector antennas that are partially covered by precipitation layers

One of the most common approached to computing the radiation characteristics of reflector antennas is based on the PO method [51, 52]. A generalization of this method onto the case of a non-perfectly reflecting surface of an antenna reflector was described in [53]. However, the authors of [53] addressed only uniform (equidistant) layers of precipitation deposits on the antenna reflector's surface. In addition, in this subsection, unlike the approach adopted in [53], we present the solution to this particular problem without breaking the fields down in two orthogonal polarizations, which is preferable, in particular given that the incidence direction of the plane EMW on the reflector's surface is close to the normal to that surface at a source point.

Let us consider a reflector antenna, the reflector of which is a circular cut from a paraboloid of revolution (figure 2.34). The antenna feed is placed at the reflector's focus

$$(z = f).$$

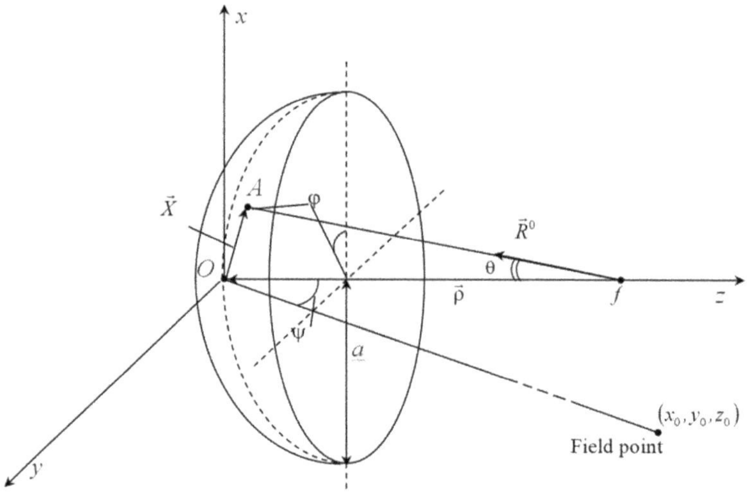

Figure 2.34. Geometry of the problem.

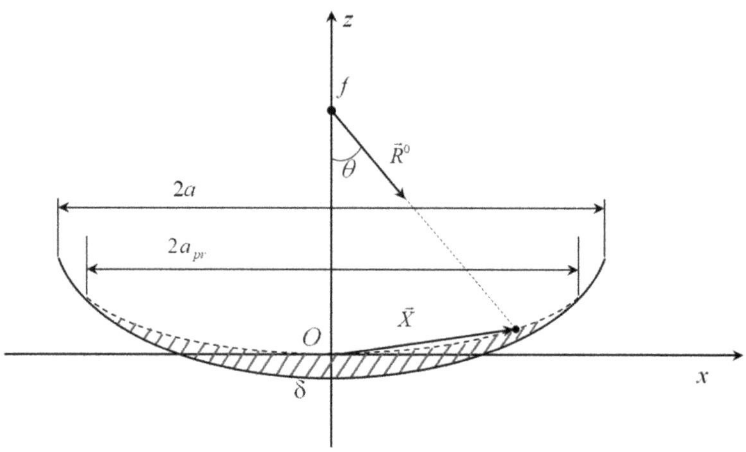

Figure 2.35. The distribution of the snow deposit on the antenna reflector's surface.

For the sake of simplicity, we introduce the following assumptions: the antenna reflector is placed in the far-field zone with respect to the antenna feed, which is usually true in practical antenna design [54]; the impact of the antenna reflector on the RP of the feed is negligible. The size of the antenna reflector is much greater than the antenna operating wavelength $2a \gg \lambda$ (a is the reflector's radius, λ is the operating wavelength of the radar), a condition which holds in most cases.

We also assume that the volume in space occupied by snow is confined between the surfaces of two paraboloids of revolution (figure 2.35) [55–57]. Furthermore, the projection of the snow footprint onto the surface Oxy is a circle of radius a_{pr}. The origin of the coordinate system is placed on the point where the reflector's axis crosses the surface that confines the snow (precipitation) deposit (figure 2.35).

To compute the DPs of the antennas considered here, we use the following integral representation of the scattered field [19]:

$$\vec{p} \cdot \vec{E}^s = -jk_0 \frac{\exp(jk_0 R)}{4\pi R} \int_S \left(\sqrt{\frac{\mu_0}{\varepsilon_0}} \left(\vec{p} \cdot \vec{H}^{\perp}\right) + \left(\vec{p} \times \vec{r}^0\right) \cdot \vec{E}^{\perp} \right) \times \exp(-jk_0(\vec{r}^0 \cdot \vec{x}))\, \mathrm{d}S \tag{2.64}$$

where S is the surface encompassing the antenna reflector and the precipitation deposit, $\vec{r}^0$ is the unit vector of radiation direction, R is the distance from the origin of coordinate system to the field point,

$$\vec{E}^{\perp} = \vec{n} \times \vec{E},\ \vec{H}^{\perp} = \vec{n} \times \vec{H},$$

$(\vec{E}, \vec{H})$ is the field excited at the surface S as resulting from the combined effect of the antenna feed and reflector, $\vec{n}$ is the external normal to the surface S, k_0 is the wavenumber in free space, ε_0 and μ_0 are the relative permittivity and permeability of free space, and $\vec{p}$ is the polarization of the wave received by the antenna.

Next, we assume that the antenna reflector is an infinitely thin screen, which has large electric size and small curvatures. Therefore, we can further assume that the field excited at the shadow part of the surface (with respect to the feed) is equal to zero, i.e. $\vec{E} = \vec{H} = 0$ (which is usually the case in the PO approximation [51, 52]). Since we are free to choose the surface S in (2.64), we set it up from considerations of convenience. Namely, it will coincide with the back of the reflector in the shadow zone, and it will follow the top of the precipitation deposit layer (if such deposit is present) or it would follow the 'illuminated' part of the reflector's surface (if precipitation deposit is absent).

Considering all the above assumptions, we can present the field radiated by the antenna feed that reached a point A at the surface of the precipitation deposit (or at the illuminated part of the reflector's surface if there is no deposit), that particular point being described by radius-vector $\vec{X}$ (figure 2.35), in the following form [54]:

$$\vec{E}^i(\vec{X}) = \frac{jk_0}{4\pi} \vec{p}\,' \frac{\exp(jk_0|\vec{X} + \vec{\rho}|)}{|\vec{X} + \vec{\rho}|} F^i(\theta, \varphi), \tag{2.65}$$

$$\vec{H}^i(\vec{X}) = \sqrt{\frac{\varepsilon_0}{\mu_0}} \frac{jk_0}{4\pi} \left(\vec{R}^0 \times \vec{p}\,'\right) \frac{\exp(jk_0|\vec{X} + \vec{\rho}|)}{|\vec{X} + \vec{\rho}|} F^i(\theta, \varphi), \tag{2.66}$$

where $\vec{R}^0$ is the unit vector of a ray pointing from the reflector's focal point f to the point A at the surface of precipitation deposit layer (figure 2.35), θ is the angle defining the orientation of vector $\vec{R}^0$ with respect to the antenna's axis, φ is the angle defining the location of point A with respect to plane xOz, $\vec{\rho}$ is the radius-vector drawn from the reflector's focus to the origin of the coordinate system,

$$\vec{p}\,' = \frac{\vec{R}^0 \times (\vec{p}^0 \times \vec{R}^0)}{|\vec{R}^0 \times (\vec{p}^0 \times \vec{R}^0)|}$$

is the unit vector characterizing the polarization of the EMW propagating in the direction $\vec{R}^0$, $\vec{p}^0$ is the unit vector depicting the direction of the antenna feed's electric field intensity, $k_0 = \omega\sqrt{\varepsilon_0\mu_0}$ is the wavenumber in free space, ε_0, μ_0 are the absolute permittivity and permeability of free space, and $F^i(\theta, \varphi)$ is the DP of the antenna feed.

This field has the form of an inhomogeneous plane EMW with polarization vector $\vec{p}\,'$ that depends on the propagation direction. In other words, field $(\vec{E}^i, \vec{H}^i)$ can be presented as follows:

$$\vec{E}^i = \Omega(\theta, \varphi)\vec{p}\,' \exp\left(jk_0(\vec{R}^0 \cdot \vec{X})\right), \tag{2.67}$$

$$\vec{H}^i = \sqrt{\frac{\varepsilon_0}{\mu_0}}\Omega(\theta, \varphi)\left(\vec{R}^0 \times \vec{p}\,'\right)\exp\left(jk_0\left(\vec{R}^0 \cdot \vec{X}\right)\right), \tag{2.68}$$

where

$$\Omega(\theta, \varphi) = \frac{jk_0}{4\pi}\frac{\exp\left(jk_0\left(\vec{R}^0 \cdot \vec{\rho}\right)\right)}{|\vec{X} + \vec{\rho}|}F^i(\theta, \varphi) \tag{2.69}$$

Considering the fact that the precipitation deposit layer has a smooth surface, we can also assume that the field excited at point A, which belongs to the surface of that deposit, is going to be the same as the field excited at the surface of a flat uniform layer of that same precipitation backed by a perfectly conducting plane, that flat uniform layer being osculant to the reflector's surface at point A. The thickness of this layer is equal to the actual thickness of the precipitation deposit along the normal to the reflector's surface at the osculating point. At the points of the illuminated reflector's surface that are free from precipitation we employ the usual physical optics approximation: $\vec{E}^\perp = 0$, $\vec{H}^\perp = 2(\vec{n} \times \vec{H}^i)$. In that case, the field $(\vec{E}^i, \vec{H}^i)$ radiated by the antenna feed is computed using expressions (2.67) and (2.68).

However, the points of the illuminated reflector's surface covered with snow are treated as those from the so-called benchmark (standard) problem, which is represented by expressions (2.67) and (2.68). In that particular benchmark problem, the field at the points on the top of the precipitation deposit is replaced by the corresponding field from the benchmark problem featuring the plane EMW scattering at the flat uniform (equidistant) layer of a dielectric backed by a perfectly conducting plane (figure 2.36).

The EMF at points $\vec{X}$ at the surface of the dielectric layer has the form [19]

$$\vec{E}(\vec{X}) = \Omega(\theta, \varphi)(\vec{p}\,' + \vec{p}^{\,s})\exp\left(jk_0\left(\vec{R}^0 \cdot \vec{X}\right)\right), \tag{2.70}$$

$$\vec{H}(\vec{X}) = \sqrt{\frac{\varepsilon_0}{\mu_0}}\Omega(\theta, \varphi)\left[\left(\vec{R}^0 \times \vec{p}\,'\right) + \left(\vec{R}^0 \times \vec{p}^{\,s}\right)\right]\exp\left(jk_0\left(\vec{R}^0 \cdot \vec{X}\right)\right). \tag{2.71}$$

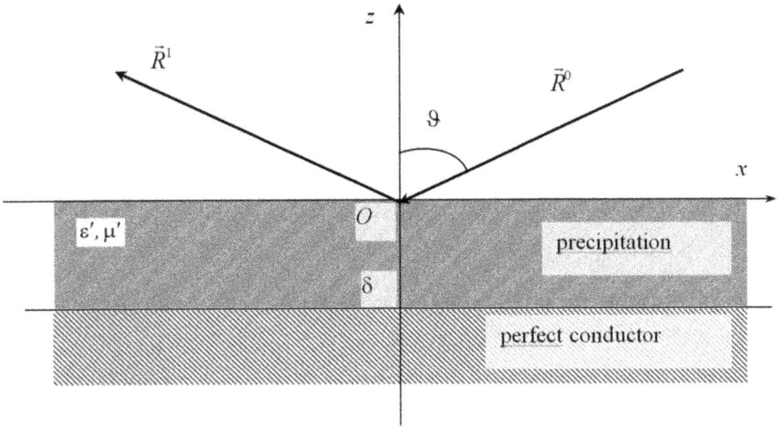

Figure 2.36. Benchmark problem of a plane EMW scattering at a flat uniform dielectric layer backed by a perfectly conducting plane.

In expressions (2.70) and (2.71), $\vec{p}^{\,s}$ is the complex vector coefficient of wave reflection that can be computed as follows [19]:

$$\vec{p}^{\,s} = \vec{p}_T^{\,s} - \vec{n}\frac{\left(\vec{p}_T^{\,s} \cdot \vec{R}^0\right)}{\cos\vartheta}, \tag{2.72}$$

where $\vec{n}$ is the unit vector of the external normal to the surface S, which means it is normal to the dielectric layer as well,

$$\vec{p}_T^{\,s} = \frac{jc\cos\vartheta + 1}{jc\cos\vartheta - 1}\vec{p}_T^{\,\prime} - \frac{2jc}{jc\cos\vartheta - 1}\left[\vec{R}_T^0\frac{\left(\vec{R}_T^0\cdot\vec{p}^{\,\prime}\right)}{jc - \cos\vartheta} + \vec{R}_\perp^0\frac{\left(\vec{R}_T^0\cdot\vec{p}^{\,\prime}\right)}{\varepsilon'\mu'\left(jc - \dfrac{\cos^2\vartheta_1}{\cos\vartheta}\right)}\right], \tag{2.73}$$

$$\vec{R}_T^0 = \vec{R}^0 - \vec{n}\left(\vec{R}^0\cdot\vec{n}\right),$$

$$\vec{R}_\perp^0 = \vec{n}\times\vec{R}^0,$$

$$\vec{p}_T^{\,\prime} = \vec{p}^{\,\prime} - \vec{n}(\vec{p}^{\,\prime}\cdot\vec{n}),$$

$$c = \sqrt{\frac{\mu'_1}{\varepsilon'_1}}\cos\vartheta_1\cdot tg\left(k_0\sqrt{\varepsilon'_1\mu'_1}\,\delta\cos\vartheta_1\right),$$

$$\cos^2\vartheta_1 = 1 - \frac{\sin^2\vartheta}{\varepsilon'_1\mu'_1},$$

δ is the depth of the dielectric layer, and ε' and μ' are the relative permittivity and permeability of the precipitation deposit layer, respectively.

Expressions (2.72) and (2.73) are applicable for computing the scattered field by expressions (2.70) and (2.70) for an arbitrary polarization of incident field and an arbitrary incidence direction with a small exception of directions that are close to the tangential ones. For $\vartheta = 0$, expressions (2.72) and (2.73) become the well-known formulas [58] for the case of normal EMW incidence on a surface.

2.4.2 Numerical simulation of the radiation characteristics of reflector antennas that are partially covered with snow

One of the most common precipitation deposits that can actually stick to an antenna reflector is snow. Thus, below we present the results of our study concerned with the impact of snow deposits on the radiation performance of a reflector antenna operated, in particular, by the MRL-1 WR [59–61]. In our computations regarding this particular radar, we assumed that a snow deposit was distributed over the surface of the antenna reflector as shown in figure 2.35. Our objective was to compute the DP $F(\psi)$ of the parabolic reflector antenna with the parameters pertaining to the MRL-1 WR. The reflector diameter was 2a = 3 m and its surface was partially covered with snow (the snow deposit footprints, i.e. its projection on the plane xOy, was shaped as circle of radius equal to 0.75 m or 1.25 m (figure 2.37)).

MRL-1 operates in two modes that differ by the signal wavelength [61]: $\lambda = 8.2$ mm and $\lambda = 32$ mm. Therefore, its feed includes two rectangular horn antennas with apertures sizes as follows: $L_E = 26$ mm and $L_H = 29$ mm for the operating wavelength $\lambda = 32$ mm; $L_E = 6.65$ mm and $L_H = 7.4$ mm for the operating wavelength $\lambda = 8.2$ mm.

We assume that the field distribution over the horn's aperture is co-phased. In this case, its DP $F^i(\theta, \varphi)$ can be presented as follows [62]:

$$F^i(\theta, \varphi) = \frac{1 + \cos\theta}{2}\frac{\pi}{4}\left[\frac{\sin\left(\psi_H - \frac{\pi}{2}\right)}{\psi_H - \frac{\pi}{2}} + \frac{\sin\left(\psi_H + \frac{\pi}{2}\right)}{\psi_H + \frac{\pi}{2}}\right]\frac{\sin\psi_E}{\psi_E}, \tag{2.74}$$

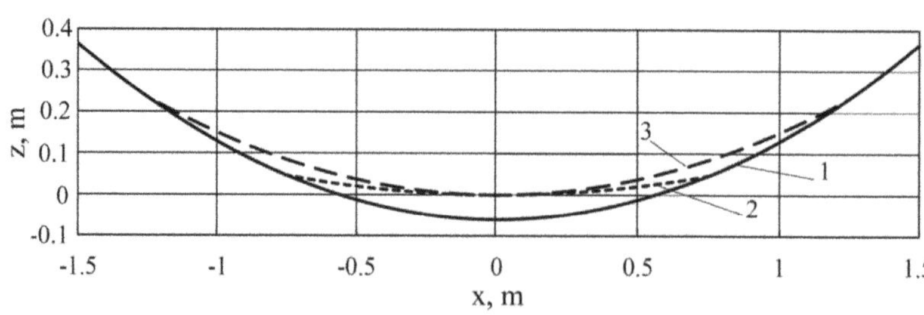

Figure 2.37. Distribution of the snow deposit over the antenna reflector: the reflector's surface (curve 1); the deposit of snow with its footprint radii of $a_{pr} = 0.75$ m (curve 2) and $a_{pr} = 1.25$ m (curve 3).

where

$$\psi_E = \frac{L_E \pi}{\lambda} \sin\theta \sin\varphi, \ \psi_H = \frac{L_H \pi}{\lambda} \sin\theta \cos\varphi,$$

and L_E and L_H are sizes of the feed in the E and H planes, respectively.

Figure 2.38 shows the DP of antenna $F(\psi)$ measured in terms of power along the H plane for the case of operating wavelength $\lambda = 8.2$ mm. In our computations, we assumed that the relative permittivity of snow was $\varepsilon' = 1.9 + j0.65$ [63].

Curve #1 presents the antenna DP for no snow deposited on its surface, curves #2 and #3 represent the cases where the antenna reflector was partially covered with snow and the radii of its footprint were equal to $a_{\text{pr}} = 0.75$ m and $a_{\text{pr}} = 1.25$ m (figure 2.37), respectively (in both latter cases the snow cover at the reflector's center was 6 cm thick).

Figure 2.39 shows similar plots for the antenna operating at the wavelength of $\lambda = 32$ mm for a relative permittivity of snow cover of $\varepsilon' = 2.8 + j0.9$. The latter value was experimentally measured using the method presented in [48].

It is obvious that in both cases the presence of snow deposited on the parabolic reflector alters the antenna DP significantly, both in terms of its shape and maximum gain. The wider the snow footprint on its surface the greater is the loss in the antenna's radiation gain.

Quantitative analysis of the data obtained above for the snow deposit with a footprint of radius $a_{\text{pr}} = 1.25$ m shows that the level of the first sidelobe of the antenna operating at the wavelength of $\lambda = 8.2$ mm increased to -6.8 dB (compared to -25.1 dB for a clear surface of the antenna reflector). For the antenna operating at a wavelength of $\lambda = 32$ mm, the snow deposit with the same footprint led to an increase of the first sidelobe level up to -4.9 dB (compare this to the sidelobe level of -26.4 dB that the antenna had for a clear reflector surface).

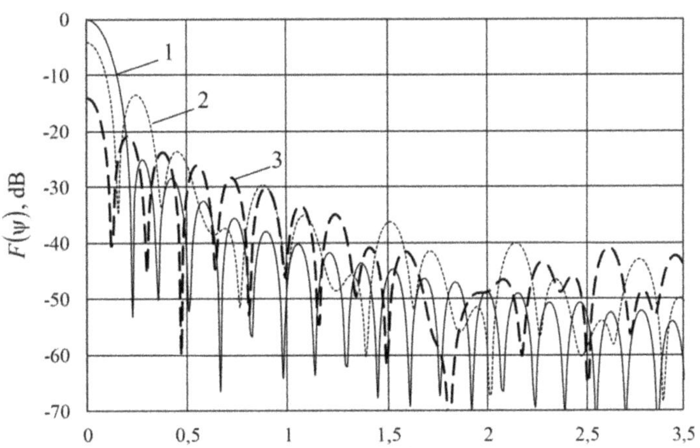

Figure 2.38. DPs of a reflector antenna operating at wavelength $\lambda = 8.2$ mm for no snow deposited on its surface (curve #1), and for the center of the antenna reflector having a circular snow deposit with the radii of their footprints equal to $a_{\text{pr}} = 0.75$ m (curve #2) and to $a_{\text{pr}} = 1.25$ m (curve #3).

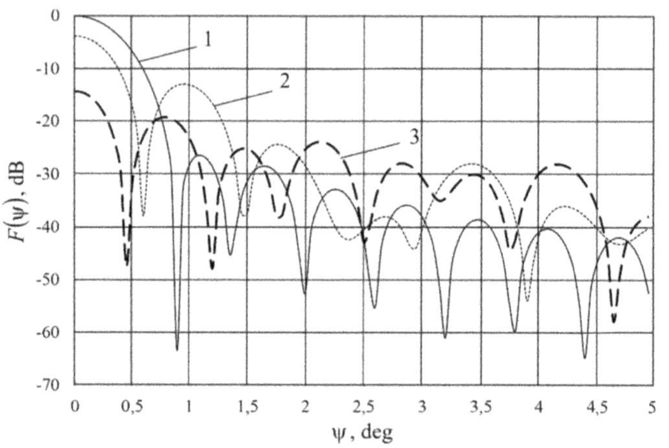

Figure 2.39. DPs of a reflector antenna operating at the wavelength $\lambda = 32$ mm for no snow deposited on its surface (curve #1), and for the center of antenna reflector having a circular snow deposit with the radii of their footprints equal to $a_{\text{pr}} = 0.75$ m (curve #2) and to $a_{\text{pr}} = 1.25$ m (curve #3).

One can also see that the main lobe of the antenna DP becomes almost twice as narrow due to the snow deposited on the antenna's reflector compared to the case of a clear antenna surface. For the antenna operating at a wavelength of $\lambda = 8.2$ mm the main antenna lobe narrowed from the value of $2\Delta\psi_0 = 0.46°$ to $2\Delta\psi_0 = 0.24°$. For the antenna operating wavelength of $\lambda = 32$ mm, the main antenna lobe narrowed from the value of $2\Delta\psi_0 = 1.8°$ to $2\Delta\psi_0 = 0.9°$. Such a decrease in the main lobe width is explained by the fact that the amplitude distribution of the field over the reflector aperture, which was significantly decaying toward the edges to begin with, came closer to the uniform distribution due to the field's (surface current's) attenuation at the center of the reflector because of the snow deposit. That field's attenuation at the center leads to a smaller antenna gain, as well as to a narrower main lobe and to greater sidelobes of its DP.

Quantitatively, the antenna gain at both operating wavelengths decreased by 14.2–14.3 dB for the snow deposit with a footprint of $2a_{\text{pr}} = 2.5$ m, this footprint meaning that almost the entire surface of the antenna reflector was covered by snow.

In order to assess the impact of the snow deposit on the radiation performance of a reflector antenna in more detail, we simulated the field radiated by such an antenna for various reflectors with relative diameters ranging as $10 \leqslant 2a/\lambda \leqslant 100$ and for various relative focal distances ranging as $0.5 \leqslant f/2a \leqslant 1$. In addition, in this case we considered a snow deposit with a flat top (figure 2.40), which can be easily verified in an experimental study.

The antenna feed was shaped as a circular horn, the radiation field of which was computed as follows [62]:

$$\overline{E}^i = \vec{e}_\theta E_\theta + \vec{e}_\varphi E_\varphi, \tag{2.75}$$

$$\overline{H}^i = \vec{e}_\theta H_\theta + \vec{e}_\varphi H_\varphi, \tag{2.76}$$

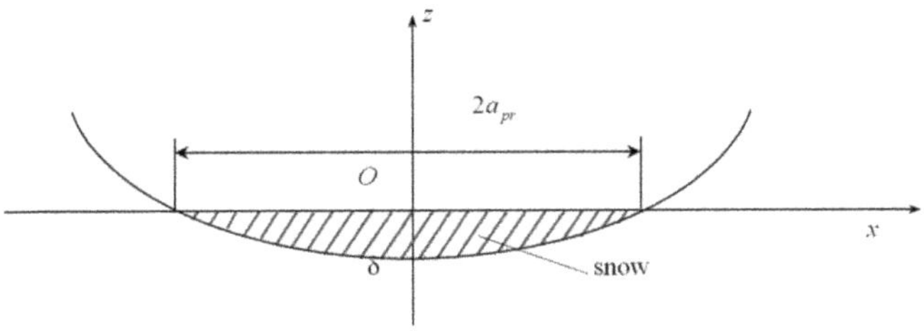

Figure 2.40. Profile of the snow deposit on the parabolic reflector's surface, that can be reproduced easily in an experiment.

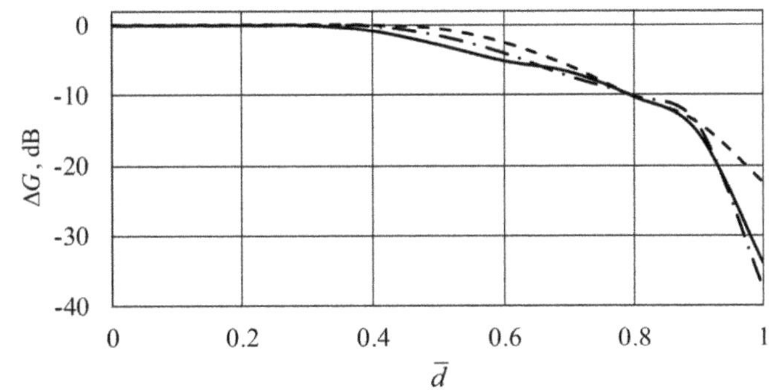

Figure 2.41. The drop in reflector antenna gain versus relative diameter of snow deposit for the relative reflector's diameter of $2a/\lambda = 10$ (the solid line corresponds to $f/2a = 0.5$, the dot-and-dash line corresponds to $f/2a = 0.75$, and the dashed line corresponds to $f/2a = 1$).

where $\vec{e}_\theta = (\cos\theta\cos\varphi,\ \cos\theta\sin\varphi,\ -\sin\theta)$, $\vec{e}_\varphi = (-\sin\varphi,\ \cos\varphi,\ 0)$,

$$E_\theta = \Phi(\overrightarrow{X})\left[1 + \sqrt{1 - \left(\frac{\lambda}{3.41a_i}\right)^2}\cos\theta\right]\frac{J_1(k_0a_i\sin\theta)}{\sin\theta}\sin\varphi, \tag{2.77}$$

$$E_\varphi = \Phi(\overrightarrow{X})k_0a_i\left[\cos\theta + \sqrt{1 - \left(\frac{\lambda}{3.41a_i}\right)^2}\right]\frac{J'_1(k_0a_i\sin\theta)}{1 - \left(\frac{2\pi a_i\sin\theta}{1.841\lambda}\right)^2}\cos\varphi, \tag{2.78}$$

$J_1(\phi)$ and $J'_1(\phi)$ are the Bessel function of first order and its derivative,

$$\Phi(\overrightarrow{X}) = \frac{\exp(jk_0|\overrightarrow{X} + \vec{\rho}\,|)}{|\overrightarrow{X} + \vec{\rho}\,|},$$

and a_i is the aperture radius of the horn feed.

Figures 2.41–2.47 show the computed values of the relative drop in antenna gain (ΔG), which is equal to the ratio of squared field amplitudes radiated in the boresight direction ($\theta = 0$) by the antenna if snow is present and if snow is absent. The latter results

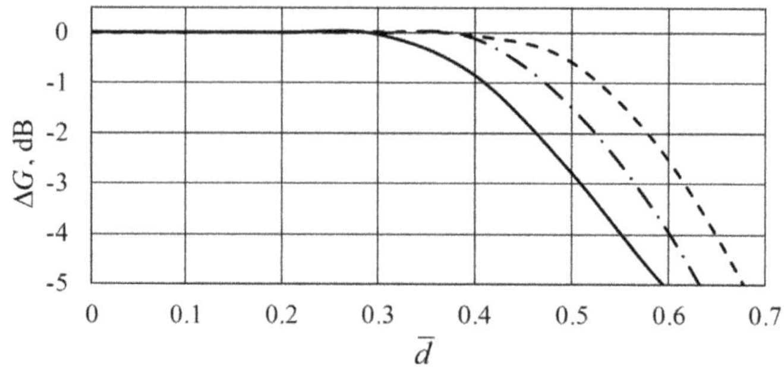

Figure 2.42. The drop in reflector antenna gain versus the relative diameter of the snow deposit for a relative reflector diameter of $2a/\lambda = 10$ (the solid line corresponds to $f/2a = 0.5$, the dot-and-dash line corresponds to $f/2a = 0.75$, and the dashed line corresponds to $f/2a = 1$).

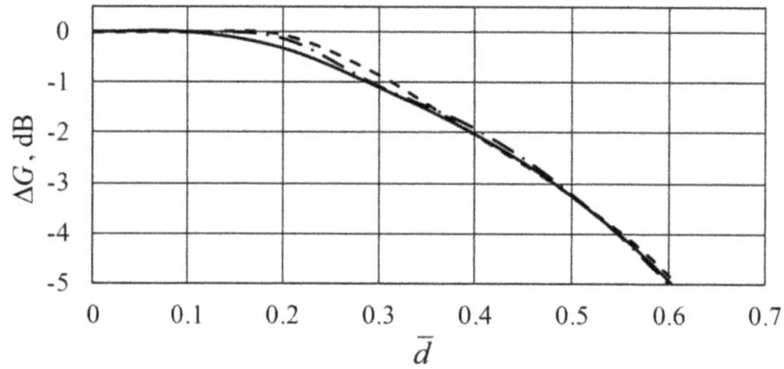

Figure 2.43. The drop in reflector antenna gain versus the relative diameter of the snow deposit for a relative reflector diameter of $2a/\lambda = 50$ (the solid line corresponds to $f/2a = 0.5$, the dot-and-dash line corresponds to $f/2a = 0.75$, and the dashed line corresponds to $f/2a = 1$).

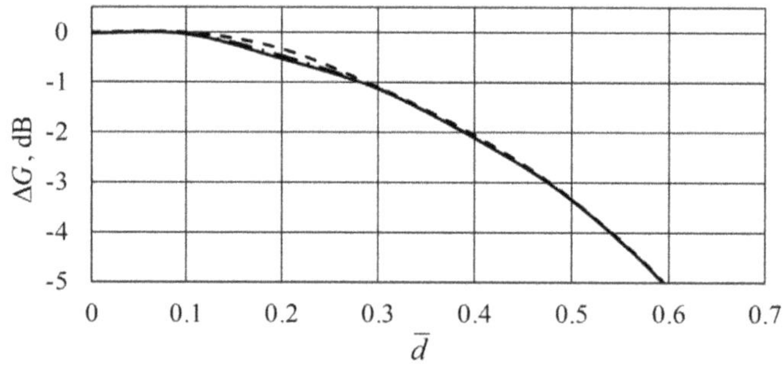

Figure 2.44. The drop in reflector antenna gain versus the relative diameter of the snow deposit for a relative reflector diameter of $2a/\lambda = 100$ (the solid line corresponds to $f/2a = 0.5$, the dot-and-dash line corresponds to $f/2a = 0.75$, and the dashed line corresponds to $f/2a = 1$).

are plotted versus relative diameter of snow deposit $\bar{d} = 2a_{\mathrm{pr}}/2a$. Here, the variables are the relative diameter of the reflector $2a/\lambda$ and the relative focal distance of its feed $f/2a$. The operating wavelength of the antenna was fixed, and it was equal to 32 mm.

The radius of the feed's aperture was set in such manner that it would result in the amplitude distribution decaying towards the reflector's edge by 9.5–9.7 dB in the E-plane. The geometrical features of the feeds adopted for the simulation are presented in table 2.1.

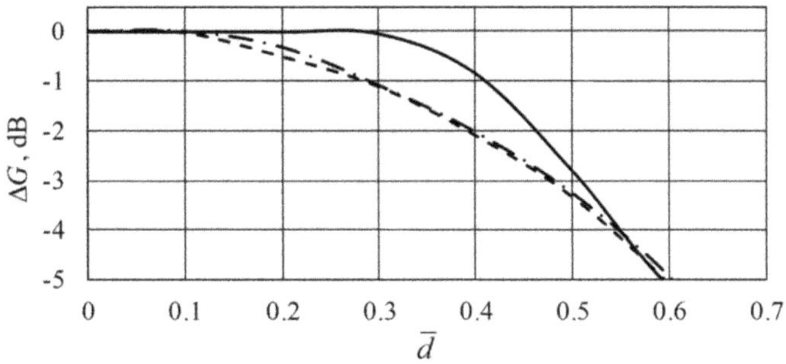

Figure 2.45. The drop in reflector antenna gain versus the relative diameter of the snow deposit for a relative focal distance of $f/2a = 0.5$ (the solid line corresponds to $2a/\lambda = 10$, the dot-and-dash line corresponds to $2a/\lambda = 50$, and the dashed line corresponds to $2a/\lambda = 100$).

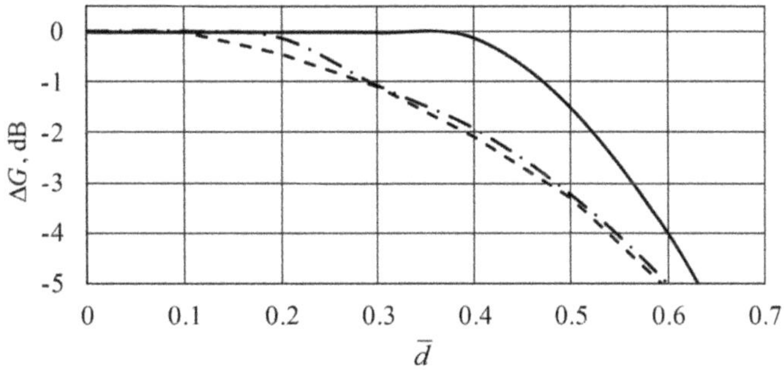

Figure 2.46. The drop in reflector antenna gain versus the relative diameter of the snow deposit for a relative focal distance of $f/2a = 0.75$ (the solid line corresponds to $2a/\lambda = 10$, the dot-and-dash line corresponds to $2a/\lambda = 50$, and the dashed line corresponds to $2a/\lambda = 100$).

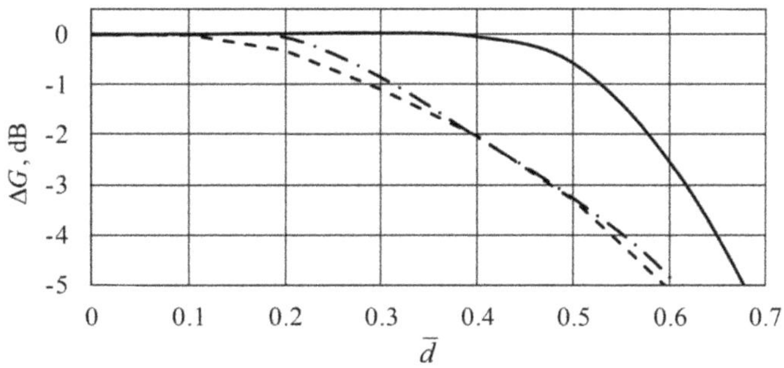

Figure 2.47. The drop in reflector antenna gain versus the relative diameter of the snow deposit for a relative focal distance of $f/2a = 1$ (the solid line corresponds to $2a/\lambda = 10$, the dot-and-dash line corresponds to $2a/\lambda = 50$, and the dashed line corresponds to $2a/\lambda = 100$).

The plots in figures 2.41–2.44 are presented for a fixed relative diameter of the reflector and varying focal relative distance, the solid line corresponding to $f/2a = 0.5$, the dot-and-dash line corresponding to $f/2a = 0.75$, and the dashed line corresponding to $f/2a = 1$.

Table 2.1. Geometrical features of the antenna feeds.

#	$f/2a$	a_i, mm
1	0.5	14
2	0.75	21
3	1	28

Figure 2.41 shows the plots for the relative reflector diameter of $2a/\lambda = 10$. The analysis of the plots obtained for various relative focal distances ($f/2a$) revealed the fact that as the relative radius of the snow deposit goes above the threshold of $\bar{d} > 0.9$ the antenna suffers a sharp drop in its gain.

Practically, the antenna can still be in operation if that its reflector has little snow deposited on its surface. Therefore, for practical reasons we present the subsequent plots of figures 2.42–2.47 for the narrower range of relative snow deposit radius of $0 \leqslant \bar{d} \leqslant 0.7$ for fixed values of the relative reflector radius (figures 2.42–2.44) and for fixed relative focal distances (figures 2.45–2.47).

The plots of figures 2.45–2.47 were computed for a fixed value of the relative focal distance and varying values of relative reflector diameter, the solid line corresponding to a relative reflector diameter of $2a/\lambda = 10$, the dash-and-dot line corresponding to a relative reflector diameter of $2a/\lambda = 50$, and the dashed line corresponding to a relative reflector diameter of $2a/\lambda = 100$.

Analysis of the simulation results demonstrates that if one assumes an antenna gain drop below 1 dB, $\Delta G \leqslant 1$, as an indicator of antenna performance degradation, then it becomes obvious that as the antenna reflector becomes larger, the impact of the snow deposit on its gain becomes more pronounced. In this regard, the antennas of greater electrical dimensions are more vulnerable to the snow depositing on the surface of their reflectors. However, the antenna performance degradation almost does not depend on the reflector's profile curvature, which is related to the feed's relative focal distance $f/2a$, across a wide range of relative antenna reflector diameters $10 \leqslant 2a/\lambda \leqslant 100$.

2.4.3 Numerical simulation of the reflector antenna radiation performance given a non-symmetrical distribution of precipitation deposit over its reflector's surface

In practical applications, symmetrical precipitation deposits occur seldom, in particular if precipitation in the form of rain is concerned. The latter can accumulate on the reflector's surface if, for example, the antenna looks in a vertical direction. Thus, below we consider the impact of a non-symmetrical water layer on the performance of a reflector antenna (figure 2.48).

Let us consider an antenna, the reflector of which is a circular cut from a paraboloid of revolution (figure 2.48), its feed being placed in the paraboloid's focus ($z = f$). We write the paraboloid's equation in the following form:

$$-\frac{x^2 + y^2}{2q} + z = 0, \tag{2.79}$$

where $q = 2f$ is the focal parameter of the reflector.

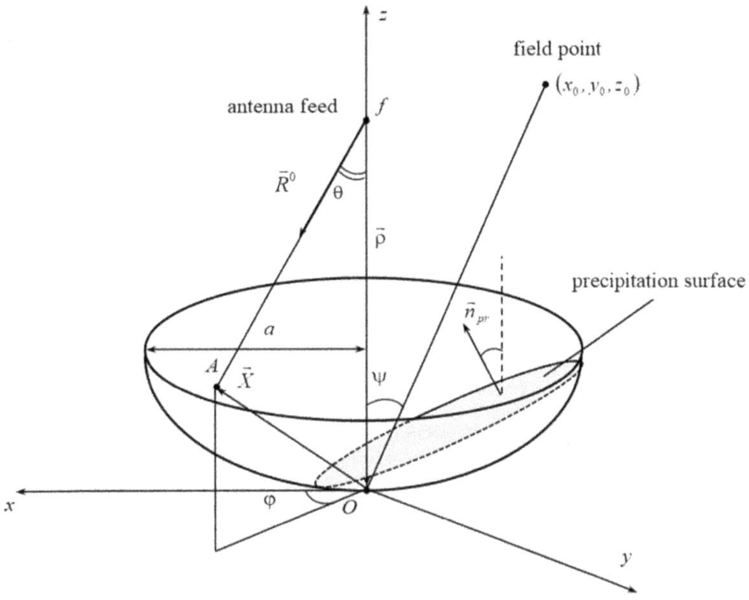

Figure 2.48. Geometry of the problem.

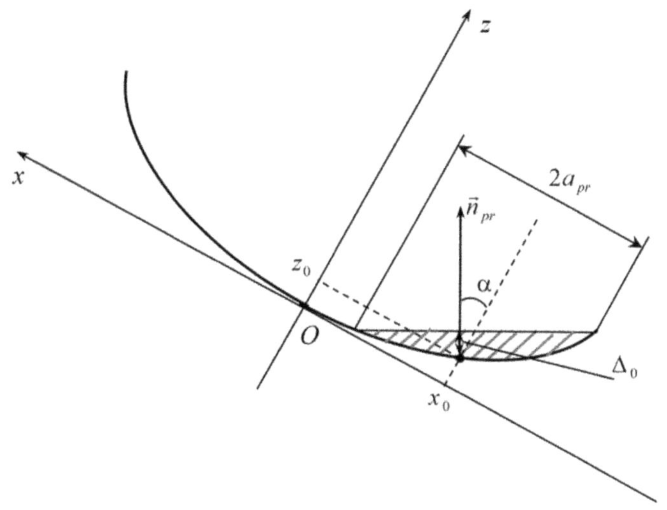

Figure 2.49 Geometry of the precipitation deposit on the reflector's surface.

Here, we consider high gain antennas, for which the condition of $2a >> \lambda$ (a is the reflector aperture radius, λ is the operating wavelength) holds.

In the general case, the antenna boresight direction is displaced off vertical by an angle α. Because of that, the layer of water settles down non-symmetrically with respect to the optical axis of the reflector (figure 2.49). In this case, the inclined surface of the reflector holds a water pool with a footprint centered around point $(x_0, 0, z_0)$ (figure 2.49). We further assume that the antenna axis is actually inclined in the plane xOz by the angle α from the vertical.

Let us find the coordinates of the water pool footprint center $(x_0,\ 0,\ z_0)$. To that end, we write the expression for the normal $\vec{n}_{\text{pr}}$ to the deposit's upper surface, which coincides with the normal to the reflector's surface at point $(x_0,\ 0,\ z_0)$:

$$\vec{n}_{\text{pr}} = \frac{\left(-\dfrac{x_0}{q},\ 0,1\right)}{\sqrt{1 + \dfrac{x_0^2}{q^2}}}. \tag{2.80}$$

Taking into account expression (2.80) and the above assumptions, we obtain the following expression for the projection of the normal $\vec{n}_{\text{pr}}$ onto the axis Oz:

$$\left(\vec{n}_{\text{pr}} \cdot \vec{e}_z\right) = \cos\alpha = \frac{1}{\sqrt{1 + \dfrac{x_0^2}{q^2}}}, \tag{2.81}$$

where $\vec{e}_z = (0,0,1)$.

Expression (2.81) yields

$$x_0 = q\ tg\ \alpha. \tag{2.82}$$

Plugging in the expression (2.81) into the reflector's equation (2.79), we land on the third coordinate of the precipitation deposit footprint:

$$z_0 = \frac{q}{2} tg^2\alpha. \tag{2.83}$$

Now we find the thickness of the deposit (depth of the pool) at point $(x_0,\ 0,\ z_0)$ (figure 2.48). To achieve that, we write the equation describing the upper surface of the water pool

$$\frac{-x_0}{q}x + z = -x\ tg\ \alpha + z = C, \tag{2.84}$$

where C is the sought for constant.

Now we determine the line along the directing vector $\vec{n}_{\text{pr}}$ that passes through the point $(x_0,\ 0,\ z_0)$. In parametric form, this line is as follows:

$$\begin{cases} x = x_0 - \dfrac{x_0 t}{q} \\ y = 0 \\ z = z_0 + t \end{cases}. \tag{2.85}$$

We find the value for parameter t_0, which corresponds to the cross point of the line (2.85) with the plane (2.84). To that end we plug expression (2.85) into (2.84):

$$t_0 = \frac{C + x_0\ tg\ \alpha - z_0}{1 + \dfrac{x_0\ tg\ \alpha}{q}}. \tag{2.86}$$

Substituting the parameter t_0 into (2.85), we receive the coordinates of cross point (x_1, y_1, z_1), we obtain the pool depth Δ_0 at point $(x_0, 0, z_0)$ as

$$\Delta_0 = \sqrt{(x_0 - x_1)^2 + (y_0 - y_1)^2 + (z_0 - z_1)^2} = t_0\sqrt{tg^2\ \alpha + 1}. \qquad (2.87)$$

Plugging the expression (2.86) into (2.87) and doing some math, we obtain the formula for computing the constant C of expression (2.84) that describes the upper surface of the water pool:

$$C = -\frac{q}{2}tg^2\ \alpha + \Delta_0\sqrt{1 + tg^2\ \alpha}. \qquad (2.88)$$

It can be shown easily that the projection of the reflector's surface cross-section by the upper plane of the water pool onto the plane xOy is a circle with radius a_{pr} (figure 2.48):

$$a_{\text{pr}} = \sqrt{2q\Delta_0\sqrt{1 + tg^2\ \alpha}}. \qquad (2.89)$$

We assume that the reflector's design presumes the water drainage hole at its center, which prevents water accumulation above a certain level given that the reflector is displaced off the strictly vertical oriented direction. Above that certain level, the water pool drains off the reflector's surface either via the drainage hole or spilling over the reflector's edge. The latter means that there is a certain range of inclination angles which make the water accumulation on top of the reflector possible. This particular range of inclination angles depends on the antenna reflector geometry. Figure 2.50 presents the limit case antenna inclination angles α, for which water accumulation is still possible, versus relative focal distance $f/2a$.

In figure 2.50 curve #1 corresponds to the maximum of water held by the reflector, while curve #2 corresponds to all the water spilling over the reflector's edge. For an antenna looking vertically up (zero displacement off the vertical, $\alpha = 0°$), all the water is drained via the drainage at the center of the reflector. Thus, for inclination angles belonging to the interval limited by the two curves of

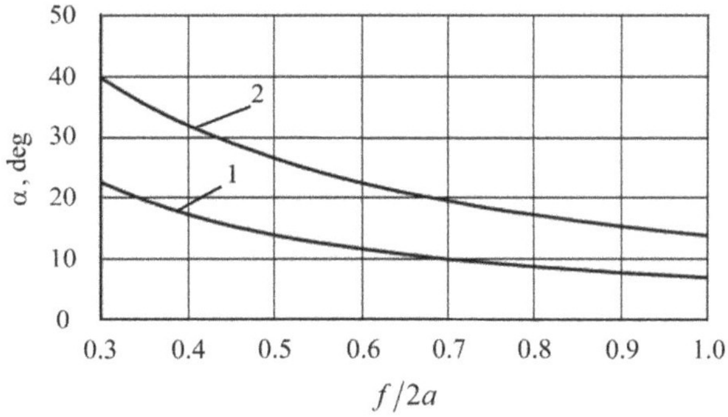

Figure 2.50. The limit case antenna inclination angles for which water accumulation is still possible, versus the relative focal distance. Curve #1 corresponds to the maximum of water held by the reflector, curve #2 corresponds to all the water spilling over the reflector's edge.

figure 2.50, there can be some water held by the reflector. Obviously, the depth of the reflector affect the possible water accumulation on its surface. Deeper antennas with smaller relative focal distances $f/2a$ can hold water in a wider range of inclination angles α. Shallower antennas with greater values of $f/2a$ are less prone to such precipitation accumulation as the angle range for the reflector holding any water becomes narrower.

For predicting the radiation performance of a parabolic antenna that has a layer of water accumulated at its surface, we use the method presented above in subsection 2.4.1. In this case, the surface S, over which integration is to be conducted in formula (2.64), will have the form depicted in figure 2.51.

Below, we present the computation results with regard to the radiation performance of parabolic antennas if the surfaces of their reflectors hold some liquid precipitation (rain) [64, 65].

The computations were carried out for an antenna feed in the form of a circular horn of two different aperture radii of 18.5 mm and of 23 mm. The electric strength vector of its radiated wave belonged to the plane yOz (the E-plane). The field amplitude distributions over the reflector's surface in two generic planes are shown in figures 2.52 and 2.53.

Figures 2.54–2.57 show the normalized DPs $F(\psi)$ of an antenna in the E- and H-planes. The plots of $F(\psi)$ were normalized to maximum antenna gain for the case that there was no precipitation deposited on the reflector's surface. All computations were conducted for a relative permittivity of water equal to $\varepsilon' = 44.43 + j41.39$ that water attains at a temperature of 0 °C [47, 66]. The plots represent situations with various quantities of water present at the antenna reflector's surface.

Curve #1 in figures 2.54–2.57 represents the situation where there is no water present on the reflector, whereas curves # 2 through #4 represent the situations where a water pool is present that has a relative footprint diameter $\bar{d} = 2a_{\text{pr}}/2a$

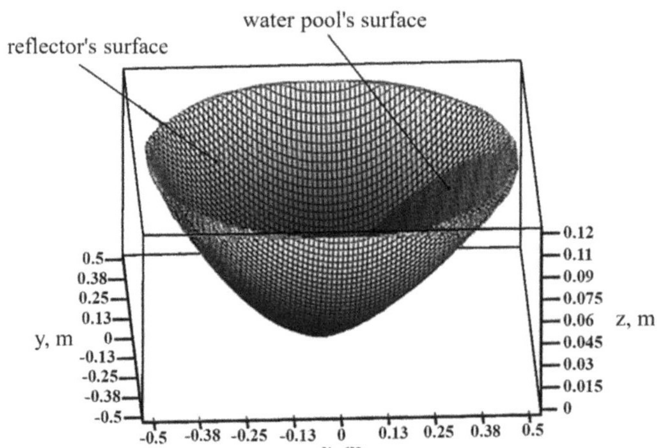

Figure 2.51. Finding the integration surface S.

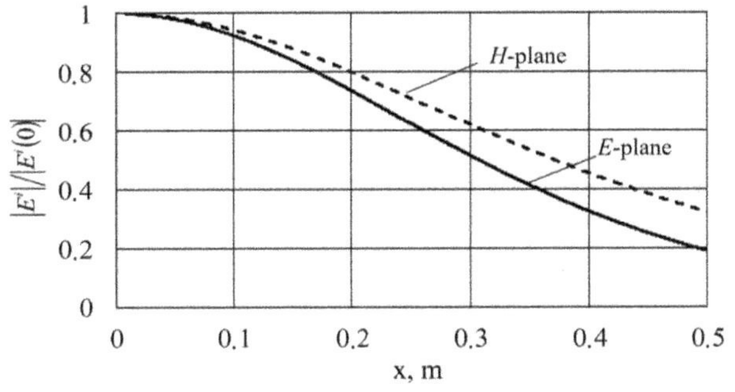

Figure 2.52. Field amplitude distributions over the reflector's surface in two generic planes (for a feed aperture radius of 18.5 mm).

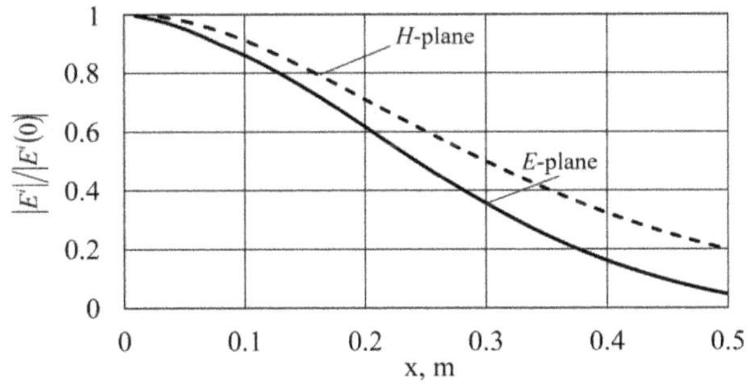

Figure 2.53. Field amplitude distributions over the reflector's surface in two generic planes (for a feed aperture radius of 23 mm).

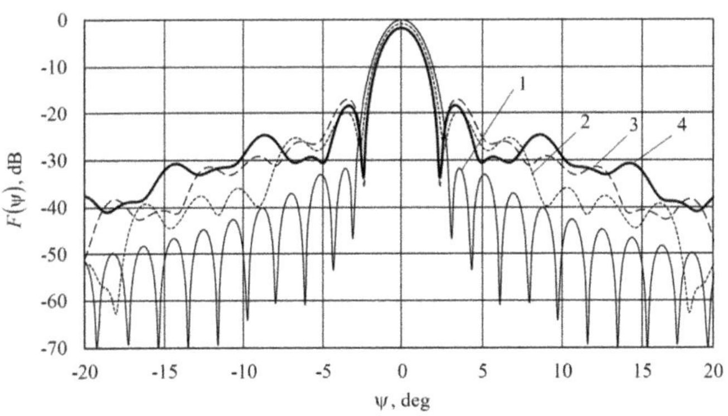

Figure 2.54. Normalized DPs $F(\psi)$ in the E-plane, the radius of the antenna feed's aperture being equal to 18.5 mm (curve #1—no water pool on the surface; curve #2—$\overline{d}$ = 0.26; curve #3—$\overline{d}$ = 0.32; and curve #4—$\overline{d}$ = 0.38).

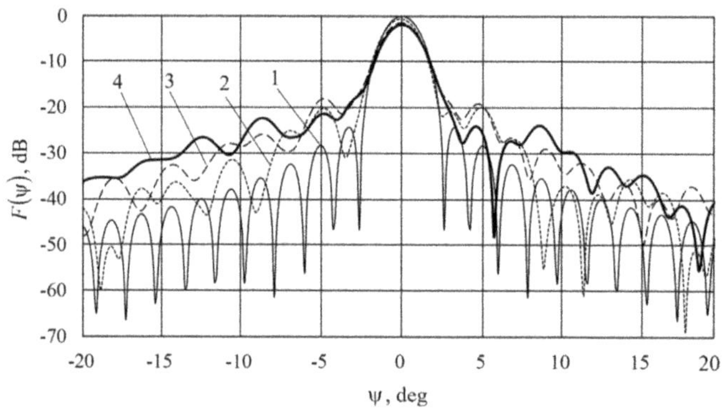

Figure 2.55. Normalized DPs $F(\psi)$ in the *H*-plane, the radius of the antenna feed's aperture being equal to 18.5 mm (curve #1—no water pool on the surface; curve #2—$\bar{d}$ = 0.26; curve #3—$\bar{d}$ = 0.32; and curve #4—$\bar{d}$ = 0.38).

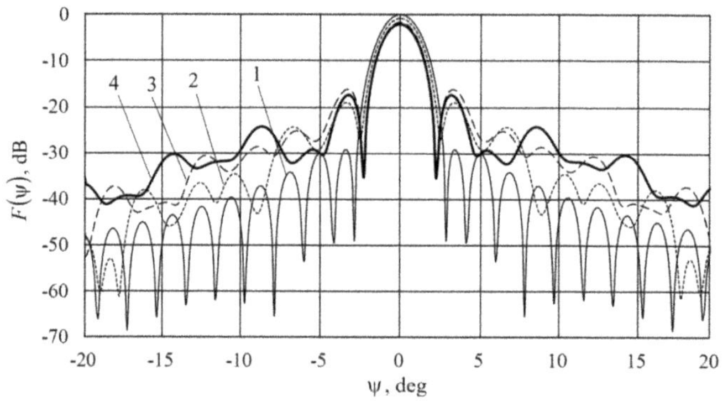

Figure 2.56. Normalized DPs $F(\psi)$ in the *E*-plane, the radius of the antenna feed's aperture being equal to 23 mm (curve #1—no water pool on the surface; curve #2—$\bar{d}$ = 0.26; curve #3—$\bar{d}$ = 0.32; and curve #4—$\bar{d}$ = 0.38).

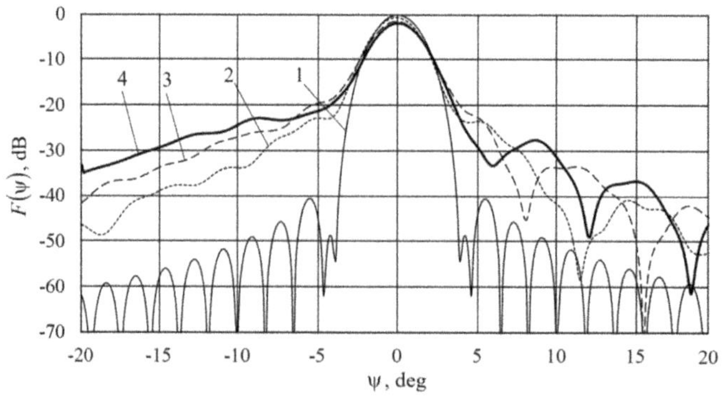

Figure 2.57. Normalized DPs $F(\psi)$ in the *H*-plane, the radius of the antenna feed's aperture being equal to 23 mm (curve #1—no water pool on the surface; curve #2—$\bar{d}$ = 0.26; curve #3—$\bar{d}$ = 0.32; and curve #4—$\bar{d}$ = 0.38).

equal to 0.26, 0.32, and 0.38, respectively. The latter footprint diameters correspond to the amount of water equal to $V_w = 0.2$ L, $V_w = 0.5$ L, and $V_w = 1$ L, respectively.

Analysis of figures 2.54–2.57 demonstrates that the presence of water on the reflector's surface affects both the DP shape and the gain of the reflector antenna. First, the level of sidelobes increases significantly. In the case of the antenna represented in figures 2.54 and 2.55, such a sidelobe increase amounts to 12.5 dB or even 15.5 dB in the *E*-plane (figure 2.54) depending on the amount of water in the pool on its surface. In the *H*-plane, such a sidelobe increase amounts to 5.5 dB or even to 7.5 dB (figure 2.55).

Second, the notches of the DP become shallow and they may disappear altogether, while its sidelobes widen significantly.

Third, in the plane of the antenna scan (*H*-plane), the presence of a water pool on the reflector's surface, as anticipated, introduces a lot of asymmetry into the antenna's DP (figures 2.55 and 2.57). In addition, such asymmetry is more pronounced in the second case (figure 2.57). That could be due to the fact that a greater portion of the reflector's surface illuminated by a higher intensity field is covered by the pool of water.

A closer look at the main lobe in figures 2.54 and 2.55 reveals that in the antenna scan plane (the *H*-plane in our case), the presence of a water pool on the reflector's surface makes the main lobe of the DP wider while distorting its symmetry. Thus, for a water pool on the surface of the reflector with a relative diameter of its footprint of $\bar{d} = 0.38$ (curve #4 in figure 2.55), the main lobe of the DP becomes 5% wider. This can be explained by the fact that the presence of precipitation affects the amplitude distribution in this plane in a stronger manner, causing the field to decay faster towards the reflector's edge.

In the *E*-plane the opposite effect takes place. The main lobe of the DP in the E-plane narrows (for a water footprint radius of $\bar{d} = 0.38$ the half-power pattern width decreases by about 6.5%). The latter could be explained by the smoothing effect of the water pool upon the amplitude distribution of the field (current) over the reflector's surface in the *E*-plane, which means that the field attenuation due to the water pool at the reflector's center makes the otherwise quickly decaying towards the edges field more uniform.

In addition to altering the shape of the DP, the impact of precipitation also manifests as decreased antenna gain, ΔG (which can be regarded as the gain loss). For example, for a water footprint radius of $\bar{d} = 0.26$, such a decrease amounts to 0.75 dB, for $\bar{d} = 0.32$ such a decrease is 1.6 dB, and for $\bar{d} = 0.38$ such an increase amounts to 1.85 dB.

In order to assess the impact of the water layer held by the reflector's surface on the antenna radiation performance, we carried out a round of simulations for antennas featuring reflectors with relative radii of $2a/\lambda = 10, \ 50, \ 100$, and relative focal distances of $f/2a = 0.5, \ 0.75, \ 1.0$.

Figures 2.58–2.63 present the results of computations regarding the antenna gain loss (ΔG) versus the relative diameter of footprint $\bar{d} = 2a_{\text{pr}}/2a$ of the water pool held by its reflector's surface. The antenna was tilted by an angle of $\alpha = 5°$ off the vertical in the *E*-plane. The diameter of a circular feed, similar to subsection 2.4.2,

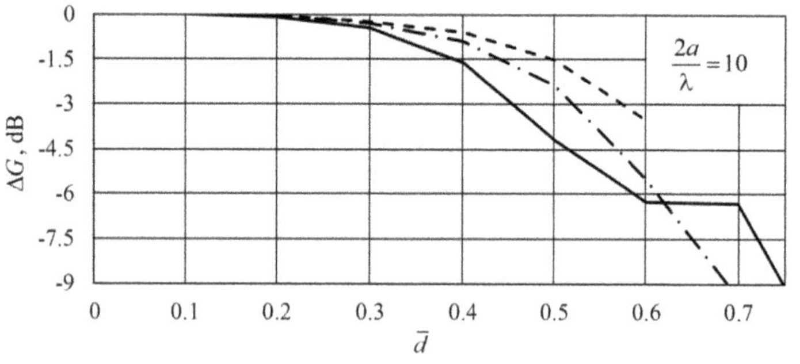

Figure 2.58. Antenna gain loss due to a non-symmetrical water layer on the surface of its parabolic reflector for a relative reflector diameter of $2a = 10\lambda$, the antenna operating at a wavelength of $\lambda = 32$ mm.

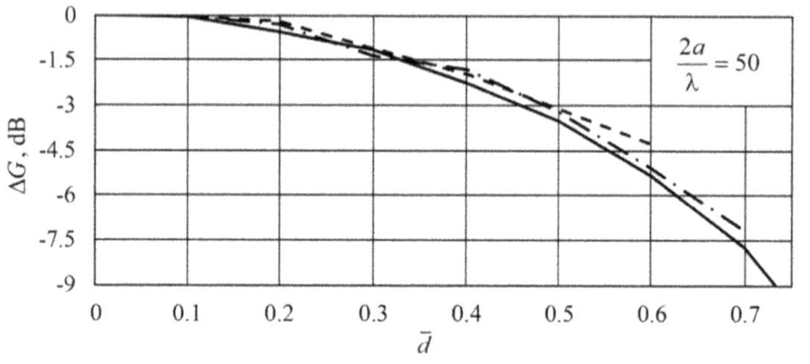

Figure 2.59. Antenna gain loss due to a non-symmetrical water layer on the surface of its parabolic reflector for a relative reflector diameter of $2a = 50\lambda$, the antenna operating at a wavelength of $\lambda = 32$ mm.

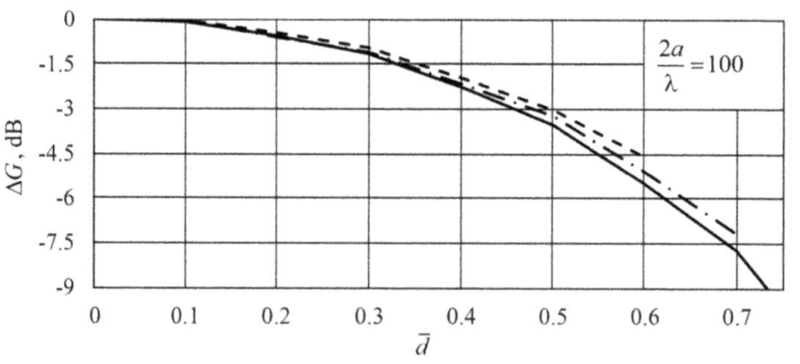

Figure 2.60. Antenna gain loss due to a non-symmetrical water layer on the surface of its parabolic reflector for a relative reflector diameter of $2a = 100\lambda$, the antenna operating at a wavelength $\lambda =$ of 32 mm.

was set in such manner that the amplitude distribution of the field the reflector was illuminated with in the E-plane would be decaying by 9.5–9.7 dB towards the edge. Under such conditions, the relative diameter of the reflector $2a/\lambda$ and the relative focal distance $f/2a$ become the parameters that cannot take arbitrary values and become characteristic of the antenna, depending on the task at hand.

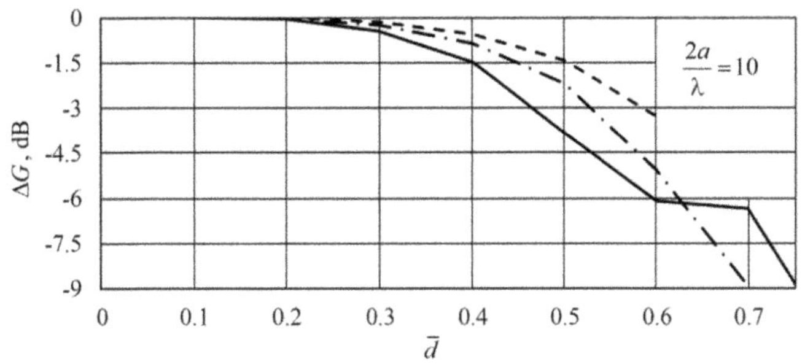

Figure 2.61. Antenna gain loss due to a non-symmetrical water layer on the surface of its parabolic reflector for a relative reflector diameter of $2a = 10\lambda$, the antenna operating at a wavelength of $\lambda = 8.2$ mm.

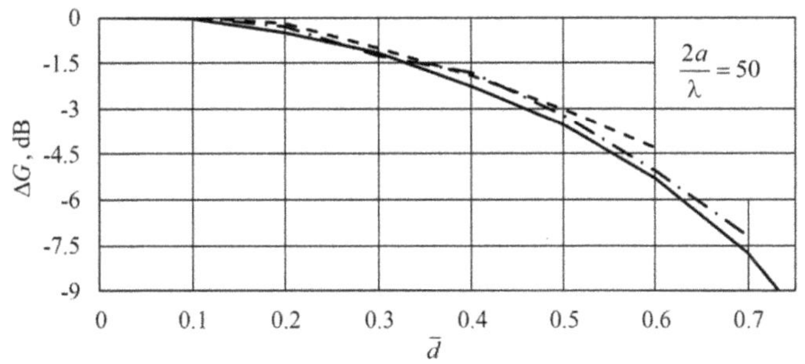

Figure 2.62. Antenna gain loss due to a non-symmetrical water layer on the surface of its parabolic reflector for a relative reflector diameter of $2a = 50\lambda$, the antenna operating at a wavelength of $\lambda = 8.2$ mm.

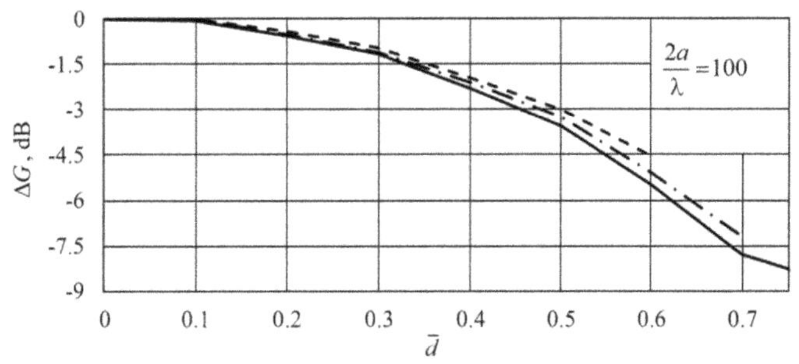

Figure 2.63. Antenna gain loss due to a non-symmetrical water layer on the surface of its parabolic reflector for a relative reflector diameter of $2a = 100\lambda$, the antenna operating at a wavelength of $\lambda = 8.2$ mm.

The plots of figures 2.58–2.60 were computed for an antenna operating wavelength $\lambda = 32$ mm, and the plots of figures 2.61–2.63 were computed for an antenna operating wavelength of $\lambda = 8.2$ mm. There, the solid lines correspond to $f/2a = 0.5$, the dot-and-dash lines correspond to $f/2a = 0.75$, and the dashed lines correspond to $f/2a = 1$. While computing the gain loss for an antenna operating frequency $\lambda =$ 8.2 mm, we assumed that the relative permittivity of water was equal to $\varepsilon' = 10.16 + j19$ for a temperature of 0 °C [47, 66].

The plots show that the gain loss due to a water layer on the reflector's surface for antenna operating frequencies of both 32 mm and 8.2 mm. As in the case of the snow deposit above (subsection 2.4.2), a significant drop in antenna gain versus the relative diameter of precipitation footprint $\bar{d}$ begin at smaller values of the latter for antennas of smaller electrical sizes, however, in case of antennas that are larger electrically, such gain loss has a more abrupt character.

2.4.4 Verification of the proposed method for predicting the radiation performance of reflector antennas whose surface was partially covered by precipitation

In order to check the reliability of the computation results obtained using the developed method against the real processes that have to do with EMWs scattered by the antenna's reflector with precipitation, we compared our results to experimental data obtained together with a group of employees of the AY Usikov Institute of Radio Physics and Electronics of the NAS of Ukraine [56, 65]. That experiment was designed and conducted under the leadership of Grygoriy Khlopov. It consisted in measuring the actual antenna gain depending on the amount of precipitation deposited on the antenna reflector.

The measurements were carried out at a wavelength of $\lambda = 32$ mm using a truss tower (figure 2.64(a)), at the footing of which there was a reflector antenna with a reflector of diameter $2a = 1$ m and a relative focal distance $f/2a = 0.52$. A circular horn with an aperture radius of 23 mm served as a feed for that antenna.

The tower height provided for the far-field zone condition to be satisfied for the antenna of interest. The antenna fixture provided an accuracy of the antenna axis positioning of not worse than 0.1°. The whole measuring rig is depicted in figure 2.64(b). The transmitter signal was generated by the G4–83 standard signal generator [67], which provided calibrated power output.

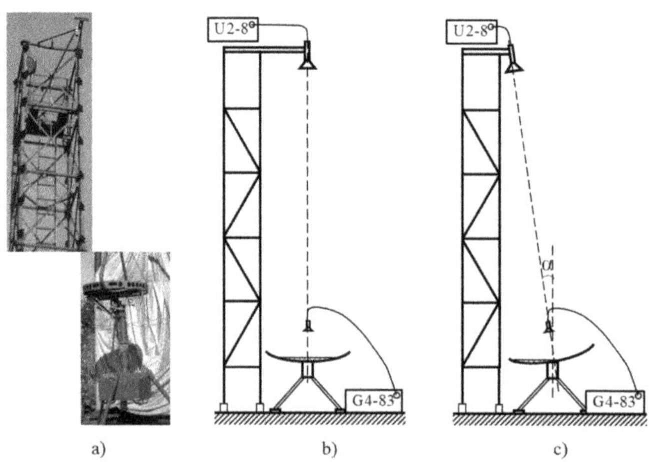

Figure 2.64. The appearance of the measurement rig (a), and its schematic diagram (b) and (c).

The generator was placed near the antenna under study, the receiver horn antenna (with an aperture of 10 × 15 cm) coupled with a detector unit being put on top of the rig and being connected to a narrowband amplifier of the U2–8 type [68]. As a result, the energy potential of the measuring rig was above 40 dB, which provided for quite stable and repeatable results.

The experiment was conducted in two variants. In the first variant, the antenna under study was directed towards the zenith (figure 2.64(b)); the central part of the antenna reflector was covered by a symmetrical snow deposit, as shown in figure 2.40. In the second variant, the antenna under study was tilted by the angle $\alpha = 11.7°$ (figure 2.64(b)); the water was poured on the reflector's surface to form a non-symmetrical pool with respect to its axis pool (figure 2.58).

Experimental study of the influence of a symmetrical (with respect to the antenna's axis) snow deposit on the antenna gain

The measurement rig was calibrated when the antenna's surface was clear of any foreign objects. After calibration, a flat top snow deposit with a certain footprint diameter was placed symmetrically on the reflector's surface (figure 2.40). The depth and the moisture content (in kg m^{-3}) of the snow deposit were kept under control. That particular round of measurements resulted in the averaged data presented in figure 2.65 in the form of antenna gain loss versus the diameter of the snow deposit footprint $2a_{\mathrm{pr}}$ (curve #1).

The radiation performance of the same antenna in terms of the gain loss was also predicted using the method presented in subsection 2.4.1. The computations were carried out for a relative permittivity of snow equal to $\varepsilon' = 2.8 + j0.9$ [63], (curve #2 in figure 2.65).

The difference between the measured and computed points on the plots are below 11%.

A study of the effect a non-symmetrical (with respect to the antenna axis) water pool on top of its reflector has on the antenna's gain

The experiment was carried out for the operating frequency λ = 32 mm at the measuring rig (figure 2.64(c)). This time the antenna axis was tilted away from a

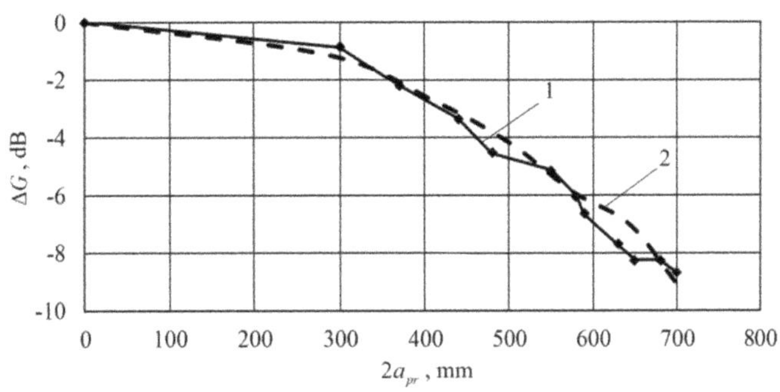

Figure 2.65. The antenna gain loss as measured (curve #1) and computed (curve #2) versus the radius of the snow deposit.

strictly zenith orientation by an angle of 11.7°. The antenna gain loss due to water present on the reflector's surface was measured by comparing the received power for two conditions, the first condition being water pool present, and the second being water pool absent. Individual datum points were averaged to obtain the value of the gain loss versus the amount of water present in the pool V_w.

The relation between the volume of water in the pool V_w, the diameter of its footprint $2a_{\text{pr}}$, and the depth of the pool Δ_0 on the reflector's surface is presented in table 2.2.

Figure 2.66 presents the results of measurements along with the data predicted using the above method with respect to the antenna gain loss versus the amount of water in the non-symmetrical pool. Predicted data points were obtained for the relative permittivity of water equal to $\varepsilon' = 44.43 + j41.39$. The measured data points (figure 2.66, curve #1) deviated from the predicted ones (curve #2) by no more than 0.27 dB (16%).

Comparing the data obtained by measurement and by computation (figures 2.65 and 2.66) for the antenna that had a layer of precipitation on the surface of its reflector, both symmetrical and non-symmetrical with respect to the antenna's axis, one can draw the conclusion that the above mathematical model and its computer representation do correspond to the real processes taking place in EMW scattering at the antenna reflector if a layer of precipitation is present on its surface.

Table 2.2. The relation between the volume of water in the pool on the reflector's surface and its footprint parameters.

V_w, liters	0.1	0.2	0.3	0.4	0.5	0.6	0.7	0.8	0.9	1	1.1
Δ_0, mm	5.42	7.66	9.38	10.8	12.1	13.27	14.33	15.32	16.25	17.13	17.97
$2a_{\text{pr}}$, mm	215	255.1	282.4	303.4	320.8	335.8	348.9	360.8	371.6	381.5	390.7

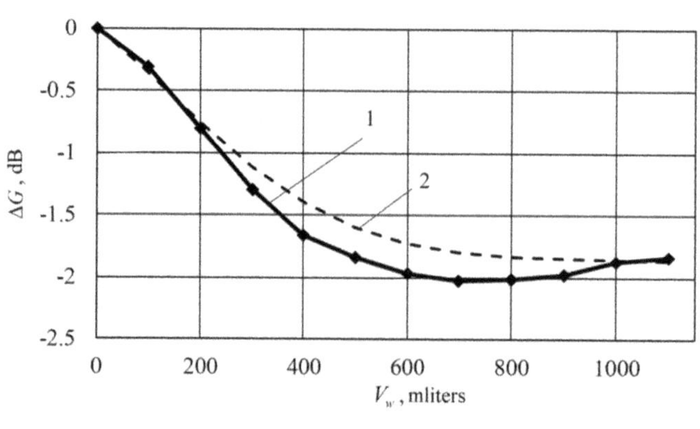

Figure 2.66. Antenna gain loss versus the amount of water in the non-symmetrical pool on the reflector's surface as measured (curve #1) and predicted (curve #2).

2.5 Near-field calculation for an electrically large reflector antenna

In a number of practical cases, there is a necessity for predicting the antenna radiation performance in the near-field zone. For instance, such a necessity appears when designing antennas to be installed on aircraft and to be placed under protective fairings, or when predicting the performance of a number of closely spaced electronic means that would be operating in overlapping frequency bands [69, 71], or when finding the locations of microwave energy concentrations around a radar that would be dangerous for personnel, or when figuring out the impact of antenna fixtures (and other elements that are usually placed in the immediate vicinity of its reflector) on the antenna's radiation pattern [70].

The computational methods based on IE solving do allow for predicting the EMF at any particular point of space (both the near- and far-field zones). However, they are very computationally extensive. In addition, this computation burden grows very fast as the electrical sizes of the antenna reflectors increase [3]. In most cases, reflector antennas are used to create pencil-like beams, which requires the use of large AS apertures. Such apertures are usually many times greater than the operating wavelength. The latter means that one can employ asymptotic high-frequency methods (AHFMs) to predict the radiation performance of reflector antennas.

The paper [71] presents a method for predicting the antenna near field by the field distribution over an offset aperture situated in the reflector's focal plane. A significant drawback of that method is its inapplicability for predicting the side and back radiation of the antenna. In addition, this method provides an acceptable accuracy only for the radiation directions around the antenna boresight (the main lobe and the first sidelobes of the antenna pattern). Other widely used methods include geometrical diffraction theory and physical diffraction theory, as well as their modifications [35, 72]. A common drawback of all these methods is that they all fail if the antenna has non-perfectly reflecting segments on its surface.

The authors of [73, 74] describe a computational approach that allows for the prediction of both the near and far-field of the antenna that may have the rim of its reflector coated with radio-absorbing material. The approach is based on an integral representation of the field scattered by the antenna reflector in the antenna near field zone. It employs the benchmark problem of plane EMW scattering at a perfectly conducting wedge, the edge of which is coated with a radio-absorbing cylinder [75], in order to evaluate the field scattered by the reflector's edge (rim).

The sources [73, 74] mentioned above considered antenna reflectors shaped as a symmetrical cut from a paraboloid of revolution, yet in many cases the need for reducing the effect of reflector shadowing by its feed calls for a non-symmetrical cut from the paraboloid (figure 2.67). Antennas featuring such reflectors are usually installed on flying platforms (figure 2.67(a)), as well as being used in naval and ground based radar (figures 2.66(b) and (c)).

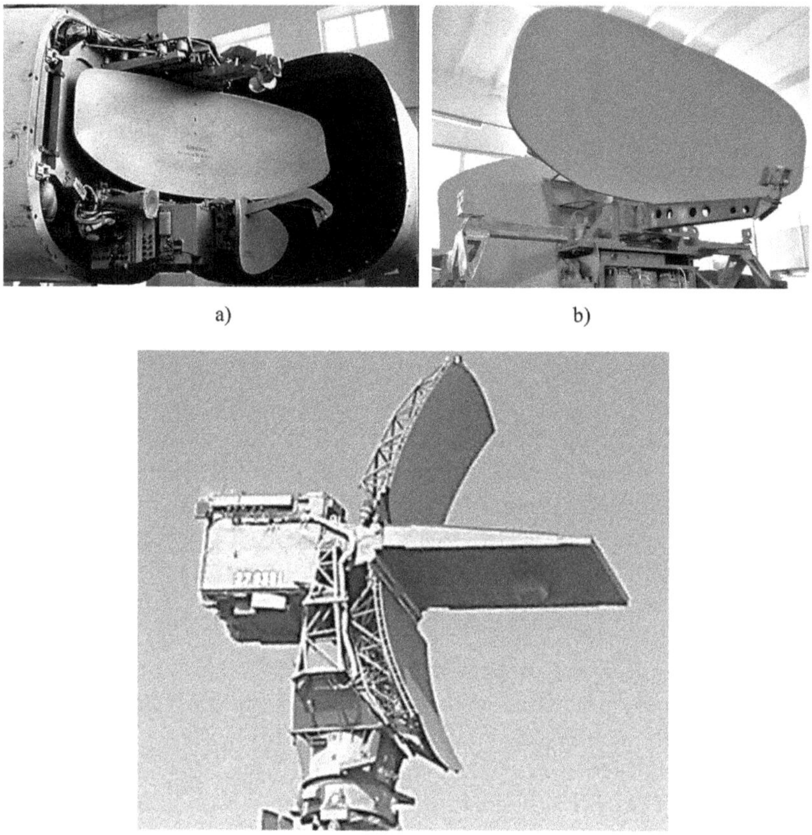

a) b)

c)

Figure 2.67. Examples of ASs shaped as non-symmetrical cuts from paraboloids: (a) the antenna of the Orion-A forward looking radar borne by Su-24M aircraft; (b) the AS of the Mineral-ME naval radar complex for target detection and target designation designed by the 'Kvant-Radiolokatsiya' research institute; and (c) the antenna of the low-altitude target detector 5N66M of S-300 air defense system.

2.5.1 The method for computing the field radiated by a reflector antenna in the near-field zone

We consider here a reflector antenna that has its reflector shaped as a non-symmetrical cut from the paraboloid of revolution shown in figure 2.68. For definiteness, we assume that the reflector parameters were as follows: $a = 3.46$ m; $b = 3.1$ m; and $f = 3.6$ m. The reflector's focal point f is occupied by the horn feed operating on the wavelength of 4.5 cm. The feed illuminates the reflector with a vertically polarized EMW, the maximum of the feed's radiation being directed towards the reflector's center (figure 2.68).

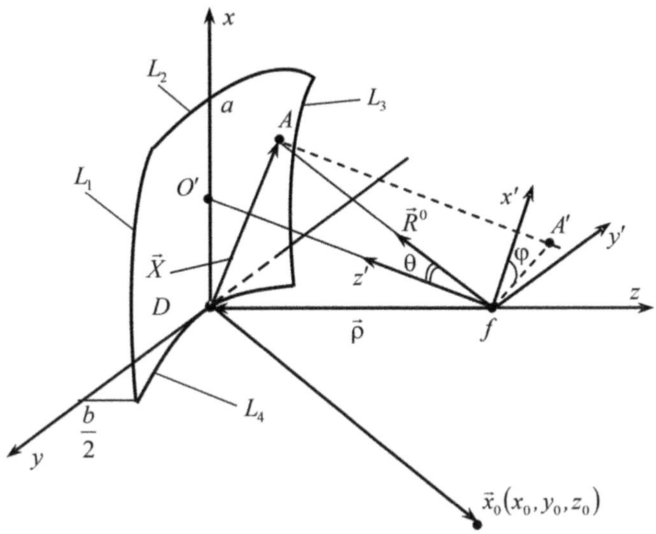

Figure 2.68. Geometry of the task at hand.

We make some additional assumptions for further analysis: the feed's DP is not altered by it being placed at the reflector's focus; and all the points belonging to the reflector's surface are in the feed's far-field zone. All the formulas are given for the time dependence of EMFs expressed as $e^{-j\omega t}$.

Now, considering the above assumptions, we can express the field excited by the feed at a point A at the reflector's surface that is described by the radius-vector $\overrightarrow{X}$ (figure 2.68), in the following form [73]:

$$\overline{E}^f(\overrightarrow{X}) = \frac{jk_0}{4\pi}\overline{p}'\frac{\exp[jk_0|\overrightarrow{X}+\overline{\rho}|]}{|\overrightarrow{X}+\overline{\rho}|}F(\theta, \varphi), \tag{2.90}$$

$$\overline{H}^f(\overrightarrow{X}) = \sqrt{\frac{\varepsilon_0}{\mu_0}}\frac{jk_0}{4\pi}\left(\overrightarrow{R}^0 \times \overline{p}'\right)\frac{\exp[jk_0|\overrightarrow{X}+\overline{\rho}|]}{|\overrightarrow{X}+\overline{\rho}|}F(\theta, \varphi), \tag{2.91}$$

where $\overrightarrow{R}^0$ is the unit vector of the wave incidence direction from the feed to field point A, $\overline{\rho}$ is the radius-vector pointing from the reflector's focus to its hump, $\overline{p}' = \dfrac{\overrightarrow{R}^0 \times \left(\overline{p}^0 \times \overrightarrow{R}^0\right)}{\left|\overrightarrow{R}^0 \times \left(\overline{p}^0 \times \overrightarrow{R}^0\right)\right|}$ is the vector describing the polarization of the field that is incident in the direction of $\overrightarrow{R}^0$ ($\overline{p}^0$ is the unit vector directed along the electric field intensity vector of the feed), $k_0 = \omega\sqrt{\varepsilon_0\mu_0}$ is the wavenumber, ε_0 and μ_0 are the permittivity and permeability of free space, respectively, and $F(\theta, \varphi)$ is the function (complex in general) that defines the amplitude and phase distribution of the field excited by the feed in its far-field zone versus polar coordinates θ and φ (figure 2.68).

Angle θ defines the angular direction of vector $\vec{R}^0$ with respect to the reflector's center O', and angle φ defines its rotation with respect to plane $x'fz'$ (figure 2.68).

The total field radiated by the antenna can be represented as a sum of two components. The first is the field excited by the antenna's feed $\vec{E}^f(\vec{X})$, $\vec{H}^f(\vec{X})$, the second one is the field scattered by the antenna reflector $\vec{E}^s(\vec{X})$, $\vec{H}^s(\vec{X})$:

$$\vec{E}(\vec{X}) = \vec{E}^f(\vec{X}) + \vec{E}^s(\vec{X}), \tag{2.92}$$

$$\vec{H}(\vec{X}) = \vec{H}^f(\vec{X}) + \vec{H}^s(\vec{X}). \tag{2.93}$$

Given that the DP of the feed $F(\theta, \varphi)$ is known, the problem is reduced to finding the field scattered by the reflector. As a basic computation expression, as was the case in [73, 74], we used the expression obtained using the Lorentz reciprocity theorem for the integral representation of the field scattered by the reflector:

$$\begin{aligned} \vec{p}\,\vec{E}^s(\vec{x}_0) &= \frac{1}{4\pi}\sqrt{\frac{\mu_0}{\varepsilon_0}} \int_{D'} \frac{e^{jk_0r}}{r}\left(\frac{2-2jk_0r}{jk_0r^2}\vec{p}^{\,||} + \left(\frac{jk_0r-1}{jk_0r^2} - jk_0\right)\vec{p}^{\,\perp}\right)\vec{K}\,\mathrm{d}s \\ &+ \frac{1}{4\pi}\int_{S'} \frac{e^{jk_0r}}{r}\sqrt{\frac{\mu_0}{\varepsilon_0}}\left(\frac{2-2jk_0r}{jk_0r^2}\vec{p}^{\,||} + \left(\frac{jk_0r-1}{jk_0r^2} - jk_0\right)\vec{p}^{\,\perp}\right)\vec{H}^{\perp}\,\mathrm{d}s \\ &+ \frac{1}{4\pi}\int_{S'} \frac{e^{jk_0r}}{r}\left(\frac{jk_0r-1}{r}\right)\vec{p}\left(\vec{E}^{\perp}\times\vec{r}_0\right)\mathrm{d}s, \end{aligned} \tag{2.94}$$

where $\vec{p}$ is the unit vector of receiving antenna polarization, S' is the toroidal surface that envelops the reflector's edge (its cross-section by the plane perpendicular to the edge is the circle S_0 of radius R (figure 2.69)), D' is the smooth portion of the reflector's surface D, i.e. its entire surface except for the edge,

$$\vec{E}^{\perp} = \vec{n} \times \vec{E},\ \vec{H}^{\perp} = \vec{n} \times \vec{H},$$

$\vec{n}$ is the internal with respect to the S' normal,

$$\vec{p}^{\,||} = (\vec{p} \cdot \vec{r}_0)\vec{r}_0,$$

$$\vec{p}^{\,\perp} = \vec{p} - (\vec{p} \cdot \vec{r}_0)\vec{r}_0,$$

$\vec{r}_0$ is the unit vector of direction from point A towards the field point,

$$r = |\vec{X} - \vec{x}_0|,$$

$\vec{x}_0$ is the field point,

$$\vec{K} = \left(\vec{H}^{\perp}\right)^{+} - \left(\vec{H}^{\perp}\right)^{-},$$

and $(\vec{H}^{\perp})^{+}$ and $(\vec{H}^{\perp})^{-}$ are the surface densities of the electric currents excited on the irradiated (by the feed) surface and on the shadow surface of the reflector, respectively. The reflector was assumed to be an infinitely thin screen.

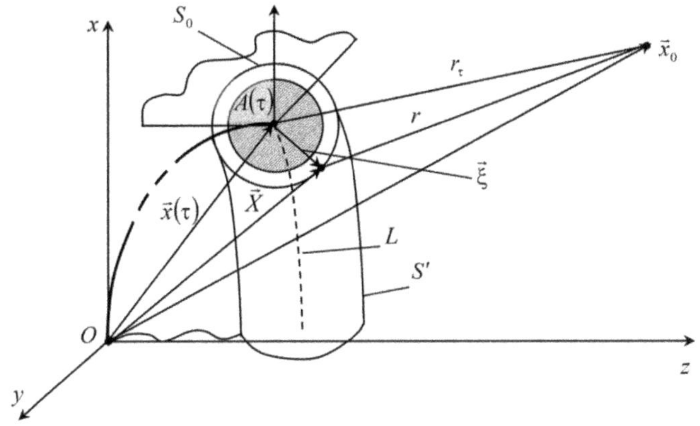

Figure 2.69. Integration over the reflector's edge.

Considering that the reflector was electrically large (a condition that holds for pencil beam antennas), we employed Kirchhoff's approximation to evaluate the integral over surface D':

$$\left(\overline{H}^{\perp}\right)^{+} = 2\left(\vec{n} \times \overline{H}^{f}\right), \quad \left(\overline{H}^{\perp}\right)^{-} = 0. \tag{2.95}$$

Computation of the EM contribution from that particular part of the reflector's surface was conducted using a special cubature formula, the design of which presumes the so-called triangulation of the surface D' where the smooth surface is actually replaced by a system of plain triangles (facets) [19].

The contribution from the edge (reflector's rim) into the field scattered by the whole reflector can be estimated as the integral over surface S' in expression (2.94) (here we denote that integral as $I_{S'}$). This integral can be rewritten as repeated integrals, the external integral being taken along the edge line L, and the internal integral being taken along the circle S_0:

$$I_{S'} = \frac{jk_0}{(4\pi)^2} \int_L e^{jk_0\Phi(\tau)} M(\tau, \vec{r}_\tau^{\,0}) \frac{F(\tau)}{|\vec{x}(\tau) + \vec{\rho}\,|} \, d\tau, \tag{2.96}$$

where

$$\begin{aligned} M(\tau, \vec{r}_\tau^{\,0}) = &\int_{S_0} \frac{e^{jk_0\vec{\xi}\cdot\vec{r}_\tau^{\,0}}}{r} \sqrt{\frac{\mu_0}{\varepsilon_0}} \left(\frac{2 - 2jk_0 r}{jk_0 r^2} \vec{p}^{\,\|} + \left(\frac{jk_0 r - 1}{jk_0 r^2} - jk_0 \right) \vec{p}^{\,\perp} \right) \overline{\vec{H}}^{\perp} \\ &(\vec{\xi}\,) \, dq \\ &+ \int_{S_0} \frac{e^{jk_0\vec{\xi}\cdot\vec{r}_\tau^{\,0}}}{r} \left(\frac{jk_0 r - 1}{r} \right) \vec{p} \cdot \left(\overline{\vec{E}}^{\perp}(\vec{\xi}\,) \times \vec{r}_\tau^{\,0} \right) dq, \end{aligned} \tag{2.97}$$

$\Phi(\tau) = |\vec{x}_0 - \vec{x}(\tau)| + \overline{R}^{0}(\tau) \cdot (\vec{x}(\tau) + \vec{\rho}\,)$ is the phase function, $\overline{R}^{0}(\tau)$ is the unit vector of wave incidence from the feed to point $A(\tau)$, $L = \sum_{i=1}^{4} L_i$ is the edge line, $\vec{r}_\tau^{\,0}$ is the unit vector of direction from point $A(\tau)$ to the field point $\vec{x}_0$, $F(\tau)$ is the function that defines the amplitude and phase distribution of the field excited by the feed in its far-field zone at the edge point τ, and $\vec{\tilde{H}}^{\perp}(\vec{\xi})$ and $\vec{\tilde{E}}^{\perp}(\vec{\xi})$ are the magnetic and electric strength vectors of the field that are excited on the surface S_0 by plane wave

$$\vec{\tilde{E}}^{0}(\vec{\xi}) = \vec{p}\exp\left[j\,k_0\left(\vec{R}^{0}\cdot\vec{\xi}\right)\right],$$

$$\vec{\tilde{H}}^{0}(\vec{\xi}) = \sqrt{\frac{\varepsilon_0}{\mu_0}}\left(\vec{R}^{0}\times\vec{p}\right)\exp\left[jk_0\left(\vec{R}^{0}\cdot\vec{\xi}\right)\right].$$

To evaluate the integral (2.96), one is required to compute the values of function $M(\tau, \vec{r}_\tau^{\,0})$ at every point of the edge line $A(\tau)$. To that end, we broke down the edge line into three segments L_i $(i = \overline{1,3})$ (figure 2.68), which are the cross-sections of the initial paraboloid of revolution with the planes that cut the reflector out of it. The segment L_4 is usually a salient line where the reflector's surface breaks to morph into a screen that separates the transmitting antenna from the receiving one (in case there is such separation). Considering that the integrand in (2.97) could be regarded as a smooth function, the integral $M(\tau, \vec{r}_\tau^{\,0})$ at every segment of the edge line was evaluated by one-dimensional numerical integration. In our computations, we used the five-point compound Gauss formula. Taking into account the large electrical dimensions of the screen and the small curvature of this edge while computing $M(\tau, \vec{r}_\tau^{\,0})$, we replaced the fields $\vec{\tilde{H}}^{\perp}(\vec{\xi})$ and $\vec{\tilde{E}}^{\perp}(\vec{\xi})$ at the actual reflector's surface by the corresponding fields at the perfectly conducting plane that included the edge and was osculant to the reflector's surface at point $A(\tau)$ (figure 2.69). The same approach was used in [73, 74, 76]. In the case of discontinuity line L_4, instead of the osculant plane we used a wedge with an external angle equal to $\pi/2$.

To compute the fields $\vec{\tilde{H}}^{\perp}(\vec{\xi})$ and $\vec{\tilde{E}}^{\perp}(\vec{\xi})$ at the surface enveloping the edge (the surface discontinuity) we employed a solution to the benchmark problem of the oblique plane wave incidence on the perfectly conducting wedge that had a toroidal radar absorbing cylinder covering its edge [4]. In the case of edge line segments L_i $(i = \overline{1,3})$, the external angle of the wedge was equal to 2π (under that condition, the edge degenerated into a half-plane).

2.5.2 Near-field computation results

In our computations, we assumed that the antenna was fed by a pyramidal horn. The feed was oriented in such way that the maximum of its DP was pointing towards the reflector's center (point O' with coordinates $\left(\frac{a}{2},\ 0,\ \frac{a^2}{16f}\right)$ in figure 2.68).

We also assumed in our computations that the polarization vector of the feed belonged to plane xOz (vertical polarization of EMW). In most cases, the phase distribution of the fields could be assumed to be uniform. Therefore, the DP of the feed can be computed as follows [62]:

$$F(\theta, \varphi) = \frac{1 + \cos\theta}{2} \frac{\pi}{4} \frac{\sin\psi_E}{\psi_E} \left[\frac{\sin(\psi_H - \pi/2)}{\psi_H - \pi/2} + \frac{\sin(\psi_H + \pi/2)}{\psi_H + \pi/2} \right], \quad (2.98)$$

where $\psi_E = \frac{L_E \pi}{\lambda} \sin\theta \sin\varphi$, $\psi_H = \frac{L_H \pi}{\lambda} \sin\theta \cos\varphi$, L_E, L_H are the sizes of horn aperture in the E- and H-planes, respectively, θ and φ are the polar angles describing the horn's DP in its local coordinate system $fx'y'z'$, and λ is the operating wavelength of the horn feed.

In this case, the horn aperture sizes were equal to $L_E = 1.1\lambda$ and $L_H = 2.9\lambda$.

Figures 2.70–2.73 below present the computation results of near fields for the AS at hand. The bold solid lines in figures 2.70 and 2.72 represent the field amplitude

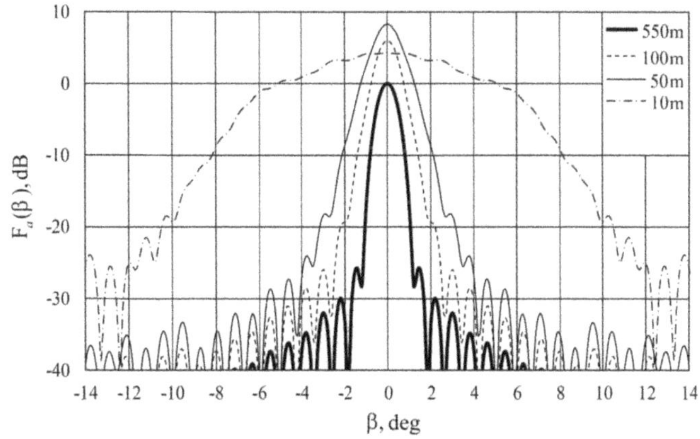

Figure 2.70. Gradual formation of the antenna DP in the azimuthal plane.

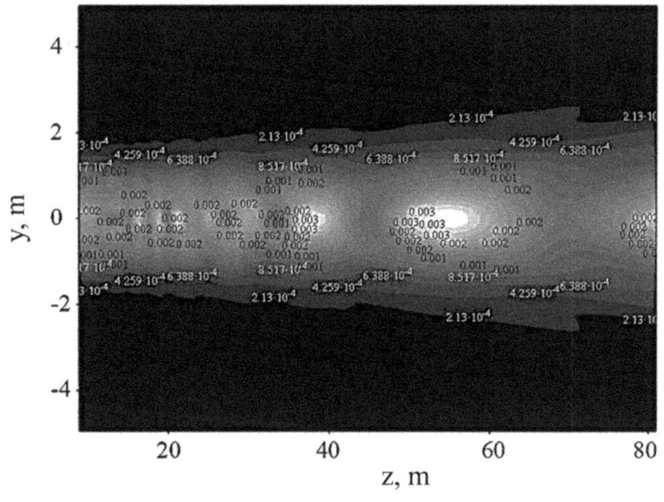

Figure 2.71. Amplitude distribution of the field $|\vec{p}^{\perp}\vec{E}^{s}(\beta)|$ radiated by the antenna in the azimuthal plane.

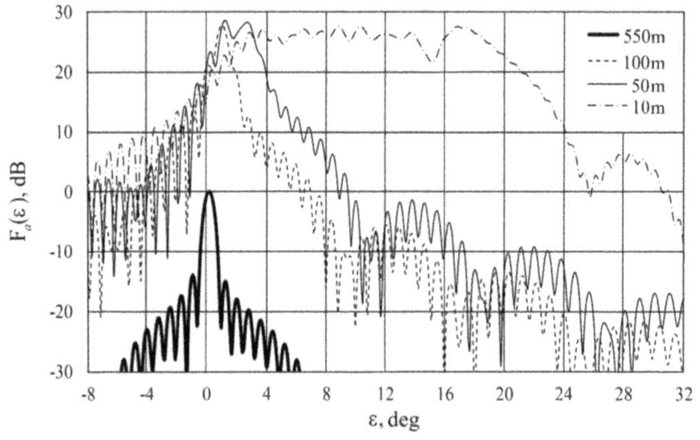

Figure 2.72. Gradual formation of the antenna DP in the elevation plane.

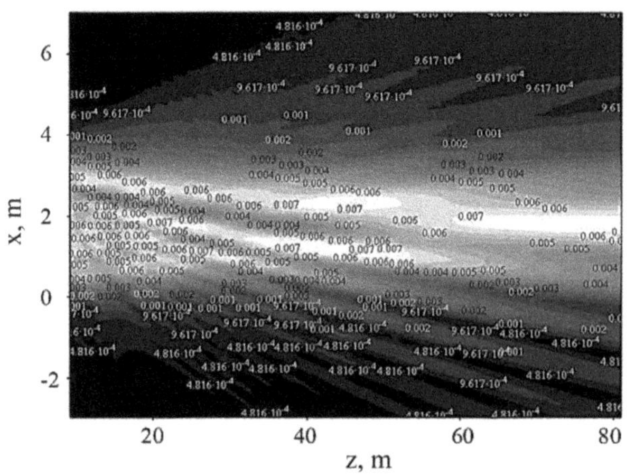

Figure 2.73. Amplitude distribution of field $|\vec{p}^{\perp}\vec{E}^{s}(\varepsilon)|$ radiated by the antenna in the elevation plane.

versus polar angle coordinates for a distance from the center of antenna of $r = 550$ m (which satisfies the far-field zone condition), the dashed line shows the same field amplitude for a distance from the antenna center $r = 100$ m, the thin solid line corresponds to the distance from the antenna $r = 50$ m, and the dash-and-dot line corresponds to the distance $r = 10$ m.

Figure 2.70 shows the computed amplitude of the perpendicular field component $F_a(\beta) = |\vec{p}^{\perp}\vec{E}^{s}(\beta)|/|\vec{p}^{\perp}\vec{E}^{s}_{550}(0°)|$ versus azimuthal direction towards the field point β (β was counted off axis Oz in plane yOz).

The plots in figure 2.70 have been normalized to the amplitude of the perpendicular field component $|\vec{p}^{\perp}\vec{E}^{s}_{550}(0°)|$ radiated by the antenna in the azimuthal direction $\beta = 0°$ as observed at the distance from the antenna reflector's center of 550 m (antenna's far-field zone).

Figure 2.71 shows the computed distribution of field $|\vec{p}^{\perp}\vec{E}^{s}(\beta)|$ radiated by the antenna in the azimuthal plane yOz.

Figure 2.72 shows the amplitude distribution of the perpendicular field component $F_a(\varepsilon) = |\vec{p}^{\perp}\vec{E}^{s}(\varepsilon)|/|\vec{p}^{\perp}\vec{E}^{s}_{550}(0.2°)|$ versus the elevation angle ε (ε is counted off the axis Oz in the plane xOz).

The plots in figure 2.72 are normalized with respect to the value $|\vec{p}^{\perp}\vec{E}^{s}_{550}(0.2°)|$, which is the perpendicular field component radiated by the antenna in the direction $\varepsilon = 0.2°$ at the distance of 550 m away from the antenna.

Figures 2.70 and 2.72 illustrate how the antenna DP gradually forms in the azimuthal and elevation planes as the field point moves away from the antenna.

Figure 2.73 shows the computed amplitude distribution of the field $|\vec{p}^{\perp}\vec{E}^{s}(\varepsilon)|$ radiated by the antenna in the elevation plane xOz.

Analysis of the amplitude distributions of the near fields radiated by the antenna in the azimuthal (figure 2.71) and elevation (figure 2.73) planes provide for finding the zones around the antenna where the EM energy is being concentrated. For example, in the particular case of the above antenna system such zones of increased energy concentration are those with coordinates (2; 0; 25); (1.5; 0; 45), (2.5; 0; 45) and so on. It is worth noting that as far as the azimuthal plane is concerned, all these zones stick to the antenna's axis and the field distribution in the azimuth is symmetrical with respect to axis Oz. The latter can be explained by the symmetry of the feed's DP in the elevation plane xOz.

A more complex amplitude distribution of the field is observed in the elevation plane, which is caused by a number of factors including the following: the reflector being non-symmetric with respect to the axis of the paraboloid it was cut from; and displacement of the feed and its main lobe off the paraboloid's axis.

Analysis of the above plots allows the formulation of recommendations regarding rational (from the viewpoint of reducing the impact on the radiation performance of antenna) placement of radar equipment in the vicinity of particular radiating antenna. The method is also suitable for finding good radar sites as well as for evaluating the levels of electromagnetic radiation around particular radar in order to ensure the safety of personnel operation on that site.

Diagrams presenting the amplitude distributions of fields in the near zone are also helpful when evaluating the EM compatibility of various electronic means that would operate on the same site.

In addition, the computation of the near fields of ASs is required at the early stages of antenna design, in particular if the antenna is to be installed on a mobile platform such as an aircraft, ship, or ground vehicle. The possibility of finding zones of increased energy concentration in the vicinity of the antenna reflector early on in the design process makes it possible to avoid errors related to the wrong placement of the components related to the antenna and its eventual carrier. In the case of ground radar, the presence of certain objects in the radar antenna vicinity may have an adverse effect on the radiation performance of the radar antenna, which needs to be mitigated or accounted for.

2.6 Electromagnetic wave scattering by thin wires placed inside dielectric shells

The majority of unmanned aerial vehicles (UAVs) are just thin dielectric or composite shells streamlined to have good aerodynamics. Inside, these shells contain metallic and non-metallic elements [77]. At the same time the measures aimed at reducing the radar observability of such UAVs are mostly concerned with the fuselage, which includes various payloads, control systems, and the engine. It should be noted that plastic UAV wings that include some wiring can significantly increase the total field scattered by the UAV as a whole [78]. Therefore, it is important to develop a method for evaluating scattering characteristics (RCS in particular) of a thin-wire conductor placed inside a dielectric shell. The purpose of the method for evaluating the UAV RCS is two-fold. It would be required both while designing the algorithms for radar detection and while figuring out the ways for reducing the radar visibility of small UAVs. Usually the problem of EM scattering from electrically thin conductors is solved using the Pocklington equation [79–85]. In addition, sections of straight and bent conductors were considered. However, the works available to us did not contain any information on studies concerned with the EM scattering of thin wires placed inside radar-transparent or dielectric shells.

Thus one would find it helpful to have a method for computing the scattering characteristics developed. Such a method would have to account for EM interactions between thin wires and the shell that houses them (UAV wing with conductors inside). In addition, in this section we also consider the impact of the wire shape and placement inside the wing on the radar visibility of such a wing as a whole.

2.6.1 Scattering of an electromagnetic wave by a thin wire housed inside a thin dielectric shell

Let us consider a plane EMW

$$\vec{E}^{0}(\vec{x}) = \vec{p}^{\,0} e^{jk_0\left(\vec{R}^{0}\cdot\vec{x}\right)},$$

$$\vec{H}^{0}(\vec{x}) = \sqrt{\frac{\varepsilon_0}{\mu_0}}\left(\vec{R}^{0} \times \vec{p}^{\,0}\right) e^{jk_0\left(\vec{R}^{0}\cdot\vec{x}\right)}, \tag{2.99}$$

which is incident from without upon a thin wire (or a group of wires) placed inside a dielectric shell (figure 2.74(a)). Here $k_0 = 2\pi/\lambda_0$ is the wavenumber in free space, λ_0 is the wavelength, $\vec{p}^{\,0}$ and $\vec{R}^{0}$ are unit vectors of polarization and illumination direction, and ε_0 and μ_0 are the permittivity and permeability of the vacuum. The cross-section of the object under consideration is shown in figure 2.74(b).

The time dependency we are going to use here is $\exp(-j\omega t)$ as before. In this case, the imaginary part of the shell's material relative to permittivity is a positive constant [86, 87].

Let $\vec{E}(\vec{x}|\vec{x}_0, \vec{p})$, $\vec{H}(\vec{x}|\vec{x}_0, \vec{p})$ be the field excited by an electrical dipole given that the object consists of a dielectric shell only (region V_2, figure 2.74(b)). Here, $\vec{x}_0$ is the radius-vector of the dipole location point, and $\vec{p}$ is the unit vector of polarization.

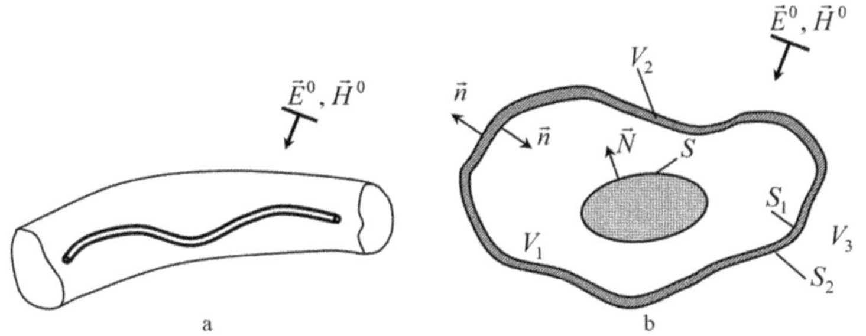

Figure 2.74. Problem layout: thin wire inside dielectric shell.

Application of the Lorentz reciprocity theorem to the sought total field ($\vec{E}$, $\vec{H}$) and auxiliary field ($\vec{E}$, $\vec{H}$) excited by the dipole in various regions of space (in particular, the regions V_1, V_2, and V_3 of figure 2.74(b)) gives us the integral representation for the vector $\vec{E}(\vec{x}_0)$ at any point of space:

$$\vec{p} \cdot \vec{E}(\vec{x}_0) = \vec{p} \cdot \vec{E}(\vec{x}_0) + \frac{1}{j\omega}\int_S \vec{J}(\vec{x}) \cdot \vec{E}(\vec{x}|\vec{x}_0, \vec{p})\ \mathrm{d}s. \tag{2.100}$$

Here $\vec{E}(\vec{x}_0)$ represents the field at point $\vec{x}_0$ that was scattered by a dielectric shell only and $\vec{J}(\vec{x})$ is the surface current density at the wire's surface S that was excited by a plane wave (2.99) propagated to the shell's inside.

The integral representation (2.100) is exact. Putting $\vec{x}_0 = -R\vec{R}^0$ into expression (2.100) and tending $R \to \infty$, we obtain the field excited by an electrical dipole $\vec{E}(\vec{x}|\vec{x}_0, \vec{p})$ transformed into the plane wave thus obtaining an asymptotical expression for the field $\vec{E}\left(-\vec{R}^0\right)$ scattered in the direction $-\vec{R}^0$ [19]:

$$\vec{p} \cdot \vec{E}\left(-\vec{R}^0\right) = \vec{p} \cdot \vec{E}\left(-\vec{R}^0\right) + \frac{1}{j\omega}\int_S \vec{J}(\vec{x}) \cdot \vec{E}\left(\vec{x}\middle|\vec{R}^0, \vec{p}\right)\ \mathrm{d}s. \tag{2.101}$$

Here $\vec{E}\left(-\vec{R}^0\right)$ is the field scattered in a backward direction by the dielectric shell alone, and $\vec{E}\left(\vec{x}|\vec{R}^0, \vec{p}\right)$ is the field at point $\vec{x}$ on the wire's surface S that was generated by plane wave (2.99) that propagated to the shell's inside. Vector $\vec{p}$ in equation (2.101) is the polarization unit vector of the receiving antenna.

The integral part of expression (2.101) corresponds to scattering by wires (from inside the shell) that takes into account the system interactions within the object itself. The dielectric shell is of large electrical size and of small curvature. The field scattered by the shell or passing through one can be calculated using the short-wave asymptotic methods. Also, since the dielectric shells of interest possess good radar transparency, we can ignore the impact of multiple wave bounces from the shell's inside onto the surface current density $\vec{J}(\vec{x})$.

Therefore, to calculate the integral part of expression (2.100) we need to obtain the wire surface current density $\vec{J}(\vec{x})$, which is excited by plane EMW (2.99) passed through the dielectric shell.

Computation of current density $\vec{J}(\vec{x})$ excited by an incoming wave on the wire's surface S

Considering the fact that the wire diameter is significantly smaller than the wavelength of the incident wave as well as its length, one can apply the Pocklington equation [79–85] to compute the surface current density over the wire's surface (assuming that the surface current flows along the wire thus having a longitudinal component only). Let us now introduce parameter ρ (figure 2.75) as follows:

$$\rho = \sqrt{|\vec{\xi} - \vec{x}|^2 + a^2}.$$

Here a is the wire radius, $\vec{x}$ is the observation point, $\vec{\xi}$ is the projection of the integration point $\vec{x}'$ (sitting on the wire's surface S) onto the wire's axis, and $\vec{l}$ and $\vec{l}^0$ are unit vectors tangential to the wire's axis at the corresponding points. Let us use the following parametrization of wire coordinates: $\vec{x} = \vec{r}(t_0)$, $\vec{\xi} = \vec{r}(t)$, and $\vec{r}(t) = (\varphi(t), \psi(t), 0)^T$ $(0 \leqslant t \leqslant T)$.

Here the assumption has been made that the whole wire belongs to a plane. Then, $\vec{l} = (\varphi'(t), \psi'(t), 0)^T$, $\vec{l}^0 = (\varphi'(t_0), \psi'(t_0), 0)^T$. Accounting for the fact that the wire is thin, one can assume that the current density is mostly concentrated along the wire's axis thus becoming a thread of current flowing along $\vec{l}$:

$$\vec{J}(\vec{x}) = \frac{I(\vec{x})}{2\pi a}\vec{l},$$

where $I(\vec{x})$ is a scalar function which depends on the parameter t only. The parameter varies along the wire's axial line $(0 \leqslant t \leqslant T)$.

At the same time, the Pocklington equation for $I(\vec{x})$ can be written in the following form

$$\frac{1}{jk_0}\sqrt{\frac{\mu_0}{\varepsilon_0}}\int_0^T \Phi(\rho)I(t)\sqrt{\varphi'^2(t) + \psi'^2(t)}\ \mathrm{d}t = F(t_0). \tag{2.102}$$

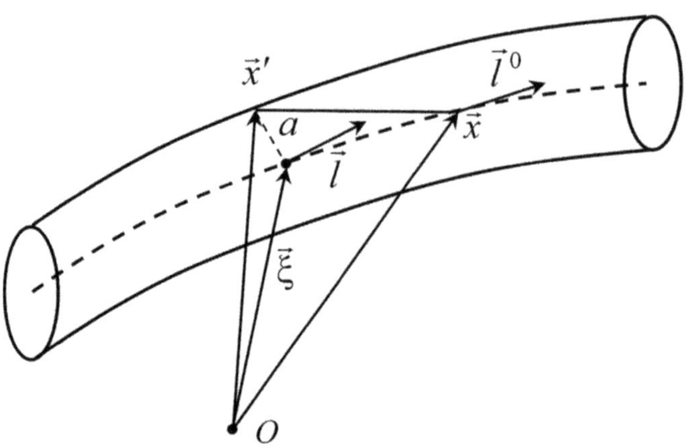

Figure 2.75. The model of a wire.

Here,

$$F(t_0) = -\vec{l}^{\,0} \cdot \vec{E}(\vec{x}(t_0)), \tag{2.103}$$

$\vec{E}(\vec{x})$ is the intensity of the field at point $\vec{x}$ (on the wire) that propagated inside the dielectric shell,

$$\Phi(\rho) = k_0^2 g(\rho)\left(\vec{l}^{\,0} \cdot \vec{l}\right) - \frac{\partial^2 g}{\partial l \partial l^0}, \quad g(\rho) = \frac{1}{4\pi}\frac{e^{jk_0\rho}}{\rho}, \tag{2.104}$$

$$\frac{\partial^2 g}{\partial l \partial l^0} = -g''_{\rho\rho}\left(\frac{\vec{\xi} - \vec{x}}{\rho} \cdot \vec{l}^{\,0}\right)\left(\frac{\vec{\xi} - \vec{x}}{\rho} \cdot \vec{l}\right) + g'_\rho \cdot D,$$

$$D = \vec{\nabla}_x\left(\frac{\vec{\xi} - \vec{x}}{\rho} \cdot \vec{l}\right) \cdot \vec{l}^{\,0}.$$

Using the midpoint rule Riemann sums to calculate the integral in equation (2.102), one can obtain the SLE with respect to values $I_i = I(t_i)$ $(i = 1, \ldots, N)$.

Having found the I_i $(i = 1, \ldots, N)$, one can compute the integral part of expression (2.101).

The calculation of the non-integral part in expression (2.101)

The non-integral part in expression (2.101) corresponds to the field scattered in the backward direction by a dielectric shell only obtained while excluding the impact of the wires inside. To calculate this particular field, we assume that there are no wires inside the shell. Thus we apply the Lorentz reciprocity theorem to the sought field in region V_3 (outside the shell, figure 2.74(b)) and to the field of an auxiliary electrical dipole in free space [19]. Moving the observation point to infinity in the direction $-\vec{R}^0$ (thus switching from equation (2.100) to equation (2.101)), we obtain

$$\vec{p} \cdot \vec{E}\left(-\vec{R}^0\right) = -jk_0\frac{e^{jk_0R}}{4\pi R}\int_{S_2}\left[\left(\vec{H}_1^{\perp}(\vec{x}) \cdot \vec{p}\right)\sqrt{\frac{\mu_0}{\varepsilon_0}} - \vec{E}_1^{\perp}(\vec{x})\left(\vec{p} \times \vec{R}^0\right)\right] \times e^{jk_0\left(\vec{R}^0 \cdot \vec{x}\right)}\, ds. \tag{2.105}$$

Here, $\vec{H}_1^{\perp}(\vec{x}) = \vec{n} \times \vec{H}_1$, $\vec{E}_1^{\perp}(\vec{x}) = \vec{n} \times \vec{E}_1$, $(\vec{E}_1, \vec{H}_1)$ is the field near (at) the shell's surface S_2 (figure 2.74(b)) if there are no wires inside, and R is the distance to the observation point.

In Kirchhoff's approximation, the field $(\vec{E}_1, \vec{H}_1)$ can be replaced by the field $\left(\vec{\overline{E}}_1, \vec{\overline{H}}_1\right)$ at the surface of a plane equidistant dielectric layer (which can be called a dielectric sheet for short), which is being tangent to the surface S_2 at every particular point $\vec{x}$ of the 'illuminated' part of this surface. Further on, we will be using the reflection coefficients $\rho_\perp$, $\rho_\parallel$ and transmission coefficients $\tau_\perp$, $\tau_\parallel$ of such a dielectric sheet at two orthogonal polarizations of the EMW for an arbitrary angle of its incidence. If the vector of electrical field strength is parallel to the plane of incidence (with respect to the plane dielectric sheet replacing the actual curved surface of the shell at a given point), we consider such situation as || polarization. Otherwise, if the vector of electrical field strength is perpendicular to the plane of incidence, we consider such a situation as $\perp$ polarization.

Let us introduce the following decomposition of the polarization unit vector of receiving (transmitting) antenna:

$$\vec{p} = p_\perp \vec{e}_\perp^{\,0} + p_{||}\vec{e}_{||}^{\,0},$$

where $\vec{e}_\perp^{\,0} = \dfrac{\vec{R}^0 \times \vec{n}}{\left|\vec{R}^0 \times \vec{n}\right|}$, $\vec{e}_{||}^{\,0} = \vec{R}^0 \times \vec{e}_\perp^{\,0}$, $p_\perp = (\vec{p} \cdot \vec{e}_\perp^{\,0})$, $p_{||} = \left(\vec{p} \cdot \vec{e}_{||}^{\,0}\right)$, and $\vec{n} = \vec{n}(\vec{x})$ is the unit vector of the normal at point $\vec{x}$ (figure 2.74(b)), the normal being external with respect to region V_2. For the latter decomposition, the field $\left(\vec{\overline{E}}_1, \vec{\overline{H}}_1\right)$ can be presented in the following form:

$$\vec{\overline{E}}_1(\vec{x}) = \left[\rho_\perp(\vec{x})p_\perp(\vec{x})\vec{e}_\perp^{\,1} + \rho_{||}(\vec{x})p_{||}(\vec{x})\vec{e}_{||}^{\,1}\right]e^{jk_0\left(\vec{R}^1\cdot\vec{x}\right)}, \tag{2.106}$$

$$\vec{\overline{H}}_1(\vec{x}) = \sqrt{\frac{\varepsilon_0}{\mu_0}}\left[\rho_\perp(\vec{x})p_\perp(\vec{x})\vec{e}_{||}^{\,1} - \rho_{||}(\vec{x})p_{||}(\vec{x})\vec{e}_\perp^{\,1}\right]e^{jk_0\left(\vec{R}^1\cdot\vec{x}\right)}. \tag{2.107}$$

Here, $\vec{e}_\perp^{\,1} = \dfrac{\vec{R}^1 \times \vec{n}}{\left|\vec{R}^1 \times \vec{n}\right|}$, $\vec{e}_{||}^{\,1} = \vec{R}^1 \times \vec{e}_\perp^{\,1}$, and $\vec{R}^1 = \vec{R}^0 - \vec{n}(\vec{x})\left(\vec{R}^0 \cdot \vec{n}(\vec{x})\right)$ is the unit vector of the propagation direction of the plane wave scattered by a dielectric sheet.

If we substitute the field expressed by (2.106) and (2.107) for the field $(\vec{E}_1, \vec{H}_1)$ in (2.105), we finally obtain the following:

$$\vec{p} \cdot \vec{E}\left(-\vec{R}^0\right) \approx -2jk_0\frac{e^{jk_0R}}{4\pi R}\int_{S_{2\text{ill}}}\left[\rho_\perp p_\perp^2 - \rho_{||}p_{||}^2\right]\left(\vec{R}^0 \cdot \vec{n}\right)e^{2jk_0\left(\vec{R}^0\cdot\vec{x}\right)}\,ds, \tag{2.108}$$

where $S_{2\text{ill}}$ is the 'illuminated' part of surface S_2.

Here, we accounted for the fact that equation

$$\left(\vec{R}^1 \cdot \vec{x}\right)_{S_{2\text{ill}}} = \left(\vec{R}^0 \cdot \vec{x}\right)_{S_{2\text{ill}}}$$

holds exactly at all points $\vec{x}$ of the 'illuminated' part of surface S_2 as long as the dielectric sheet, which is being substituted for the actual surface of the dielectric shell, is tangent to S_2 at every particular point $\vec{x}$.

It is worth noting that the 'illuminated' part of the shell's surface is determined using the ray tracing method [88].

The reflection and transmission coefficients can be computed by the geometrical-optics method or obtained directly by finding solutions to Maxwell's equations with particular boundary conditions [89, 90]. The latter method can be used also in the case when the dielectric shell consists of more than one layer of dielectric.

The field $\vec{E}(\vec{x})$ (the field that propagated to the wire after passing through the shell's surface, which is used in equation (2.103)) can be expressed using the transmission coefficients as follows:

$$\vec{E}(\vec{x}) = \left[\tau_{\perp}p_{\perp}^{0}\vec{e}_{\perp}^{0} + \tau_{\|}p_{\|}^{0}\vec{e}_{\|}^{0}\right]e^{jk_0\left(\vec{R}^0\cdot\vec{x}\right)} \tag{2.109}$$

where $p_{\perp}^{0} = (\vec{p}^{0}\cdot\vec{e}_{\perp}^{0})$, $p_{\|}^{0} = \left(\vec{p}^{0}\cdot\vec{e}_{\|}^{0}\right)$.

The field $\vec{E}\left(\vec{x}|\vec{R}^{0}, \vec{p}\right)$ (that appears in expression (2.101)) can be computed in the same manner.

Thus, having computed the current density $\vec{J}(\vec{x})$ excited at the wire's surface by means of the Pocklington equation (2.102) and having obtained the expression for the field scattered by the dielectric shell if there was no wire inside it (using expression (2.105)), we find ourselves ready to compute the field scattered by the wire (or wires) placed inside the dielectric shell. That latter field can be used, in particular, for computing the radar cross-section of thin-wire conductors housed inside the dielectric shell.

2.6.2 Verification of the method and the results of computation

In order to cross check our method for computing the EMW scattering by thin wires inside the dielectric shells detailed above, we used the EM simulation software tool called FEKO [91]. Figure 2.76 shows the results of computing the RCS of a straight perfectly conducting wire that was 0.5 m long and had a radius of 0.178 mm. There was no dielectric shell surrounding it. The wavelength of the incident EMW was 0.03 m. The wire's azimuth aspect angles of 0° and 180° corresponded to the wire being probed along the line connecting its ends. The wire's azimuth aspect angle of 90° corresponded to the wire being probed across the line connecting its ends. The solid black line in figure 2.76 shows the RCS of a straight wire calculated using our

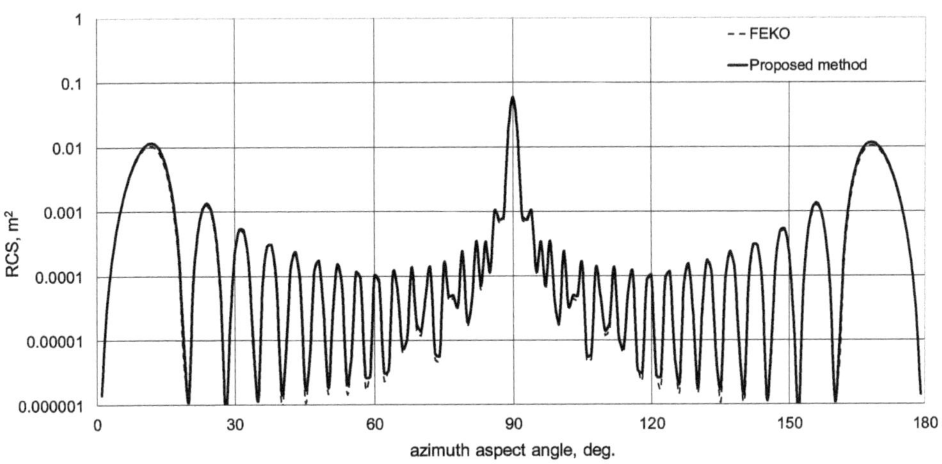

Figure 2.76. The RCS of a straight 0.5 m long wire versus the azimuth aspect of probing.

approach, whereas the dashed line shows the RCS of the same wire computed using the FEKO software pack using the method of moments (MoM).

One can see that there is a good match between the two curves. The latter means that the method detailed above yields results which are close enough to those obtained on the basis of the exact computation using MoM that directly solves the IE set. The greatest difference between the peaks of the two plots is less than 9%.

In the same manner, we also checked the accuracy of the proposed method against FEKO for the case where the wire length was comparable to the wavelength of the incident wave (0.03 m). Figure 2.77 shows the computed RCS of a straight perfectly conducting wire that was 0.06 m long (which constituted only two wavelengths of the incident wave), the radius of its cross-section being 0.066 mm. The results matched very closely, the difference between the peaks of the plots obtained by the two approaches turned out to be less than 3 percent. It is worth noting that the results of both methods (proposed here and FEKO) demonstrated higher disagreement with the results obtained by P Ufimtsev, in his classic book [51], for cylinders of the same proportions as our wires. The shape of the RCS curves in figure 2.77 are mostly in agreement with the curves presented by P Ufimtsev, but the amplitudes of the peaks are different.

To study the impact of the way the wire is laid through the dielectric wing of a small UAV, we needed to describe the various shapes of wires inside the dielectric shell. As a first approximation, we assumed that the wires would be shaped as plane segments of the harmonic function. Then the various shapes could be characterized by just a few parameters. For example, the wave hump elevation h (the amplitude of the harmonic wave), the distance between the electrical connection points a, the actual wire length, and the relative wire elongation (see figures in table 2.3), which is a dimensionless parameter equal to the ratio of the actual wire length to the distance between its two ends a. While studying the impact of the wire's shape on its RCS, we

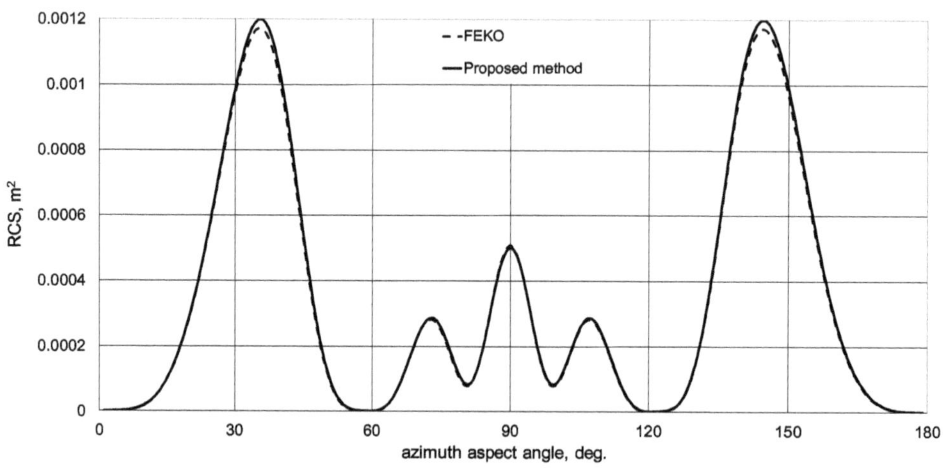

Figure 2.77. The RCS of a straight 0.06 m long wire versus the azimuth aspect of probing.

Table 2.3. Simplified shapes of wires inside a dielectric wing.

Wire shape	Distance a, m	Wave hump elevation h, m	Wire length, m	Wire elongation
Straight line a	0.5	0	0.5	1
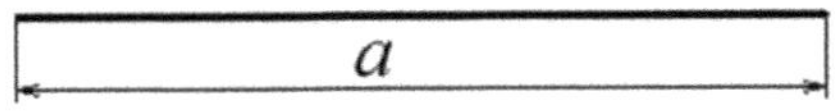				
One half-period of the harmonic function sin(x)	0.5	0.050	0.510	1.020
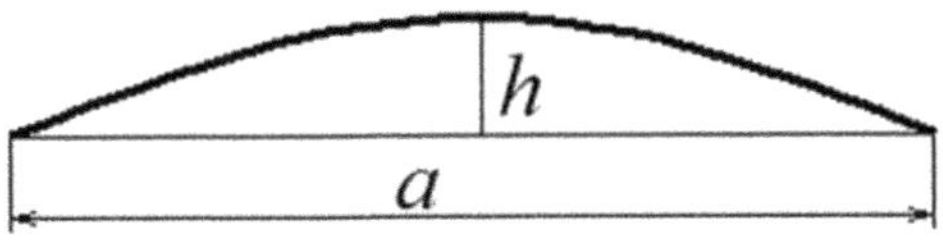	0.5	0.100	0.550	1.090
	0.5	0.150	0.590	1.190
	0.5	0.200	0.660	1.320
Two half-periods of the harmonic function sin(x)	0.5	0.010	0.502	1.004
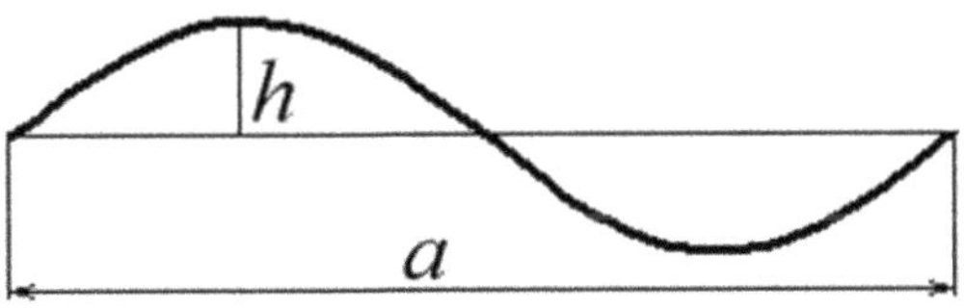	0.5	0.025	0.510	1.020
	0.5	0.050	0.550	1.090
	0.5	0.075	0.590	1.190
	0.5	0.100	0.660	1.320
	0.5	0.150	0.810	1.620
	0.5	0.200	0.970	1.950
Three half-periods of the harmonic function sin(x)	0.5	0.010	0.504	1.009
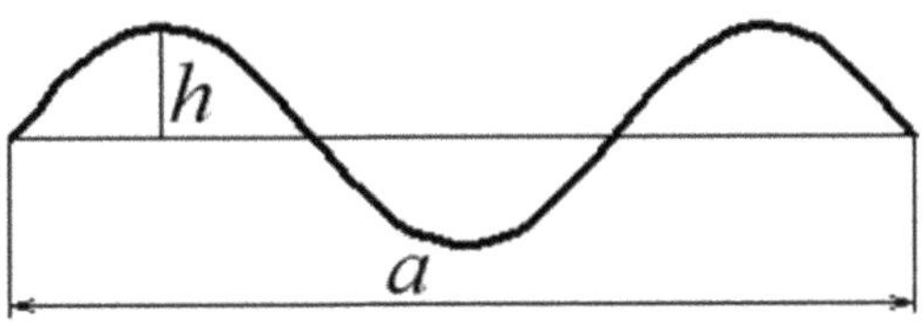	0.5	0.025	0.530	1.050
	0.5	0.050	0.590	1.190
	0.5	0.075	0.690	1.390
	0.5	0.100	0.810	1.620
	0.5	0.150	1.060	2.130
	0.5	0.200	1.330	2.670

assumed that the wires were hanging in free space, meaning that there was no dielectric shell (i.e. wing) around it.

The impact of the interference between two wires on their RCS was studied if the dielectric shell (wing) was absent too. All calculations were carried out under the assumption that the wires were routed in such manner that all their various shapes belonged to a plane. At the end of the study, we put all the pieces together and present RCS estimates for the whole wing that included both the dielectric shell and the wires inside. All results were obtained for the wavelength of probing wave $\lambda_0 = 0.03$ m.

Let us introduce a method for describing various options of wire routing inside the dielectric shell. Again, we need some simplifying assumptions to be made; otherwise there is no way to account for all possible variants of wire routing. Let us assume that a wire is to connect two points that are 0.5 m away from each other. Our objective is to compute the wire's RCS versus the illumination aspect angle for various options of wire routing. Various variants of wire routing represented by simplified shapes and the corresponding parameters they are described with are shown in table 2.3.

In calculations, we considered wires with a round cross-section of area equal to 0.1 mm^2. Such an area translates to the wire cross-section radius of 0.178 mm. A wire of the same radius was used before when we checked our computations against FEKO.

To describe the wire's attitude with respect to the radar probing direction, we introduced the reference plane, which is the plane to which the wire belongs. Thus, from now on, the wire's attitude could be represented by two angles: the elevation aspect and the azimuth aspect. In our computations, we chose three fixed values of the elevation aspect: 0° (which meant that radar probing direction vector belonged to the reference plane), 10°, and 20°, and we varied the azimuth aspect from 0° to 180°.

The azimuth aspects of 0° and 180° corresponded to probing along the line connecting the two wire's ends. The azimuth aspect of 90° corresponded to wire probing across that same line.

Figure 2.78 presents the RCS of the wire versus its aspect azimuth for a few fixed values of the hump elevation h for the wire shaped as a 'half-period of the harmonic wave' (see table 2.3). The elevation aspect was equal to 0°, meaning that the probing direction belonged to the reference plane. For the sake of convenience, we plotted the RCS of the straight wire using a thin black line. The dashed line corresponds to the curved wire with a wave hump elevation of $h = 5$ cm, whereas the bold gray line corresponds to a curved wire with a hump elevation of $h = 10$ cm.

The same RCS plots for the wire shaped as 'two half-periods of the harmonic wave' are presented in figure 2.79.

Analysis of the data in figures 2.78 and 2.79 reveals a significant drop in the maximum RCS if the wire was bent. The maximum RCS of a bent wire can be ten to fifteen times lower compared to that of the straight wire. In addition, the angle spread of the azimuth aspects, over which the RCS of the bent wire attains its peak values, becomes wider. This is particularly true for the wire shaped as 'two

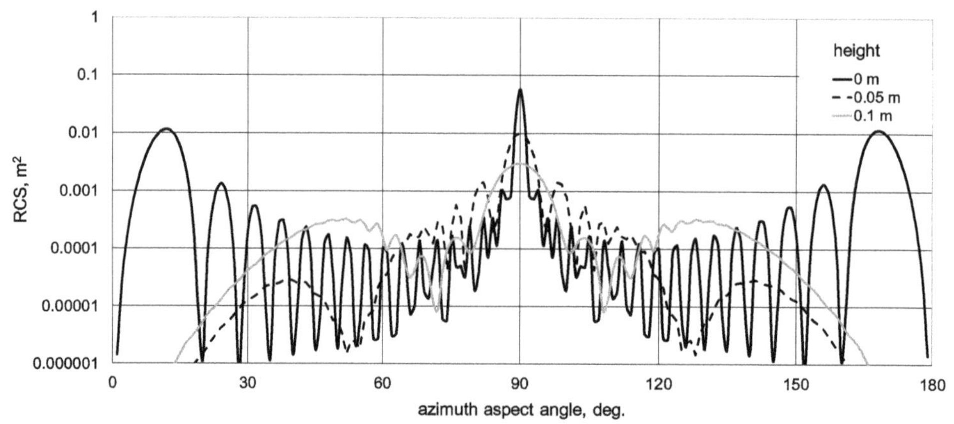

Figure 2.78. The RCS versus azimuth aspect for a wire shaped as 'one half-period of the harmonic function'.

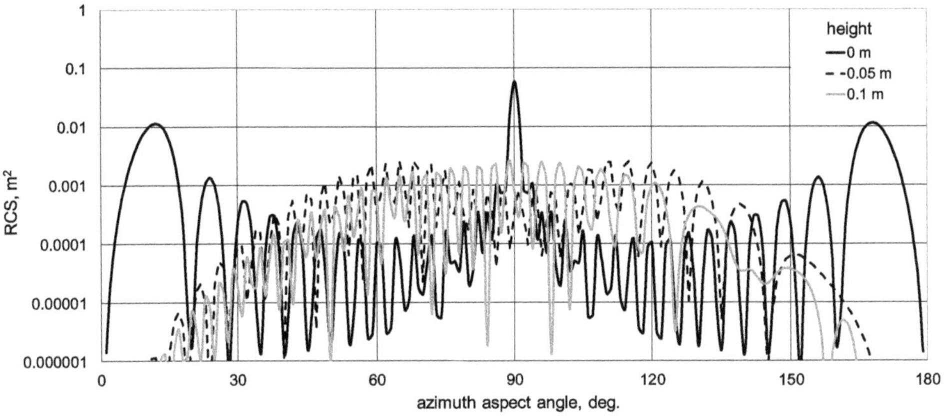

Figure 2.79. The RCS versus azimuth aspect for the wire shaped as 'two half-periods of the harmonic function'.

half-periods of the harmonic function'. Moreover, the RCS plot of that wire with two bends is not symmetrical with respect to the azimuth aspect of 90°. All these factors create difficulties while taking steps towards reducing the RCS of dielectric objects containing wires. For the wire with more bends (the wire shaped as three half-periods of the harmonic wave), the RCS behavior versus azimuth aspect becomes even more complicated. Thus, in order to avoid unnecessary complication of the analysis, from now on we will consider the scattering characteristics of only wires with a single bend, in particular those shaped as one half-period of the harmonic function.

Changing the elevation angle attitude to 20° with respect to the reference plane does not change the RCS of the wire significantly.

The scattering characteristics of a system consisting of two wires were also considered. Again, to make the study feasible, we assumed the two wires were straight and parallel at first and then they are bent simultaneously to acquire the shape of 'one half-period of the harmonic function'. In the latter case, there is an additional 'degree of freedom', namely the distance between the two wires.

Figure 2.80 shows the RCSs versus azimuth aspect of two wires, each having a cross-section area equal to 0.1 mm^2, and the distance between the wires being 0.0075 m. The direction of radar probing was held within the reference plane (the zero elevation aspect). The thin black curve in figure 2.80 corresponds to the RCS of two straight wires. The dashed line corresponds to the bent wires with a wave hump elevation value of $h = 10$ cm, and the solid line corresponds to the bent wires with a wave hump elevation of $h = 20$ cm.

The RCS versus azimuth aspect of two wires for a distance between wires of 0.015 m is shown in figure 2.81.

The analysis of the results in figures 2.80 and 2.81 shows that the maximum RCS value decreases by a factor ranging from five to ten in the case of bent wires compared to straight wires. This drop in magnitude is lower than that for a single wire. At the same time, the azimuth range featuring the higher RCS magnitudes is

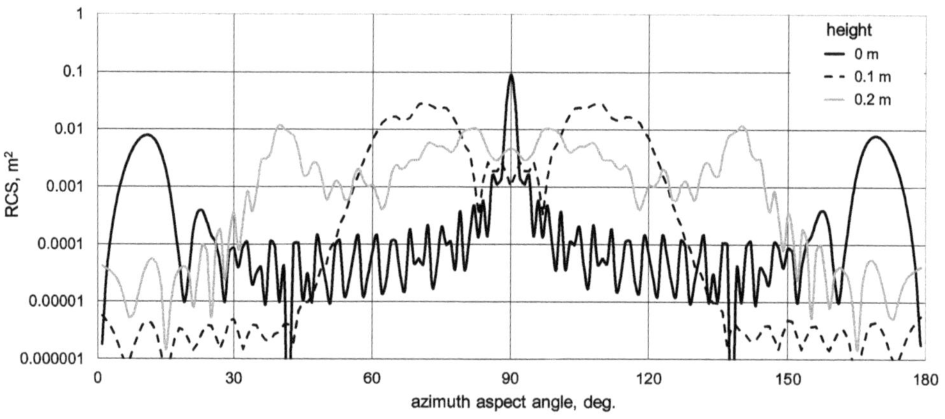

Figure 2.80. The RCS versus azimuth aspect of two wires shaped as 'one half-period of the harmonic function' for a distance between wires of 0.0075 m.

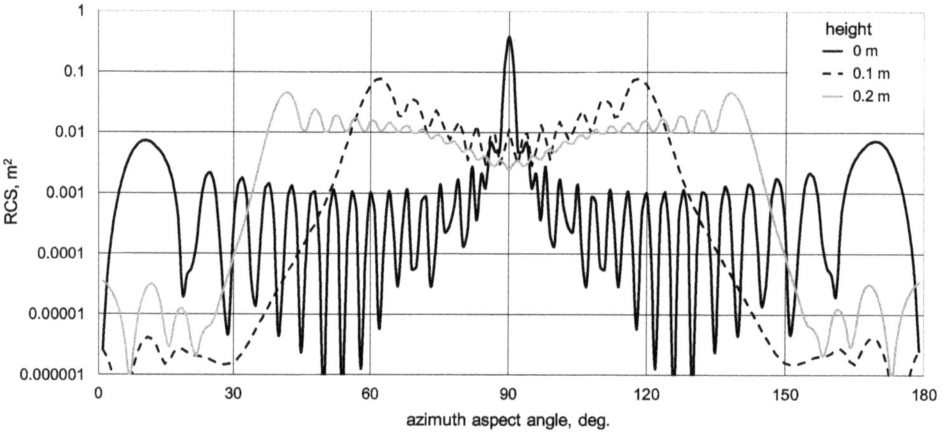

Figure 2.81. The RCS versus azimuth aspect of two wires shaped as 'one half-period of the harmonic function' for a distance between wires of 0.015 m.

wider in the case of curved wires than that in the case of straight wires. In this case, the following should be noted. If the UAV designer's objective is to keep the RCS of the wires inside a small plastic UAV below a certain level given the entire range of its aspects, then the designer needs to stick to bent wires. On the other hand, if a UAV is to include another dominant scatterer and the magnitude of its RCS versus azimuth aspect cannot be kept below the desired level, then the designer can route straight wires in such a way that the wire RCS would go up at the same azimuth aspects as the RCS of the dominant scatterer.

Now, we conclude our study by bringing together all the individual pieces we have considered so far, as we are about to obtain the RCS of a more complicated object. Let us place the wires inside a plastic dielectric shell, say the wing of a UAV.

A model for the UAV wing is shown in figure 2.82. The length of the wing was 0.94 m, its width was 0.25 m, and its thickness was 0.03 m. We assumed that there was a necessity to run a wire along the wing's trailing edge to connect a load at a distance of 0.5 m from the wing's root. Figure 2.80 shows an example of such a wing model that includes a straight wire (white line) running along the wing's trailing edge.

The simplified wing model consisted of skin only (no ribs or spars inside), which was a single-layer dielectric 0.873 mm thick, its relative permittivity being $\varepsilon = 3.2 + j0.1$, and its relative permeability being $\mu = 1$.

Figure 2.83 shows the RCS versus azimuth aspect for the model of a wing if there were no wires inside it. Here and below, the direction of wing probing was held within the wing's plane. The azimuth aspect of 90° corresponded to radar probing along the wing's leading edge.

The wing model's RCS reaches its maximum of 0.012 m^2 for an azimuth aspect of 0° (radar probing right across the wing's leading edge). For the rest of the azimuth aspects the wing model demonstrated an RCS which would not go higher than 0.0001 m^2.

Figure 2.84 shows the RCS of the wing model containing the wire inside (solid line). Here and below we assumed that the wire's cross-section area was 0.1 mm^2. The wire ran the half span of the wing's trailing edge (see figure 2.80). As a reference, in the same plot, we presented the RCS of that same wire alone (thin line).

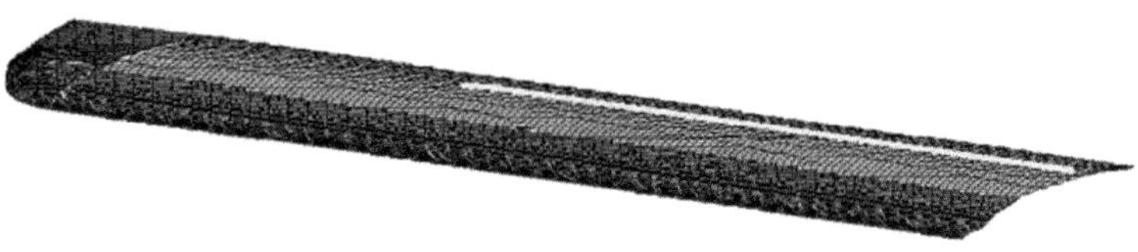

Figure 2.82. The model of a wing including a straight wire running along the wing's trailing edge.

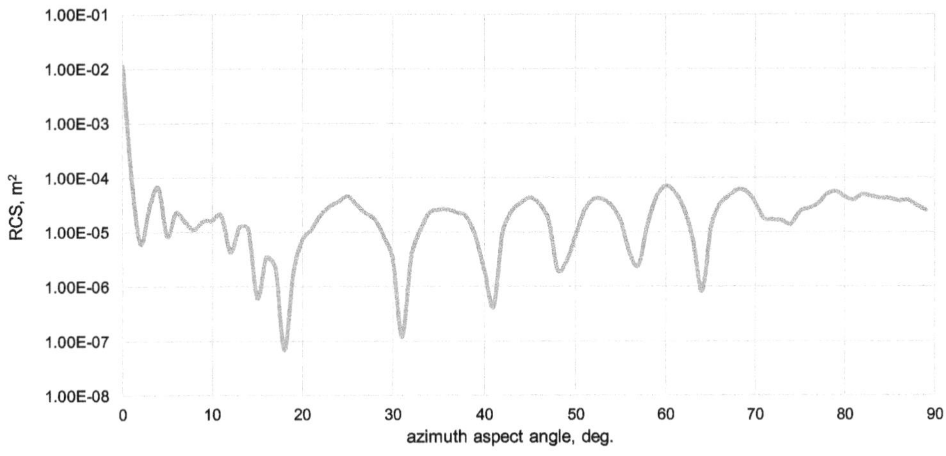

Figure 2.83. The RCS versus azimuth aspect for the wing model containing no wires inside.

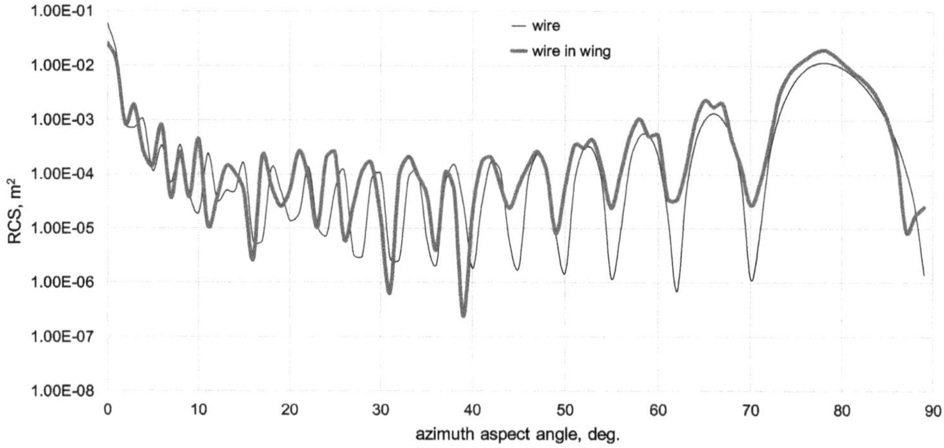

Figure 2.84. The RCS versus azimuth aspect of the wing model containing a straight wire running along the wing's trailing edge.

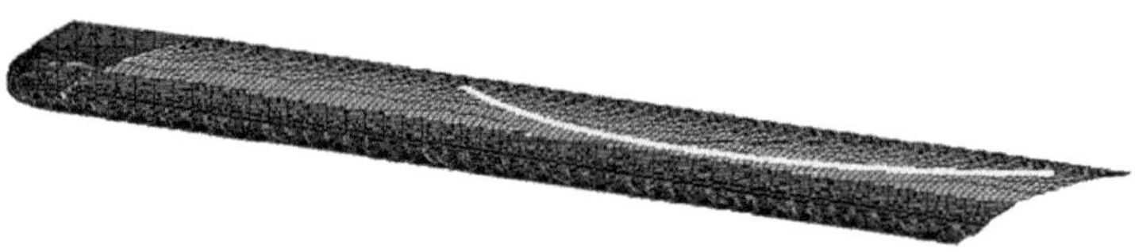

Figure 2.85. The model of a wing containing a bent wire running along the wing's trailing edge.

In this case, the RCS of the wire inside the wing differs a bit from the RCS of a standalone wire (with no shell around it). With the exception of the zero azimuth vicinity, the RCS of the wire placed inside the wing was not significantly different from the RCS of a standalone wire. At zero aspects, the leading edge of the wing's shell had a significant impact on the RCS of the entire object. However, given this particular frequency range, this contribution was less than the RCS of the wire itself.

Figure 2.85 presents a wing containing a bent wire. The wire was shaped as 'one half-period of the harmonic function', it was 0.59 m long, and it had a wave hump elevation of $h = 0.1$ m. The ends of the wire were placed at distance of 0.03 m from the wing's trailing edge, the linear distance between the ends of the wire being 0.5 m (the same distance as in the previous case of a straight wire).

Figure 2.86 shows the RCS versus azimuth aspect of the bent wire for the two cases. In the first case the wire was placed inside the wing, and in the second case the wire was a standalone one.

As can be seen in figure 2.86, the RCS of the wire dominates the RCS of the whole object for the majority of aspect angles. However, there are two exceptions. The first is the neighborhood of the zero azimuth aspect angle. The RCS of the wire at those aspects goes below that of the shell. That is because the wire was not straight. Thus, in that particular case, the object's RCS was dominated by that of the wing shell's leading edge. The second exception is the sector of azimuth aspect angles beyond 65°.

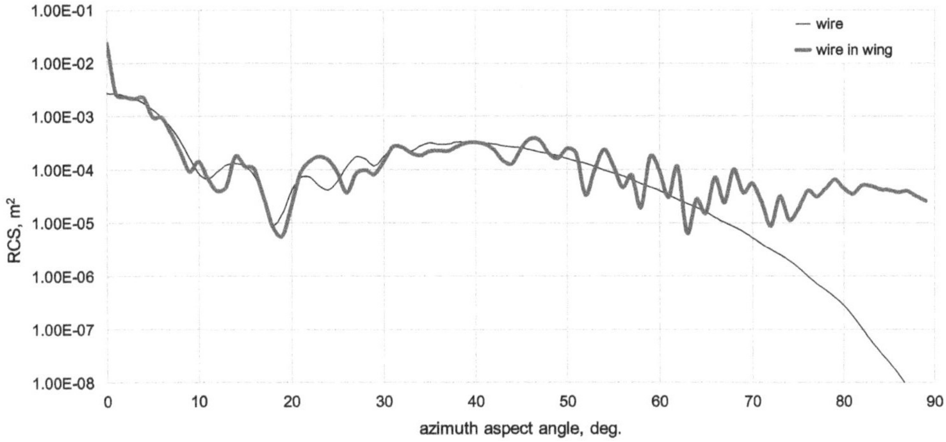

Figure 2.86. The RCS versus azimuth aspect of the wing model containing a bent wire running along the wing's trailing edge, as well as the RCS of the same wire standing alone.

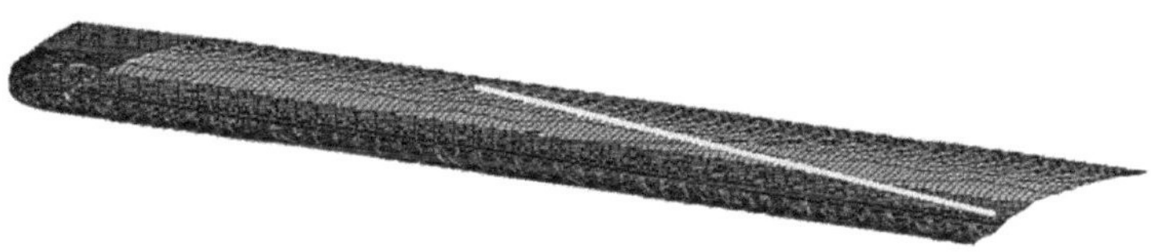

Figure 2.87. The model of a wing containing a straight wire inside running obliquely from the leading to the trailing wing's edge.

At those aspect azimuths, the wire's RCS tends to zero yet the dielectric shell with a rounded wingtip retains the RCS at a level of 0.00001–0.0001 m^2 (see figure 2.83). Our computation data demonstrate that one can reduce the RCS of a wing including the bent wire running along the wing's trailing edge either to the level of the wing's shell with no wire inside or to a value of about 0.0003 m^2 (for azimuth aspects of 30°–50°). However, the latter RCS is a quite a low level in any case.

Figure 2.87 shows the model of a wing with a straight wire inside. The wire was 0.52 m long. The wire connected two points: one end of the wire (the one further away from the wing's root) was located at a distance of 0.03 m from the wing's trailing edge; the other end of the wire (the one closer to the wing's root) was located at a distance of 0.18 m from the wing's trailing edge. The distance between the wire and the wing's root was 0.5 m (the same distance as in the two previous cases featuring straight and bent wires running along the wing's trailing edge).

Figure 2.88 shows the RCS of the wing model for a wire running inside it as shown in figure 2.87.

The plots in figure 2.88 show that the latter design featured two major RCS peaks. The first one was around the zero azimuth aspect. It corresponded to scattering from the leading edge of the dielectric wing shell. The second was around the azimuth aspect of 17°, it corresponded to the peak scattering from the wire inside the shell. That particular azimuth aspect happened to be exactly perpendicular to the wire.

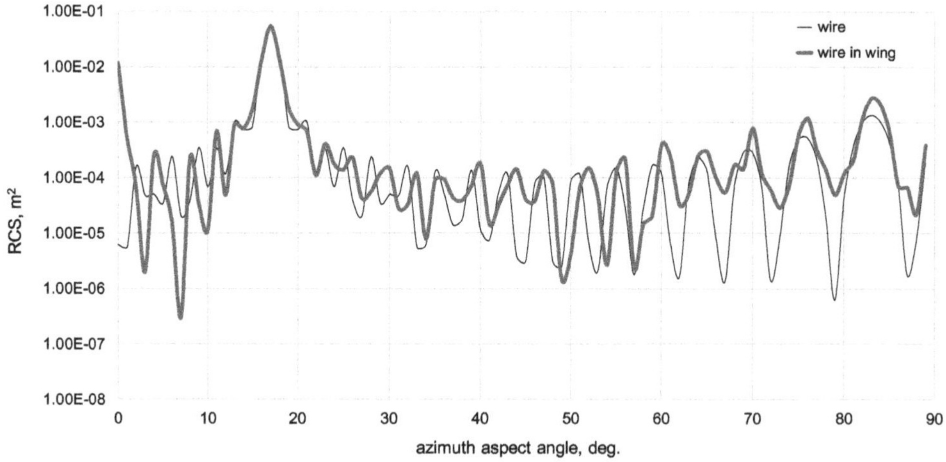

Figure 2.88. The RCS versus azimuth aspect of the wing model containing a straight wire inside running obliquely from the leading to the trailing edge.

Figure 2.89. The model of a wing containing the bent wire running from the leading to the trailing wing's edge.

A situation like that (two major peaks at the RCS plot) is not acceptable to a UAV designer. These peaks need to be reduced as much as possible. The latter can be done by avoiding straight wires inside the wing.

Figure 2.89 shows a wing featuring a bent wire inside. The wire was shaped as 'one half-period of the harmonic function' and was 0.60 m long, its wave hump elevation being $h = 0.1$ m. The wire was routed in such a way that it started around the wing's center at a distance of 0.03 m from the wing's trailing edge, and it ended in the vicinity of the wing's root at a distance of 0.18 m from the wing's trailing edge. The distance between the wire and the wing's root was 0.5 m (the same distance as in all the previous cases). The distance between the ends of the wire was equal to 0.52 m.

In figure 2.90 the wing RCSs are represented for this wire. The latter variant of the wire routing inside the wing led to an increased wing RCS in the sector of azimuth aspects from 15° through 22°, which was due to scattering from the bent wire itself (the shell contributing negligibly to the total RCS figure).

Our brief analysis reveals that wires inside the plastic wing of a UAV contribute significantly to the RCS of the whole wing. The shape and routing of the wire inside the wing's shell also matter. This is particularly true for UAVs designed as load bearing dielectric shells. Thus if the designer's objective is to bring the UAV's RCS to as low a level as possible, then special attention has to be paid to the shape and

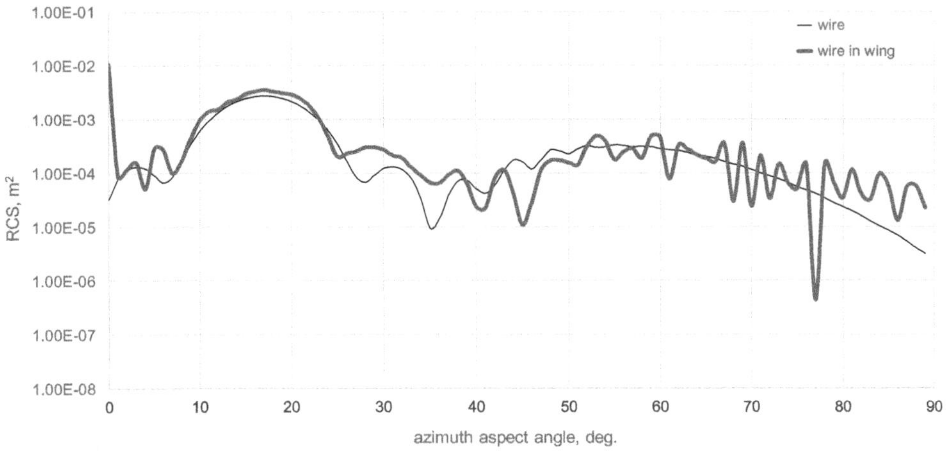

Figure 2.90. The RCS versus azimuth aspect of the wing model containing a bent wire running from its leading edge to its trailing edge.

routing of the wire. In that latter case, the designer should avoid using straight wires or placing the wires too close to the wing's leading edge (the best way to achieve that is to run the wires as shown, for instance, in figure 2.85). On the one hand, the latter design of wire routing brings the RCS of the object of interest to a minimum for zero azimuth aspects. On the other hand, such a design does not increase the RCS of the whole significantly given the other aspects of radar probing.

This subsection details the original method for computing the scattering characteristics of a dielectric shell with wires inside. A practical example of such a shell is, for example, the wing of a small plastic UAV.

A brief analysis of the effects of the wire shape and location on the RCS of the whole wing has also been carried out. Possible interference between the object's elements was also studied. The RCS computations were based on integral representations of the scattered field obtained using the Lorentz reciprocity theorem and the Pocklington equation for thin-wire antennas. Some remarks were provided regarding an optimal (from the object's RCS minimization perspective) shape and placement of the wire inside the UAV's wing.

2.7 Backscattering of an inflatable dielectric lifting–turning device designed for ground measurement of radar object scattering characteristics

In order to reduce the radar visibility of aerial objects one needs to know the parameters of these objects' scattering. In particular, one of these parameters is the object's RCS. The values of various objects' RCS can be obtained by means of radar measurements at testing grounds. One can use both the objects themselves or their scaled models for that purpose. Despite rapid development of computational electrodynamics [19, 51, 90, 92–96], which can provide information regarding the

RCS and which can even substitute for actual radar measurements, engineers still need the RCS to be measured at testing grounds, particularly for air vehicles of complex shapes. That is where computational electrodynamics cannot replace the physical measurements altogether. In addition, actual RCS measurements are necessary to verify and upgrade contemporary methods of electrodynamic computations.

The complete set of equipment required for measuring the RCS of an object includes a fixture for positioning and holding the object at a certain height and attitude with respect to the incident EMF. The scattering characteristics of this fixture, which is known as a lifting–turning device, need to be kept as low and stable as possible. Otherwise, the impact of backscattering from the device itself would alter the results of the RCS measurements for the objects under test in an unpredictable way.

Currently, there are two different approaches to physical RCS measurements that can be singled out from the whole range of known options [92, 94, 97, 98]:

- testing range measurements of the RCS of real objects (in particular, measurement systems by the Microwave Vision Group (MVG) [99], and by the Howland Company [100]), and
- anechoic chamber measurement of the RCS of scale models [99–102].

Obviously, testing range measurements are more straightforward. The RCS of real objects are characterized by a wide dynamic range. For this reason, the fixtures used to position and hold real objects in precision measurements need to meet very stringent requirements as far as their RCS are concerned [94, 97]. In addition, ground measurements of the airborne object RCS must be arranged in such a manner that the far-field zone condition would be held for the object under test as if the object was in the air. The latter means that one needs to reduce all secondary effects such as radar echoes from the underlying surface and from the lifting–turning device (LTD) itself. Usually, the object would be held at some height above the ground by means of special pillars shaped in a certain way to reduce their backscattering [94, 97, 99, 100], the object can also be suspended by cords [94, 97]. Radar absorbing coatings are also applied to reduce backscattering effect of the LTD.

Inflatable balloons or thin dielectric shells (TDS) can be offered as the LTD for measuring the RCS of objects that are relatively heavy (for example, missiles or small aircraft). The basic advantages of TDS are considered to be their light weight, simplicity of object handling, and mobility. In this section, we consider a method for finding a preliminary estimate of the scattering characteristics for such an LTD. To that end, we have created a method for evaluating the RCS of the inflatable lifting–turning device, which is an air-inflated electrically thin dielectric shell shaped as a torus, which is electrically large. The method developed here is based on a PO approach [19, 51, 90, 94].

Moreover, it turned out that for this particular LTD we had to generalize the concept of RCS a bit. That was because of the fact that the integral effect of the radar reflectivity of the whole LTD was not represented well by its regular RCS concept, which is applicable to point objects in the far-field zone. However, the test range measurements are carried out at much smaller distances so that one can no longer neglect the influence of the radar antenna DP on the radar reflectivity of the LTD. The latter means that even though the far-field zone condition is held for the entire measurement set-up, the impact of the DP of radar antenna, which is in relative proximity to the LTD, causes the situation where different LTD parts are irradiated by plane waves of different amplitudes, thus making the radar reflectivity of the entire fixture dependent on the amplitude of the incident field. Thus, instead of regular RCS we had to employ the concept of the so-called generalized RCS [103] in order to include this particular yet practical case.

We present the results of RCS computations for the aforementioned LTD for the centimeter wavelength range. Even though the method presented below provides only a coarse RCS estimate, yet it is very helpful since the prompt generation of the result allows one to take measures towards reducing the backscattering of such LTDs.

2.7.1 Method for computing the radar cross section of a thin dielectric shell of large electric sizes

Let us consider an object which is a thin dielectric shell with a wall thickness δ (figure 2.91). The object is located in free space, both its relative permittivity ε and relative permeability μ being equal to unity. The same medium fills the inside of the shell. The relative permittivity of the shell's wall is $\varepsilon_s = \varepsilon'_s + j\varepsilon''_s$ and its relative permeability is $\mu_s = 1$. The characteristic geometric dimensions of the shell are much greater than the probing signal wavelength. Moreover, all the elements that the shell consists of have small curvatures (the whole object is smooth).

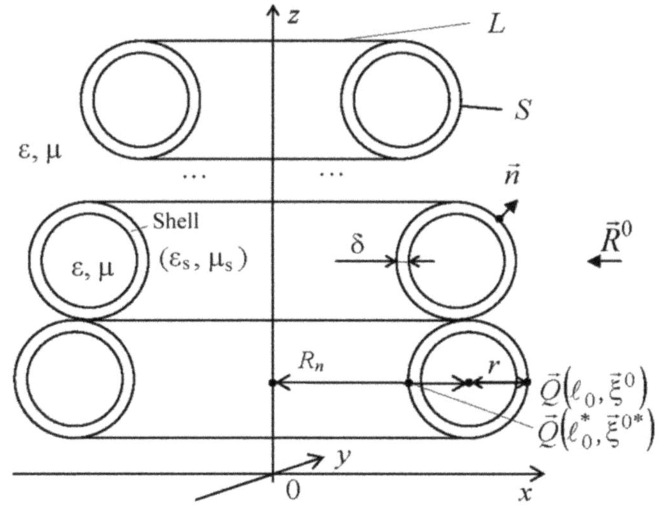

Figure 2.91. Section of the inflatable TDS model.

We use an integral representation of the Stretton–Chu type [3, 94, 104] for an EMF scattered by an arbitrary dielectric object, which is electrically large, in the far-field zone given a monostatic radar layout:

$$\begin{aligned}\vec{p}\cdot\vec{E}^{s}\left(\vec{R}^{0}\right) &\approx -jk_0\frac{\exp\left(jk_0D\right)}{D}\\ &\times\int_S\left[\left(\vec{J}^{e}\cdot\vec{p}\right)\sqrt{\frac{\mu_0}{\varepsilon_0}}-\vec{J}^{m}\left(\vec{p}\times\vec{R}^{0}\right)\right]\exp\left(jk_0\left(\vec{R}^{0}\cdot\vec{Q}\right)\right)\,\mathrm{d}s_Q,\end{aligned}\tag{2.110}$$

where $\vec{p}$ is the polarization unit vector of the receiving antenna, $\vec{E}^{s}$ is the vector of the electric field scattered by the object, $\vec{R}^{0}$ is the unit vector of the incident wave propagation direction (the radar probing direction), $k_0 = 2\pi/\lambda_0$ is the wavenumber in free space, λ_0 is the wavelength of the illumination (primary) wave, D is the distance from the object's center to a point of reception (the field point), $\vec{J}^{e} = (\vec{n}\times\vec{H}')$ and $\vec{J}^{m} = (\vec{n}\times\vec{E}')$ are the densities of the equivalent electric and magnetic currents on the object's surface S, $\vec{E}'$ and $\vec{H}'$ are the electric and magnetic field vectors on the object's surface S, respectively, $\vec{n}$ is the unit vector of the external normal to surface S, ε_0 and μ_0 are the permittivity and permeability of free space (vacuum), respectively, and $\vec{Q}$ is a point on the object's surface S (the source point).

Let us further assume that the main contributor to the echo backscattered in the direction $-\vec{R}^{0}$ is the 'illuminated' part of the outer shell. The incident wave that actually gets inside the shell has to bounce back from the inner shell's wall and has to propagate through the outer wall once again. Such a backscattered wave loses its power significantly. The latter means that the backscattering impact of such a wave backscattered from the shell's inside can be considered as a second order effect compared to the first-order one, which is backscattering from the outer shell's wall. The second order backscattering effects can be neglected when computing a coarse RCS estimate of the LTD in question.

So, we can use expression (2.110) to obtain the field at the receiver's input in the far-field zone given Kirchhoff's approach [51]:

$$\begin{aligned}\vec{p}\cdot\vec{E}^{s}\left(\vec{R}^{0}\right) &\approx -jk_0\frac{\exp\left(jk_0D\right)}{D}\int_{S_{ill}}\left[\vec{p}\cdot\left(\vec{n}\times\vec{H}'\left(\vec{Q}\right)\right)\right.\\ &\left.+\vec{E}'(\vec{x})\left(\vec{n}\times\left(\vec{p}\times\vec{R}^{0}\right)\right)\right]\exp\left(jk_0\left(\vec{R}^{0}\cdot\vec{Q}\right)\right)\,\mathrm{d}s_Q.\end{aligned}\tag{2.111}$$

At every point of the surface S, polarization vector $\vec{p}$ can be decomposed into two orthogonal components $p_{\perp}$ and $p_{||}$:

$$\vec{p} = p_{\perp}\vec{e}_{\perp}^{\,0} + p_{||}(\vec{R}^{0}\times\vec{e}_{\perp}^{\,0}).$$

The field $\vec{E}'$, $\vec{H}'$ on surface S (or in its vicinity) for Kirchhoff's approximation can be represented as follows:

$$\vec{E}'(\vec{Q}) = \left[\rho_\perp(\vec{Q})p_\perp(\vec{Q})\vec{e}_\perp^1 + \rho_{||}(\vec{Q})p_{||}(\vec{Q})\vec{e}_{||}^1\right]\exp\left(jk_0\left(\vec{R}^1 \cdot \vec{Q}\right)\right), \tag{2.112}$$

$$\vec{H}'(\vec{x}) = \frac{1}{j\omega\mu_0}\vec{\nabla} \times \vec{E}'\left(\vec{Q}\right), \tag{2.113}$$

where $\rho_\perp$, $\rho_{||}$ are reflection coefficients for the waves at two orthogonal polarizations introduced above,

$$\vec{R}^1 = \vec{R}^0 - 2\vec{n}\left(\vec{R}^0 \cdot \vec{n}\right),$$

$$\vec{e}_\perp^m = \frac{\vec{R}^m \times \vec{n}}{|\vec{R}^m \times \vec{n}|}, \ (m = 0, 1),$$

$$\vec{e}_{||}^m = \left(\vec{R}^m \times \vec{e}_\perp^m\right).$$

Let us then make another assumption, which is commonly used for Kirchhoff's approximation [51]. Taking into account the fact that the object is electrically large and its curvatures are small, the field excited at (source) point $\vec{Q}$ on the surface S will be deemed as the field excited on a planar equidistant layer of a dielectric (dielectric sheet) of thickness δ, which is osculant to the surface S at point $\vec{Q}$. The latter means that the point $\vec{Q}$ now also belongs to the osculating sheet of the dielectric. Thus, now the values of the reflection coefficients $\rho_\perp$, $\rho_{||}$ become constant and they can be calculated as follows:

$$\rho_{\perp,\,||} = \frac{j(c^2 - 1)\sin(k_s\delta)}{2c\cos(k_s\delta) - j(c^2 + 1)\sin(k_s\delta)},$$

where $c = \dfrac{\sqrt{\varepsilon_s - \sin^2(\theta)}}{\beta\cos(\theta)}$, θ is the incident angle,

$$\beta = \begin{cases} 1, & \text{for } \rho_\perp, \\ \varepsilon_s, & \text{for } \rho_{||}, \end{cases}$$

$$\cos(\theta) = \left|\vec{R}^0 \cdot \vec{n}\right|;$$

$$\sin^2(\theta) = 1 - \left(\vec{R}^0 \cdot \vec{n}\right)^2,$$

$$k_s = k_0\sqrt{\varepsilon_s - \sin^2(\theta)}.$$

Substituting expressions (2.112) and (2.113) in (2.111) and performing simple transformations we arrive at

$$\begin{aligned}\vec{p} \cdot \vec{E}^s\left(\vec{R}^0\right) &\approx -jk_0\frac{\exp(jk_0D)}{4\pi D}\int_{S_{\text{ill}}}\left[\rho_\perp\left(\vec{Q}\right)p_\perp^2\left(\vec{Q}\right) - \rho_{||}\left(\vec{Q}\right)p_{||}^2\left(\vec{Q}\right)\right]\left(\vec{R}^0 \cdot \vec{n}\right) \\ &\times \exp\left(2jk_0\left(\vec{R}^0 \cdot \vec{Q}\right)\right)\,ds_Q = -2jk_0\frac{\exp(jk_0D)}{4\pi D}I\left(\vec{R}^0\right).\end{aligned} \tag{2.114}$$

While obtaining equation (2.114), we took into account the fact that $\left(\vec{R}^1 \cdot \vec{Q}\right)\Big|_S = \left(\vec{R}^0 \cdot \vec{Q}\right)\Big|_S$, which was made possible by replacing the field at every point of the actual shell surface by the field on a corresponding osculating dielectric sheet, and we used the relation $\vec{R}^1 = \vec{R}^0 - 2\vec{n}\left(\vec{R}^0 \cdot \vec{n}\right)$.

Now, we introduce coordinate system $Oxyz$ so that the Oz axis is directed along the object's construction line. The basic element of the LTD is the torus that is 'set on' the Oz axis (figure 2.91). Let us consider EMW scattering from such a torus. Let us denote the major radius of the nth torus as R_n, ($n = \overline{1,\ 7}$), whereas r will stand for the minor radius, which is the radius of the torus's ring cross-section. Let us write the double integral $I\left(\vec{R}^0\right)$ in the form of an iterated one:

$$I\left(\vec{R}^0\right) = \int_L \exp\left(2jk_0\left(\vec{R}^0 \cdot \vec{Q}(\ell)\right)\right)M(\ell)\ \mathrm{d}\ell, \tag{2.115}$$

where

$$M(\ell) = \int_{S_0} \exp\left(2jk_0\left(\vec{R}^0 \cdot \vec{Q}(\vec{\xi})\right)\right)\left[\rho_\perp\left(\vec{Q}(\vec{\xi})\right)p_\perp^2\left(\vec{Q}(\vec{\xi})\right)\right.$$

$$\left.-\rho_{||}(\vec{Q}(\vec{\xi}))p_{||}^2(\vec{Q}(\vec{\xi}))\right](\vec{R}^0 \cdot \vec{n})\ \mathrm{d}s_\xi, \tag{2.116}$$

$\mathrm{d}s_\xi = r_0\ \mathrm{d}\varphi$ is an arc element, S_0 is the 'illuminated' part of the vertical torus's ring cross-section, which is perpendicular to contour L (center line of the torus),

and r_0 and φ are polar coordinates of a point within the cross-section S_0.

Taking into account that $R_n,\ r \gg \lambda_0$, and applying the stationary phase method [19] consequentially to integrals (2.115), (2.116), we obtain the following:

$$I\left(\vec{R}^0\right) \approx \exp\left(2jk_0\left(\vec{R}^0 \cdot \vec{Q}(\ell_0)\right) + j\frac{\pi}{4}\right)M(\ell_0)\sqrt{\frac{\pi}{k_0\chi(\ell_0)|\vec{R}^0 \cdot \vec{\nu}(\ell_0)|}}, \tag{2.117}$$

where ℓ_0 is the point on the contour L that is closer to the field point (point of reception), i.e. $\vec{R}^0 \cdot \vec{Q}(\ell_0) = -R_n$, $\chi(\ell_0) = \dfrac{1}{R_n}$ is the curvature of the contour L at point ℓ_0, $\vec{\nu}(\ell_0)$ is the unit vector of principal normal at point ℓ_0, i.e. $\left|\vec{R}^0 \cdot \vec{\nu}(\ell_0)\right| = 1$,

$$M(\ell_0) \approx \exp\left(2jk_0\left(\vec{R}^0 \cdot \vec{\xi}^0\right) + j\frac{\pi}{4}\right)\rho\left(\vec{R}^0 \cdot \vec{n}^0\right)\sqrt{\frac{\pi}{k_0\chi_1(\vec{\xi}^0)|\vec{R}^0 \cdot \vec{\nu}(\vec{\xi}^0)|}},$$

$\vec{\xi}^0$ is the radius-vector of a point within the cross-section S_0, which includes the point ℓ_0 of the contour L, this radius-vector being defined in the local coordinate system originating at point ℓ_0 that belongs to contour L,

$\chi_1\left(\vec{\xi}^0\right) = \dfrac{1}{r}$ is the curvature of cross-section S_0, and

$$\left|\vec{R}^0 \cdot \vec{\nu}\left(\vec{\xi}^0\right)\right| = 1.$$

Thus $\left(\vec{R}^0 \cdot \vec{\xi}^0\right) = -r$ and $\left(\vec{R}^0 \cdot \vec{n}^0\right) = -1$. Also, we take into account the fact that at the point $\left(\vec{\xi}^0, \ell_0\right)$ the osculating dielectric sheet (which substitutes for the actual toroid shell's wall at this particular point) is orthogonal to $\vec{R}^0$ (i.e. $\theta = 0$) [95] to arrive at

$$\rho = \rho_{\perp}(0) = -\rho_{\parallel}(0) = \frac{j(\varepsilon_s - 1)\sin(k_s\delta)}{2\sqrt{\varepsilon_s}\cos(k_s\delta) - j(\varepsilon_s + 1)\sin(k_s\delta)}. \tag{2.118}$$

Combining expressions (2.115)–(2.118), we finally obtain

$$I(\vec{R}^0) \approx -\frac{j\pi\rho}{k_0}\sqrt{R_n r}\exp(-2jk_0(R_n + r)), \tag{2.119}$$

and, hence,

$$\vec{p} \cdot \vec{E}^s\left(\vec{R}^0\right) \approx -\frac{\exp(jk_0 D)}{2D}\rho\sqrt{R_n r}\exp(-2jk_0(R_n + r)). \tag{2.120}$$

The torus RCS is

$$\sigma_n = \lim_{D\to\infty} 4\pi D^2 \left|\vec{p} \cdot \vec{E}^s\right|^2 \approx 4\pi D^2\frac{1}{4D^2}\rho^2 R_n r \approx \pi\rho^2 R_n r. \tag{2.121}$$

Given the LTD, which consists of N tori differing by their major radii, we obtain the following expression:

$$\vec{p} \cdot \vec{E}^s\left(\vec{R}^0\right) \approx -\frac{\exp(jk_0 D)}{2D}\rho\sum_{n=1}^{N}\sqrt{R_n r}\exp(-2jk_0(R_n + r)), \tag{2.122}$$

and the resulted RCS is

$$\sigma_{\text{LTD}} \approx \pi\left|\rho\sum_{n=1}^{N}\sqrt{R_n r}\exp(-2jk_0(R_n + r))\right|^2. \tag{2.123}$$

In expression (2.123), the value $R_n + r$ in the exponent is the distance from the axis Oz to such a point on the surface of the nth torus, which has the least distance to the receiver. Thus, the RCS of the LTD can be computed as a sum of the EMFs scattered by individual tori given that the initial phase values of such fields were taken into account.

Expression (2.123) was obtained in the assumption that the EMF propagated inside every torus did not contribute to the whole object's RCS. However, this formula can be refined easily by taking into account the backscattering contributions from the inner sides of each torus constituting the LTD. In this case, the main contribution will again be provided by the stationary phase point belonging to the inner surface of the torus $\left(\ell_0^*, \vec{\xi}^{0*}\right)$, which is diametrically opposite to a 'specular point' lying on the 'illuminated' part of the outer torus surface. Thus, at that particular point, the following conditions will hold: $\vec{R}^0 \cdot \vec{Q}(\ell_0^*) = -R_n$, and $\vec{R}^0 \cdot \vec{\xi}^{0*}=r$. It is necessary to take into account the fact that the stationary phase point lying on the inner surface of the torus is of hyperbolic type. That particular fact accounts for the appearance of a $-j$ multiplier in the corresponding term [8]. Here we can take into account the fact that the field, which actually propagated as far as the inner surface of the torus, suffers additional attenuation since it needs to pass through the shell's wall again on its way back. Thus, while designing an additional term for the field backscattered from the inner shell's wall according to formula (2.120), we need to make the following substitutions: $\rho \to \tau^2\rho$, and $\exp(-2jk_0(R_n + r)) \to \exp(-2jk_0(R_n - r))$. Now, the formulas (2.122) and (2.123) that would take into account the curvature of the inner surface in the neighborhood of the specular point $\left(\ell_0^*, \vec{\xi}^{0*}\right)$ [8] take the following form:

$$\vec{p} \cdot \vec{E}^s\left(\vec{R}^0\right) \approx -\frac{\exp(jk_0 D)}{2D}\rho \sum_{n=1}^{N} \exp(-2jk_0 R_n) \times \sqrt{R_n r}\left[\exp(-2jk_0 r) - j\tau^2 \exp(2jk_0 r)\right], \tag{2.124}$$

$$\sigma_{\mathrm{LTD}} \approx \pi |\rho \sum_{n=1}^{N} \exp(-2jk_0 R_n)\sqrt{R_n r} \times \left[\exp(-2jk_0 r) - j\tau^2 \exp(2jk_0 r)\right]|^2. \tag{2.125}$$

In formulas (2.124) and (2.125), the complex transmission coefficient τ has the form [19]

$$\tau = \exp(-jk_0\delta)\left[\cos(k_s\delta) + \frac{j}{\sqrt{\varepsilon_s}}\sin(k_s\delta) + \rho\left(\cos(k_s\delta) - \frac{j}{\sqrt{\varepsilon_s}}\sin(k_s\delta)\right)\right],$$

where ρ is defined by formula (2.118).

Expression (2.125) represents a refined formula for the RCS of the LTD that takes into account the following waves:

- the one scattered from the 'illuminated' part of the shell's surface;
- the one propagated inside the shell and was scattered from the 'back' wall.

Below, we present the results of RCS computations for the LTD in question.

2.7.2 Computation of the radar cross section for the inflatable lifting–turning device

Using the method developed above, we have computed the values of generalized RCS of an inflatable LTD. The generalized RCS values are shown in figure 2.92. The LTD consisted of a main load bearing unit that included seven tori (LTD1) and a smaller device for an incremental change of elevation angle for the object under research (LTD2). LTD2 had the same inflatable design (the same as the main load

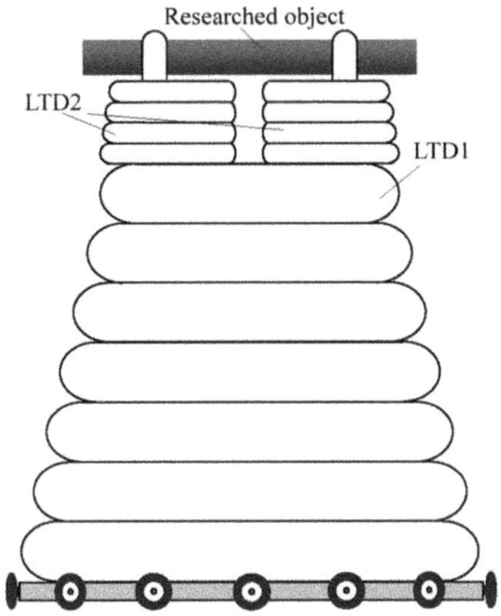

Figure 2.92. LTD set-up for the RCS measurement of an object.

bearing LTD1), however, this one consisted of toroidal sectors. The elevation angle of the object could be changed by inflating (deflating) a certain number of sections of the LTD2.

In our calculations we used the following set-up: both LTD1 and LTD2 were put on a rotating wheeled platform, but to reduce the impact of that platform onto the RCS measurements we assumed that the null of the instrumentation antenna DP in the vertical plane was directed towards the wheeled platform. In addition, the antenna's boresight axis was lifted to the height of $z = h_{max} = 19$ m and was held perpendicular to the construction line of the LTD. Thus, such arrangements provided for an almost linear variation of the incident wave amplitude along the vertical axis all the way up to the object under test. The instrumentation antenna's DP width measured by field amplitude at the level of '0.707' was $\Delta\varepsilon_{0.707} \approx 0.89°$. The bottom side of the LTD was used as a coordinate reference, i.e. $z = 0$.

LTD1 had the following features:

- its height was $h_1 = 8.2337$ m;
- the smaller diameter of all the tori was identical and it was equal to $d_1 = 1.176$ m (the minor radius being $r = r_1 = 0.588$ m, figure 2.91).
- the major radii of the tori varied according to the expression (figure 2.91)

$$R_n = R_{\text{LTD1}\ n} = R_{\text{LTD1}\ 1} + \Delta_1(n - 1), \tag{2.126}$$

  where $n = \overline{1, N}$ is the serial number of a torus (the upper one was numbered as $n = 1$), total number of tori in the LTD1 was $N = 7$ and $\Delta_1 = 0.26$ m;
- the greater diameter of the top-most torus was

$$D_{\min\ \text{LTD1}} = 2(R_{\text{LTD1}\ 1} + r_1) = 5.82 \text{ m};$$

- the greater diameter of the bottom-most one was

$$D_{\max\ \mathrm{LTD1}} = 2(R_{\mathrm{LTD1}\ 7} + r_1) = 8.9375\ \mathrm{m}.$$

LTD2 had the following features:

- its height was $h = 1.668$ m;
- the smaller diameter of torus sector (figure 2.92) $d_2 = 0.417$ m (its minor radius being $r = r_2 = 0.208$ m);
- the major radii of torus sectors varied according to the following expression

$$R_n = R_{\mathrm{LTD2}\ n} = R_{\mathrm{LTD2}\ 1} + \Delta_2(n - 1), \tag{2.127}$$

where $n = \overline{1, 4}$, $\Delta_2 = 0.087$ m, so,

$$R_{\min\ \mathrm{LTD}\ 2} = (R_{\mathrm{LTD2}\ 1} + r_2) = 2.65\ \mathrm{m},$$

$$R_{\max\ \mathrm{LTD}\ 2} = (R_{\mathrm{LTD2}\ 4} + r_2) = 2.91\ \mathrm{m}.$$

Now we need to specify the electrical parameters of which the material the LTD's shells were made. To showcase our computation method, we used an example LTD that included individual shells made of a TE62 tarpaulin of thickness $\delta = 5 \times 10^{-4}$ m. The measured value of real part of the shell's permittivity was $\varepsilon'_s = 2.5$, and its dielectric dissipation factor was $\varepsilon''_s/\varepsilon'_s = 0.05$.

We assumed that a non-uniform plane wave was incident onto the LTD. The amplitude distribution of the incident field is shown in figure 2.93.

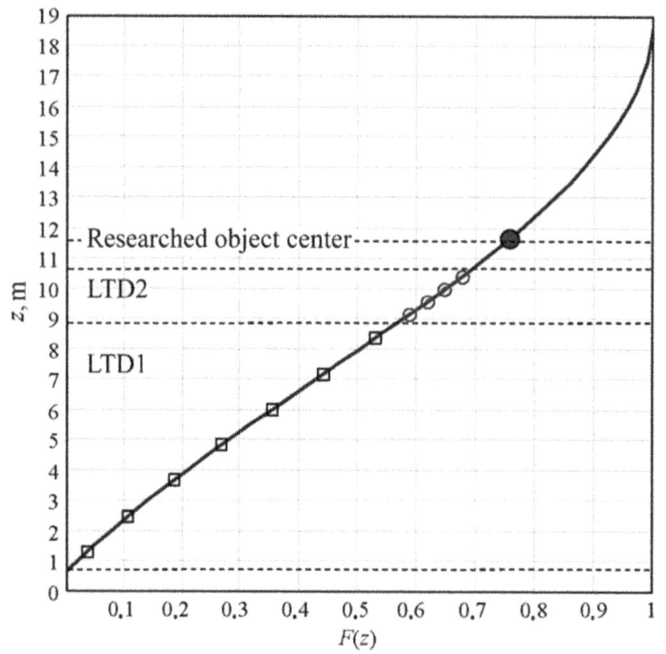

Figure 2.93. Amplitude distribution of the incident field versus height.

The function

$$F(z) \equiv F(\varepsilon_{rp}) = \frac{\sin\left(\frac{\pi L_\varepsilon}{\lambda_0}\sin(\alpha)\right)}{\frac{\pi L_\varepsilon}{\lambda_0}\sin(\alpha)}$$

represents the amplitude distribution of the field that was incident on the object.

Here,

$$\alpha = \text{arctg}\left(\frac{z}{D}\right),$$

z is the current height (see figure 2.91), D is the distance from the geometrical center of the antenna to the construction line of the LTD1 (for an illumination signal wavelength of $\lambda_0 = 0.03$ m, this distance was set to $D = 1100$ m; and for a wavelength of $\lambda_0 = 0.06$ m, this distance was set to $D = 750$ m), and $L_\varepsilon = 18.3$ m is the height corresponding to the instrumentation antenna boresight axis height minus the height of the rotating wheeled platform, which was 0.7 m high.

The dependence of the incident field amplitude can be approximated by a set of step functions, each step being equal to the diameter of the torus section (its smaller diameter). In this case, we needed to use the concept of a generalized RCS for a non-uniform incident plane wave. The generalized RCS concept becomes useful when one starts comparing the RCSs of the LTD and of the object under test, since we needed to take into account the varying amplitude of the incident field. An expression for the generalized RCS of the LTD has the following form:

$$\sigma_{\text{LTD1(2)}} \approx \pi \left|\rho \sum_{n=1}^{N} \exp(-2jk_0R_n)F_n\sqrt{R_n r}\,[\exp(-2jk_0 r) - j\tau^2 \exp(2jk_0 r)]\right|^2, \quad (2.128)$$

where F_n is the set of step functions representing the amplitude of the incident field, which was assumed to be constant within the height occupied by one torus and changed in a step-like manner for other tori. The difference in square brackets under the sum sign in expression (2.128) shows that the RCS of both LTDs, namely σ_{LTD1} and σ_{LTD2}, fluctuate if the two exponents vary. This is the case, in particular, when one changes the radius of the torus section or the carrier frequency of the radar probing signal. Below we present the results of our computations regarding the generalized RCS of the LTD1 given that the minor radius of the torus was $r = 0.588 \pm 0.01$ m. We suggest here that the maximum value of the generalized RCS for the LTD1 would be used as the estimate for σ_{LTD1} if the radius of the torus section varied within the range specified in advance.

Now the impact of changing the carrier frequency of the radar probing signal on the LTD1's radar reflectivity manifested itself as varying values of its generalized RCS for differing wavelengths of the radar probing signal. Our estimate was as follows:

$$\sigma_{\text{LTD1}} = 0.010 \text{ m}^2 \text{ for } \lambda_0 = 0.03 \text{ m};$$

$$\sigma_{\text{LTD1}} = 2.489 \times 10^{-3} \text{ m}^2 \text{ for } \lambda_0 = 0.06 \text{ m}.$$

In the case of LTD2, we also estimated the impact of changing the minor radius of the torus section on the LTD2's radar reflectivity. In particular, we varied the minor radius within the range of $r = 0.208 \pm 0.01$ m. As a result, we obtained the following figures for the generalized RCS of LTD2:

$$\sigma_{\text{LTD2}} = 4.\ 69 \times 10^{-3}\ \text{m}^2 \text{ for } \lambda_0 = \ 0.03 \text{ m};$$

$$\sigma_{\text{LTD2}} = 0.034\ \text{m}^2 \text{ for } \lambda_0 = 0.06 \text{ m}.$$

The value σ_{LTD2} may exceed that of σ_{LTD1} due to the oscillatory nature of the generalized RCS and due to the fact that the upper shell was irradiated by a wave of greater amplitude (figure 2.93).

While computing the EMF backscattered from the entire system of inflatable modules (LTD1 and LTD2 taken as a whole), we tried to find the maximum RCS of the whole system. This maximum was considered a ceiling RCS value estimate, which is usually one of the key parameters that characterizes the RCS of the fixture used to hold the object. To find the maximum, we varied the EMF scattered by LTD1 by changing its minor torus radius within the range of $r = 0.588 \pm 0.01$ m. We also maximized the field scattered by elements of LTD2 by setting its minor torus section radius to $r_2 = 0.212$ m, which provided for the strongest backscattering from the LTD2 ($\sigma_{\text{LTD2}} = 4.69 \times 10^{-3}$ m^2 for $\lambda_0 = 0.03$ m; and $\sigma_{\text{LTD2}} = 0.034$ m^2 for $\lambda_0 = 0.06$ m). In that last case, the generalized RCS of the entire system (LTD1+LTD2) was:

$$\sigma_{\text{LTD}} = 0.025\ \text{m}^2 \text{ for } \lambda_0 = 0.03 \text{ m};$$

$$\sigma_{\text{LTD}} = 0.049\ \text{m}^2 \text{ for } \lambda_0 = 0.06 \text{ m}.$$

In order to avoid the appearance of constructive interference of the EMFs scattered from the individual elements of LTDs, the sizes of the individual tori (their curvature radii) must differ. Our research resulted in the following value: a decreased level of backscattering from the LTDs considered above, one should pick the value of Δ_1 that enters expression (2.126) from the range from 0.25 m through 0.29 m, while the value of Δ_2 that enters expression (2.127) should be ranging from 0.085 m through 0.09 m.

2.7.3 Possible measures for reducing parasitic scattering from lifting–turning device

It would be a good idea to use this kind of inflatable LTD for measuring the radar reflectivity of objects, the RCS of which would greatly exceed that of the LTD. The provisional estimates presented above show that the generalized RCS of the inflatable dielectric LTD of the kind that was considered above can be as high as $\sigma_{\text{LTD}} = 0.025$ m^2 for a radar signal wavelength $\lambda_0 = 0.03$ m (or even $\sigma_{\text{LTD}} = 0.049$ m^2 for a radar signal wavelength $\lambda_0 = 0.06$ m). Thus, in real test range measurements the impact of the signals backscattered from the LTD can alter the RCS of the object under test. The latter means that radar echoes from the LTD should be regarded as parasitic radar returns. They need to be suppressed as much as possible. The suppression of parasitic radar returns could begin as early as the LTD design stage.

Based on provisional estimates for this particular LTD's radar reflectivity, one can employ the following technical solutions to reduce its parasitic radar returns:

1. The upper LTD, which would be used for incrementally changing the object's elevation angle attitude (LTD2) towards the instrumentation radar, can be screened off using flexible radar absorbers. The same radar absorbing sheets could be used to screen off the wheeled platforms the lower LTD rests on (figure 2.92) from the instrumentation radar set-up.

2. The instrumentation radar can be provided with a spatial signal processing kit capable of canceling parasitic radar returns from the bottom part of the LTD in question by placing the receiving antenna DP notch towards the wheeled platform. The latter could be achieved by employing an additional antenna, the maximum of its directivity pattern being directed towards the bottom part of the LTD where the wheeled platform is. Then, the parasitic waves scattered from the platform could be received by that antenna, which output would include mostly parasitic returns. Finally, the parasitic returns could be subtracted coherently from the signal received by the main antenna thus reducing their impact on the RCS measurement of the object under test.

3. One could also measure the parasitic scattering from the LTD when the object under test was absent. The result of that particular measurement could be recorded for future reference. Later it could be subtracted from the measurements involving the object under test thus correcting the resulting RCS.

4. Our computation of LTD backscattering shows that even a small change in the carrier frequency of the radar probing signal causes a significant change in the radar scattering by the LTD alone. Thus, before starting actual RCS measurements for the object under test, one could find the frequency at which the LTD returns minimal echo while the object under test is not there yet. Then one could account for those radar returns while carrying out the measurements for the object under test.

The methods listed above would allow one to reduce the LTD backscattering values by 16–20 dB.

A lifting–turning device or LTD is an element of the instrumentation set-up used for measuring the scattering characteristics of real-life objects at testing grounds. One of the possible designs for such an LTD involve the use of inflatable balloons that could be used to lift and hold quite large objects and put them at various attitudes with respect to instrumentation radar. There is a need to have a quick method for predicting the backscattering figures of such an LTD. In this section, a method was developed for the case of an inflatable LTD that consists of a set of balloons, every balloon in the set being an electrically large toroid. The method applies the PO approach to evaluate backscattering from balloons, which are nothing more than thin dielectric shells. We also took into account the fact that both the LTD and the object under test were probed by a non-uniform plane wave. It is shown that the RCS of the dielectric shell depends significantly on the carrier frequency of the radar probing signal. Our results demonstrate that the RCS of the LTD in question can go as high as a hundredth of square meter. We also suggest a few possible technical solutions that could additionally decrease the RCS of this particular kind of LTD.

2.8 A computation method for electromagnetic wave scattering by dielectric toroid formations

In this section, we discuss one of the possible physical mechanisms behind the so-called 'angel-echoes' (AEs), a radar phenomenon of actual radar returns received from objects that are not visually observable. Such AEs can be caused by vortex updrafts carrying a volume of air with certain radio-physical parameters into a region with different radio-physical parameters. Toroidal vortices (vortex rings) may exist for quite a while and may cover considerable distances without collapsing. The calculation of the radar backscattering diagrams from such vortex rings using contemporary software packages that are aimed at solving general electrodynamic problems and that are based on the method of moments may turn out to be very burdensome in terms of computer resources and time. The fast asymptotic method for evaluating the EMW scattering by dielectric toroidal formations of large radius been developed and described below might be helpful in many applications. The comparison of simulation results against the renowned Altair Feko software demonstrated a good performance of the method presented in this section. Below, we also present the RCSs of vortex rings that appear in many practically interesting cases for both monostatic and bistatic radar layouts. In addition, we check the results of our simulations against the results of experimental measurements.

2.8.1 Introduction to electromagnetic scattering by meteorological formations

The electromagnetic wave propagation (EWP) medium is an integral, yet not frequently emphasized, part of any electronic system (ES). The impact of EWP conditions on the performance of an ES can be assessed using various radio-physical models that describe how the electrical properties of the medium vary in space and in time. Depending on the assignment of the ES, these models can include certain parts of the atmosphere (ionosphere, troposphere, or planetary boundary layer). Since the atmosphere is a gas-like bubble around the Earth and the intrinsic permittivity of atmospheric gases is very close to unity, even though it varies somewhat depending on local terrestrial conditions, it is commonly referred to as the dielectric constant of the air ε. In general, it is a complex value determined stochastically. Its particular value depends on the time and spatial coordinates, but it can be characterized by a quasi-steady mean and a variable component. The mean value of the dielectric constant of air depends on the averaging scale, both in time and in space. The scale of averaging, in its turn, depends on whether it is its local or global atmospheric movement that is to be of concern in the model [105]. The dynamic structure of the atmosphere makes the space–time structure of its radio-physical properties heterogeneous. Vertical air flows and air stratification in the planetary boundary layer may create conditions for the so-called tropospheric ducting that would increase the effective range of the ES. However, active air mixing in the troposphere reduces the variation in the electrical properties of air, thus decreasing refraction with elevation, which, in contrast, would decrease the effective range of the ES [106].

A particular problem in radar is when coherent discrete echoes come from no apparent source (objects that cannot be spotted visually, thus the name

AEs) [107]. The classification of received echoes is still quite difficult, even with the contemporary state of radar sophistication [108]. A great number of experimental studies regarding the echoes that appear from non-evident sources and volumes of concomitant sources (season, time of the day, temperature of the ground) resulted in the following. Part of the AEs was attributed to coming from larger birds or flocks of smaller birds, and another part of the AEs was attributed to swarms of insects. However, a large portion of these echoes could be attributed to local variations in the index of atmospheric refraction. Such variations appear as a result of large volumes of air with certain electrophysical parameters entering a region where the air has significantly differing characteristics. A candidate mechanism for the latter is an upflow of warm air that is heated by the ground, such an upflow creating an ascending flow of air with a vortex ring on top of it (thermals). Sometimes, local variations in atmospheric refractivity are caused by differences in the velocities of air flowing in neighboring strata or flowing around an obstacle [107, 109, 110].

Atmospheric vortices can vary in sizes considerably from large ones (such as hurricanes and tornados) to quite small ones (such as the wake created by a bird's wings). Yet another form of turbulent air movement is a toroidal vortex or vortex ring (VR), which is formed when a linear vortex becomes closed on itself rather than on a solid (underlying) surface [111].

A VR contains a well pronounced practically homogeneous spinning core and the vortex ring 'atmosphere'. The appearance of a vortex ring 'colored' with smoke is shown in two projections in figure 2.94 [112]. VRs can exist for a long time and, moving through their entire existence, can cover considerable distances [112].

There is a whole range of meteorological formations in the atmosphere that may affect radar performance in the atmosphere. They can have various sizes and shapes, such as spherical and elongated raindrops, ice crystals, snowflakes, and multi-layer hailstones, either covered with a layer of water or not. The electrical properties of these formations vary in a wide range as well, depending on the ambient air conditions, which may include a static charge on the surface of raindrops, ionized cloud regions, and lightning channels. Therefore, the RCSs of meteorological formations are computed using a whole variety of methods [113–118].

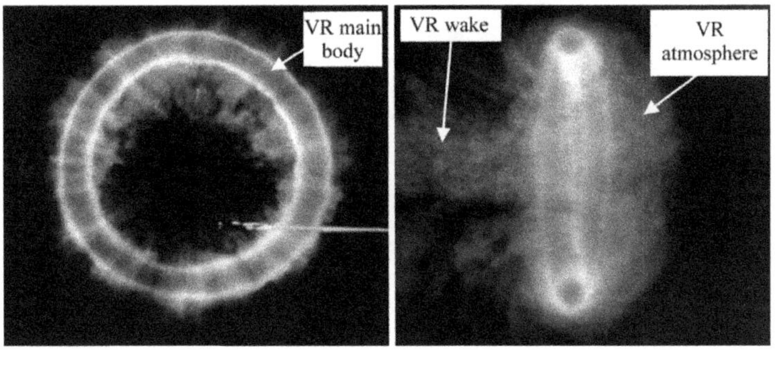

Figure 2.94. Two projections of a vortex ring 'colored' with smoke.

As far as linear vortices are concerned, their RCSs are computed using the Bourne approximation for a vortex represented by a dielectric cylinder [119]. One group studied EMW scattering at a perfectly conducting torus using numerical methods [120]. In the case of dielectric toroidal surfaces, the use of numerical computation methods would also seem reasonable. Contemporary software packages aimed at solving electrodynamics problems (Ansys HFSS, Altair Feko) are helpful because of their universal nature. However, the versatility of such software provided by the use of the MoM comes at the cost of computational redundancy in some cases, which may result in a heavy burden imposed on the computer that runs the software. As soon as the electrical dimensions of the objects under study increase, the EMW carrier frequency increases, or the dielectric constant of the material decreases, the number of discrete elements that represent the object's surface increases substantially. If the whole problem does not 'fit' into the 'fast' computer RAM and requires an additional virtual space on a 'slow' hard disk, the computation time increases drastically.

Therefore, below we develop an asymptotical method for computing the EMW scattering diagram of toroid vortices of large radius.

2.8.2 A method for computing the electromagnetic waves scattering by dielectric torus

Let us consider a homogeneous dielectric torus (figure 2.95) that is illuminated by a plane EMW:

$$\vec{E}^0(\vec{X}) = \vec{p}^0 \exp\left(-jk_0\left(\vec{R}^0 \cdot \vec{X}\right)\right),$$
$$\vec{H}^0(\vec{X}) = \sqrt{\frac{\varepsilon_0}{\mu_0}}\left(\vec{p}^0 \times \vec{R}^0\right)\exp\left(-jk_0\left(\vec{R}^0 \cdot \vec{X}\right)\right). \tag{2.129}$$

The direction of incidence is described by the unit vector $-\vec{R}^0 = (\cos\vartheta,\ 0,\ \sin\vartheta)$ and the direction of scattering is described by the unit vector $\vec{r}^0 = (\cos\psi,\ 0,\ \sin\psi)$.

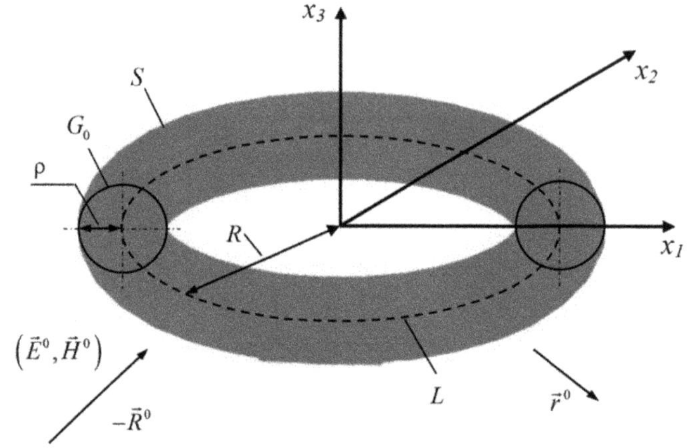

Figure 2.95. Homogeneous dielectric torus.

The latter means that both the illumination and reception directions belong to the plane Ox_1x_3, which passes through the torus axis. The polarization of the incident wave $\vec{p}^0$ is either horizontal ($\vec{p}^0 = \vec{e}_2$) or vertical ($\vec{p}^0 = -\sin\vartheta \cdot \vec{e}_1 + \cos\vartheta \cdot \vec{e}_3$). The polarization of the scattered wave is also either horizontal ($\vec{p} = \vec{e}_2$) or vertical ($\vec{p} = -\sin\psi \cdot \vec{e}_1 + \cos\psi \cdot \vec{e}_3$). Here, $\vec{e}_i$ ($i = 1,2,3$) are the unit vectors of the Cartesian coordinate system (figure 2.95). The monostatic radar reception is obtained for $\psi = \pi + \vartheta$.

Using the Lorentz reciprocity theorem [121], one can present the far-field scattered by the torus as follows:

$$\vec{p} \cdot \vec{E}^{\text{sc}}(\vec{X}) = -jk_0 \frac{\mathrm{e}^{jk_0 r}}{4\pi r} \cdot \int_S \left[(\vec{p} \cdot \vec{J}^e) \sqrt{\frac{\mu_0}{\varepsilon_0}} + \vec{J}^m \cdot (\vec{p} \times \vec{r}^0) \right] \times \mathrm{e}^{-jk_0(\vec{r}^0 \cdot \vec{x})} \, \mathrm{d}S, \tag{2.130}$$

where $\vec{J}^e = (\vec{n} \times \vec{H})$, $\vec{J}^m = (\vec{n} \times \vec{E})$ is the equivalent current density on the surface of the torus S, $\vec{n}$ is the internal normal to surface S, r is the distance from the torus center to the field point, $k_0 = \dfrac{2\pi}{\lambda_0}$ is the wavenumber in free space, λ_0 is the wavelength of the incident wave, and ε_0 and μ_0 are the absolute permittivity and permeability of the free space.

Let us introduce a Cartesian coordinate system with the origin at the torus center line L:

$$\vec{X} = \vec{x}(l) + \vec{\zeta}(\rho, \varphi).$$

Here, $\vec{x}(l)$ is the radius-vector of a point on L that has the arc coordinate l and $\vec{\zeta}(\rho, \varphi)$ is the vector orthogonal to the torus surface at that point, its length being equal to ρ (the smaller torus radius) and its direction being defined by the angle φ ($0 \leqslant \varphi \leqslant 2\pi$).

Considering the linearity of the problem at hand, it is legitimate to assume that quantities $\vec{J}^e$ and $\vec{J}^m$ on surface S at point $(l, \vec{\zeta})$ can be presented as

$$\begin{pmatrix} \vec{J}^e(\vec{X}) \\ \vec{J}^m(\vec{X}) \end{pmatrix} = \begin{pmatrix} \tilde{\vec{J}}^e(\vec{\zeta}) \\ \tilde{\vec{J}}^m(\vec{\zeta}) \end{pmatrix} \cdot \exp(-jk_0 \vec{R}^0 \cdot \vec{x}(l)), \tag{2.131}$$

where $\tilde{\vec{J}}^e(\vec{\zeta})$, $\tilde{\vec{J}}^m(\vec{\zeta})$ are the equivalent surface current densities excited on surface S by the following plane wave

$$\tilde{\vec{E}}^0(\vec{\zeta}) = \vec{p}^0 \exp\left(-jk_0\left(\vec{R}^0 \cdot \vec{\zeta}\right)\right),$$

$$\tilde{\vec{H}}^0(\vec{\zeta}) = \sqrt{\frac{\varepsilon_0}{\mu_0}} \left(\vec{p}^0 \times \vec{R}^0\right) \exp\left(-jk_0\left(\vec{R}^0 \cdot \vec{\zeta}\right)\right).$$

Now, taking into account expression (2.131), we can write:

$$\vec{p} \cdot \vec{E}^{\text{sc}}(\vec{r}^0) = -jk_0 \frac{\mathrm{e}^{-jk_0 r}}{4\pi r} I(\vec{r}^0), \tag{2.132}$$

where

$$I(\vec{r}^{\,0}) = \int_L e^{-jk_0\left(\vec{x}(l)\cdot\left(\vec{R}^0+\vec{r}^{\,0}\right)\right)} M(l)\ \mathrm{d}l, \tag{2.133}$$

$$M(l) = \int_{G_0(l)} e^{-jk_0(\vec{r}^{\,0},\vec{\zeta})}\left[\left(\vec{p}\cdot\tilde{\vec{J}}^{e}\right)\sqrt{\frac{\varepsilon_0}{\mu_0}} + \tilde{\vec{J}}^{m}\cdot\left(\vec{p}\times\vec{r}^{\,0}\right)\right]\mathrm{d}g. \tag{2.134}$$

Here, $G_0(l)$ is the ring line bounding the torus cross-section, which is perpendicular to the torus center line L at the point with arc coordinate l, and $\mathrm{d}g = \rho\ \mathrm{d}\varphi$.

Given that $R > >\lambda_0$, it is also legitimate to evaluate integral (2.133) using the stationary phase method [122]:

$$\begin{aligned} I(\vec{r}^{\,0}) &\exp\left(-jk_0\left(\left(\vec{R}^0+\vec{r}^{\,0}\right)\cdot\vec{x}(l_0)\right)+j\frac{\pi}{4}\delta\right)M(l_0)\\ &\times\sqrt{\frac{2\pi}{k_0\chi(l_0)\left|\left(\vec{R}^0+\vec{r}^{\,0}\right)\cdot\vec{\nu}(l_0)\right|}}. \end{aligned} \tag{2.135}$$

Here, l_0 is the point on L which is the closest to (furthest from) the observer. In our case there are two such points. To obtain the scattered field according to expression (2.132), one needs to sum up the electrodynamic contributions of both points into $I(\vec{r}^{\,0})$.

The quantities entering expression (2.135) are as follows:

$$-\left(\vec{R}^0+\vec{r}^{\,0}\right)\cdot\vec{x}(l_0) = \mp(\cos\vartheta - \cos\psi)R,$$

$\chi(l_0) = \dfrac{1}{R}$ is the curvature of line L at point l_0,

$$\delta = \mp\mathrm{sgn}\left(\left(\vec{R}^0+\vec{r}^{\,0}\right)\cdot\vec{\nu}(l_0)\right) = \pm\mathrm{sgn}\,(\cos\vartheta - \cos\psi)$$

$\left|\left(\vec{R}^0+\vec{r}^{\,0}\right)\cdot\vec{\nu}(l_0)\right| = |\cos\vartheta - \cos\psi|$, $\vec{\nu}(l_0)$ is the unit vector of principal normal to line L at point l_0. The signs $\pm$ belong to the two points of the stationary phase.

Considering that $R > >\lambda_0$, we can asymptotically assume that the equivalent surface current densities $\tilde{\vec{J}}^{e}$ and $\tilde{\vec{J}}^{m}$ at points belonging to line $G_0(l)$ are equal to the values that would be excited on the surface of an infinite dielectric cylinder that would contain the cross-section line $G_0(l)$ as its own cross-section, the radius of the cylinder and its dielectric constant coinciding with those of the torus, ρ and ε.

Below, we limit ourselves to only two polarizations of incident wave that are orthogonal to each other. In particular, in the case of horizontal polarization the vector $\vec{E}$ is parallel to the cylinder's axis, whereas in the case of vertical polarization the vector $\vec{H}$ is parallel to it. For the sake of convenience we can introduce the same notation for both orthogonal polarizations. Now, in the case of horizontal polarization (with respect to the cylinder that would be substituted for the torus in every one of its cross-sections across the smaller radius), we have:

$$u = \vec{e}_2 \cdot \vec{E};\ \widetilde{\vec{J}}^e = -\frac{1}{j\omega\mu_0}\frac{\partial u}{\partial n}\vec{e}_2;\ \widetilde{\vec{J}}^m = (\vec{n} \times \vec{e}_2)u.$$

In the case of vertical polarization (with respect to the substituting cylinder), we have,

$$u = \left(\vec{e}_2 \cdot \vec{H}\right)\sqrt{\frac{\mu_0}{\varepsilon_0}};\ \widetilde{\vec{J}}^e = (\vec{n} \times \vec{e}_2)\sqrt{\frac{\varepsilon_0}{\mu_0}}u;\ \widetilde{\vec{J}}^m = \frac{1}{j\omega\varepsilon_0}\frac{\partial u}{\partial n}\vec{e}_2\sqrt{\frac{\varepsilon_0}{\mu_0}}.$$

Having transformed expression (2.134), we obtain the following:

$$jk_0 M(l_0) = \widetilde{M}(l_0) = \int_0^{2\pi} \mathrm{e}^{-jk_0\rho\cos(\varphi+\vartheta-\psi)}\left[jk_0 u\cos(\varphi + \vartheta - \psi) - \frac{\partial u}{\partial n}\right]\rho\, \mathrm{d}\varphi. \quad (2.136)$$

In addition, expression (2.136) holds for the two polarizations when both transmitted and received waves are concerned, the only difference being in the sense and particular expressions for u and $\frac{\partial u}{\partial n}$.

The particular expressions for u and $\frac{\partial u}{\partial n}$ can be found as solutions to two-dimensional problems concerned with a dielectric cylinder illuminated by the following wave:

$$U^0 = \mathrm{e}^{jk_0 t\cos\varphi} = \sum_{n=0}^{\infty} j^n \varepsilon_n J_n(k_0 t)\cos n\varphi, \qquad (0 \leqslant \varphi \leqslant 2\pi), \quad (2.137)$$

where $\varepsilon_n = \begin{cases} 1, & n = 0 \\ 2, & n \neq 0 \end{cases}$.

Let us find the above solutions in two different forms. Those obtained within the cylinder ($t \leqslant \rho$) will have the form

$$u^- = \sum_{n=0}^{\infty} j^n \varepsilon_n B_n J_n(k_i t)\cos n\varphi, \quad (2.138)$$

and those obtained without the cylinder ($t > \rho$) will have the form

$$u^+ = \sum_{n=0}^{\infty} j^n \varepsilon_n C_n H_n^{(1)}(k_0 t)\cos n\varphi. \quad (2.139)$$

Here, $J_n(z)$ and $H_n^{(1)}(z)$ are the Bessel and Hankel functions, respectively, $k_i = k_0\sqrt{\varepsilon'}$, and ε' is the cylinder's dielectric constant.

We can express the boundary conditions at the cylinder's surface ($t = \rho$) as follows:

$$\begin{cases} u^- = u^+ + U^0, \\ \dfrac{1}{\beta}\dfrac{\partial u^-}{\partial n} = \dfrac{\partial u^+}{\partial n} + \dfrac{\partial U^0}{\partial n}, \end{cases} \quad (2.140)$$

where $\beta = 1$ given the horizontal polarization of the incident wave and $\beta = \varepsilon'$ for vertical polarization of the incident wave.

Plugging the expressions (2.137), (2.138), and (2.139) into (2.140), and solving the SLE, we land at the following expression:

$$C_n = \frac{J'_n(k_i\rho)J_n(k_0\rho) - \dfrac{k_0\beta}{k_i}J'_n(k_0\rho)J_n(k_i\rho)}{J'_n(k_i\rho)H_n^{(1)}(k_0\rho) - \dfrac{k_0\beta}{k_i}J_n(k_i\rho)H_n^{(1)\prime}(k_0\rho)},$$

where $J'_n(z)$ and $H_n^{(1)'}(z)$ are derivatives of the Bessel and Hankel functions, respectively.

Substituting the above expression for C_n into (2.139), we obtain $u = u^+$. Now,

$$\frac{\partial u}{\mathrm{d}n} = -\frac{\partial u^+}{\mathrm{d}t}\bigg|_{t=\rho} = -\sum_{n=0}^{\infty} j^n \varepsilon_n C_n k_0 H_n^{(1)\prime}(k_0\rho)\cos n\varphi. \qquad (2.141)$$

Let us plug expressions (2.139) and (2.141) into (2.136) and change the order of integration and summation to obtain the following:

$$\widetilde{M}(l_0) = k_0\rho\sum_{n=0}^{\infty} j^n \varepsilon_n C_n[jH_n^{(1)}(k_0\rho)\cdot I_1 + H_n^{(1)\prime}(k_0\rho)\cdot I_2],$$

where I_1 and I_2 are integrals that can be reduced to the table ones [7].

After some transformations we obtain the following:

$$I_1 = \pi e^{-j(n-1)\frac{\pi}{2}}[J_{n-1}(k_0\rho) - J_{n+1}(k_0\rho)]\cos(n\alpha),$$

$$I_2 = 2\pi e^{-jn\frac{\pi}{2}}J_n(k_0\rho)\cos(n\alpha).$$

Finally, using known functional relations that exist for Bessel functions [19], we land at the following formula:

$$\widetilde{M}(l_0) = 4\sum_{n=0}^{\infty} j^{n+1}\varepsilon_n C_n e^{-jn\frac{\pi}{2}}\cos(n\alpha),$$

where $\alpha = \vartheta - \psi$.

Now, using expressions (2.132) through (2.135), we can obtain the expression for a bistatic RCS of the torus:

$$\begin{aligned}\sigma &= 4\pi r^2\,|\vec{p}\,\vec{E}^{\,p}(\vec{r}^{\,0})|^2\\ &\approx \frac{\lambda R\cdot|\widetilde{M}(l_0)|^2}{\pi|\cos\vartheta - \cos\psi|}\cdot\cos^2\left[k_0R(\cos\vartheta - \cos\psi) - \frac{\pi}{4}\,\mathrm{sgn}\,(\cos\vartheta - \cos\psi)\right].\end{aligned} \qquad (2.142)$$

Expression (2.142) can be directly used for computing the RCS for an entire range of incidence and scattering directions ϑ and ψ except for a narrow angular sector where $\cos\vartheta = \cos\psi$. However, as it turned out, as soon as one abstains from the latter values leaving a gap of few degrees and interpolating the missing values, one can obtain a reasonably good idea of the dielectric torus RCS even in those 'problematic' points.

2.8.3 Results of simulating the radar cross section of the dielectric torus

In order to check the method presented above we carried out the same computation with regard to the dielectric tori of various electrical sizes using the Altair FEKO software. The initial data adopted for the simulation are presented in table 2.4, and the results are plotted in figures 2.96–2.98.

Table 2.4. Initial data adopted for the dielectric torus RCS computation by two methods.

Quantity	R, m	ρ, m	ε	f_0, GHz (λ_0, m)	Angle of wave incidence ϑ, degree	Polarization
Value	1	0.2	1.01	1 (0.3)	30	horizontal, vertical

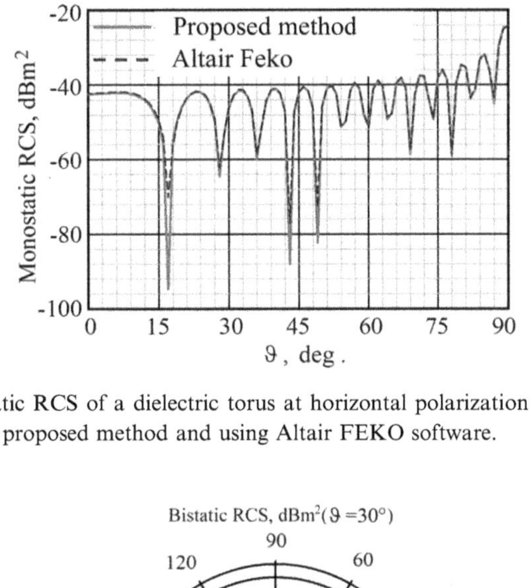

Figure 2.96. The monostatic RCS of a dielectric torus at horizontal polarization versus the wave reception angle computed using the proposed method and using Altair FEKO software.

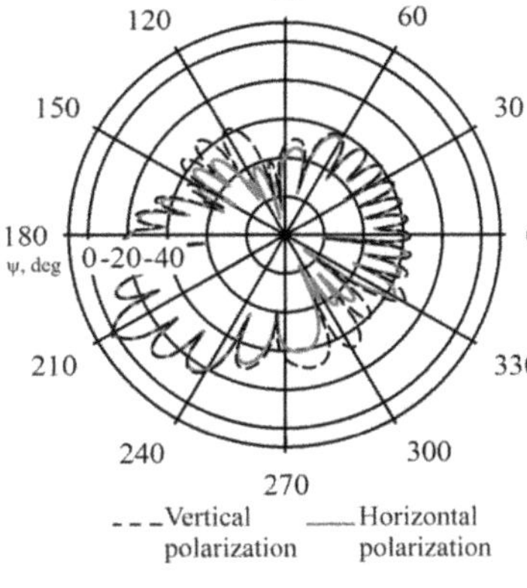

Figure 2.97. Bistatic RCS of a dielectric torus versus the wave reception angle in the case of both polarizations.

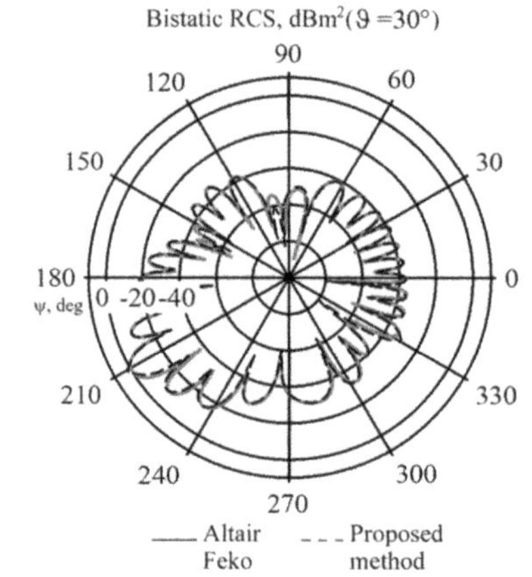

Figure 2.98. Bistatic RCS of the dielectric torus versus the reception direction using different calculation methods.

Figure 2.96 presents the plots of the monostatic RCS of the dielectric torus at horizontal polarization versus the wave reception angle computed using the above method and using Altair FEKO software. In the case of the bistatic RCS of the torus, both the incidence and reception directions belonged to the plane Ox_1x_3, which included the torus axis. We do not show the corresponding plots of the RCS for vertical polarization here because the differences between the above method and Altair FEKO for vertical polarization were minuscule.

It becomes evident from the plots of figure 2.96 that both methods generate similar results.

Figure 2.97 shows the bistatic RCS of the dielectric torus versus the wave reception angle in the case of both vertical and horizontal polarization, the angle of EMW incidence was $\vartheta = 30°$. There are well pronounced humps around the angles of reception around 210° (forward scatter) and around 330° (mirror scatter with respect to the torus plane). Also, there are evident differences between the RCS at vertical and horizontal polarizations in the directions that are approximately perpendicular to the direction of wave incidence.

Figure 2.98 shows the plots of the bistatic RCS of the dielectric torus versus the reception direction that were computed for horizontal polarization of the incident wave using the above method and using the Altair FEKO software. Again, it is evident that both methods generate nearly the same result. As was the case with figure 2.96 above, we do not present the corresponding plots computed for vertical polarization of the incident wave since there was almost no difference between the two methods.

Now, it is obvious from figures 2.96 and 2.98 that both the monostatic and bistatic RCS values computed using the above method and using the universal software like Altair FEKO are basically the same. However, there is a big difference in the computation load imposed on a computer. In our particular case, we used a regular personal computer (Dell G3 3579, CPU Intel Core i5 8300H, RAM DDR4–2666 32GB) rather than a dedicated workstation. Thus our computer was able to yield results during the time interval that was three orders of magnitude shorter than that required by the universal software running on the same computer. It is worth noting that the quantity of discrete elements representing the object's surface was set in such manner that the entire computing task would fit into the computer's RAM.

Moreover, when Altair FEKO software was used on a personal computer featuring the above specs, we observed a spike in computation errors every time as the carrier frequency of the incident wave went above $f_0 > 1$ GHz. Our attempts aimed at reducing those errors by setting increased numbers of discrete elements for describing the surface of the torus led to an unacceptable increase in computation time. The latter means that given a practically interesting case of radars operating on frequencies around $f_0 = 3$ GHz, the computation of the RCS with regard to thermals was not feasible using the universal software packages. Therefore, the RCS of a dielectric meteorological formation of large radius, which is characteristic of thermals [123] ($R = 30$ m, $\rho = 6$ m) was computed using the asymptotic method presented above. In addition, considering the fact that thermals appear in the atmosphere and not in a vacuum, we took into account the contrast of the thermal and its ambient air in terms of the dielectric constant. Thus the dielectric constant ε_1 was used for the air and dielectric constant ε_2 was used for the thermal. The method presented above does not require any changes in order for the latter contrast to be accounted for. The absolute values of ε_1 and ε_2 do not matter as much as their ratio. In the case of standard atmosphere [106] and for a thermal ascended to a height of 1000 m, we assumed that ratio to be equal to $\frac{\varepsilon_2}{\varepsilon_1} = 1.000022$.

Figure 2.99 shows the plots of monostatic RCS versus the angle of reception for the case of the above thermals given their illumination by a wave at horizontal and vertical polarizations, the carrier frequency being $f_0 = 3$ GHz. The two plots in figure 2.99 practically coincide, which is usual for dielectric inhomogeneities of turbulent nature [107].

The values of the monostatic RCS of thermals obtained by our computation for a small deviation of illumination angles from the axis of the thermal were pretty close to the RCSs of the turbulent tropospheric inhomogeneities obtained by measurement [107].

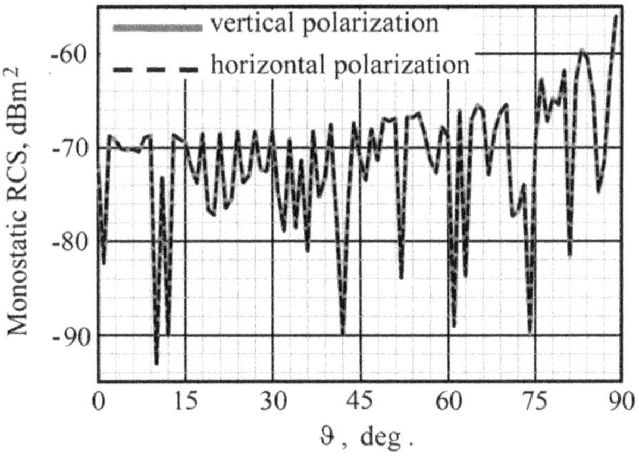

Figure 2.99. Monostatic RCS versus the angle of reception for the case of the above thermals for their illumination by a wave at horizontal and vertical polarizations, the carrier frequency being $f_0 = 3$ GHz.

Figures 2.100 and 2.101 present the bistatic RCS of thermals versus reception angle (scattering diagram) for their illumination angle $\vartheta = 30°$ that were obtained by measurements. The scattering diagrams are highly irregular since the dielectric torus (vertex ring) was electrically large. The directions of the increased scattering that correspond to the forward scatter (210°) and mirror scatter (330°) are easily visible in figures 2.100 and 2.101. As anticipated, as the electrical dimensions of the torus increase, the lobes of its RCS in those directions increase in amplitude and become narrower. In the case of a horizontally polarized incident wave, the vertex ring scattering in the directions approximately perpendicular to the incidence direction is about 20 dB stronger than that in case of a vertically polarized incident wave.

It should also be noted that the accuracy of the above method would only increase if the frequency of the incident wave went up.

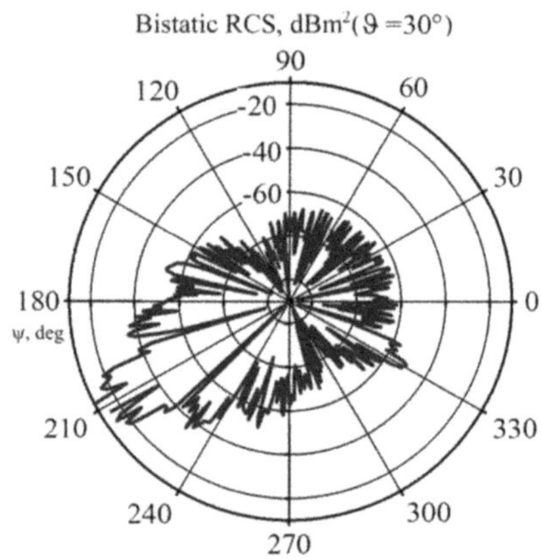

Figure 2.100. The bistatic RCS of thermals versus the reception angle at vertical polarization, the carrier frequency being f_0 = 3 GHz.

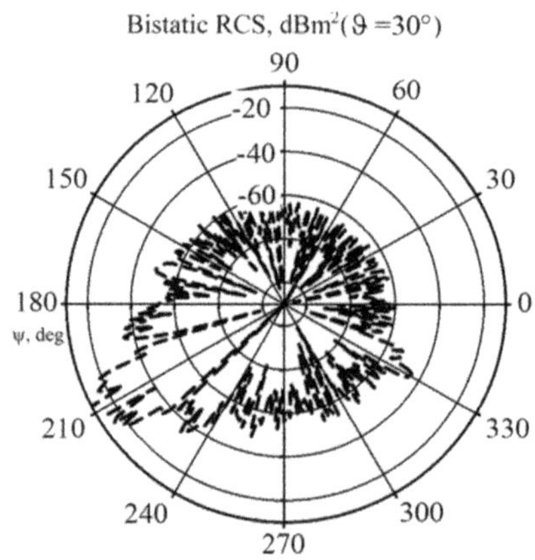

Figure 2.101. The bistatic RCS of thermals versus the reception angle at horizontal polarization, the carrier frequency being f_0 = 3 GHz.

2.9 A method for camouflaging ground vehicles to prevent their detection by radar

The method presented in [19] of computing the EMW scattering by ground objects (even those covered with a radar absorbing coating) is used here to develop an optimization technique aimed at finding some scheme for applying the radar absorber selectively onto the object's surface. As result the mean RCS of the object would be as close as possible to the mean RCS of the underlying surface shadowed by the object if the aspect angles of the object illumination are confined within certain angular sectors set in advance. We refer to this scheme as 'camouflage'. The main purpose of such camouflage is to reduce the radar contrast of the objects with respect to their surroundings thus making the object 'invisible' to radar surveillance. Such a technique can be a part of an automated tool for finding the radar absorber application scheme for any particular tactical scenario.

Given a model of a tank, we present the corresponding computations of its RCS for various sectors of the angular aspects of illumination and for various types of underlying surface that have different degrees of roughness.

Ensuring the survivability of ground vehicles on contemporary battlefields is crucial for operational success. An important aspect of such survivability is the reduction of the radar observability of such objects [94, 124–128]. In addition, in contrast to airborne objects, for which radar visibility is proportional to the object's RCS, reducing the visibility of ground objects has to take into account the surface on which the objects stand. Radar returns from ground objects are observed against the background of returns from their surroundings, which are usually characterized by some specific RCS (RCS per square meter) [129]. Thus, if one manages to make the object perfectly invisible (producing no radar returns), the object will 'punch' a hole in the radar returns from the background, thus providing a contrast with its surroundings that would give away its location.

The key benchmark problem that is to be solved is the problem of finding the RCS of a ground object with a non-perfectly reflecting surface. In particular, such non-perfect reflection of EMWs from the object's surface may be caused by the presence of a radar absorbing coating. The optimization technique presented below is based on the method presented in [19], which is concerned with predicting EMW scattering from ground objects (even those covered with a radar absorbing coating).

2.9.1 Basic relations for the radar cross section of ground objects

According to [19], one can use the integral representation of the field scattered by the ground object described using the surface S in order to solve the key benchmark problem:

$$\vec{p}\cdot\vec{E}(x_0) = \frac{1}{j\omega}\int_S \left[\vec{H}^{\perp}(x)\cdot\vec{\mathcal{E}}(x|x_0,\vec{p}\,) + \vec{E}^{\perp}(x)\cdot\vec{\mathcal{H}}(x|x_0,\vec{p}\,)\right]\mathrm{d}S_x, \quad (2.143)$$

where $\vec{\mathcal{E}}, \vec{\mathcal{H}}(x|x_0,\vec{p}\,)$ is the field of the point source situated at the point x_0 characterized by the vector-moment $\vec{p}$ for an interface with underlying surface D

present, $(\vec{E}^{\perp}, \vec{H}^{\perp})$ are tangential components of the total EM field at the object's surface S, which are rotated here by 90°.

We note here that $\vec{E}^{\perp}$and $\vec{H}^{\perp}$ are the equivalent surface densities of the electric and magnetic current, respectively.

Let us assume that the object is irradiated with two waves: the first is a plane wave propagating along the unit vector $\vec{R}^{0}$, its polarization being defined by the unit vector $\vec{P}^{0}$, and the second wave is the one reflected from the underlying surface and propagating in the direction set by $\vec{R}^{1}$. We also assume, at first, that the underlying surface has a flat interface boundary. Then,

$$\vec{R}^{1} = \vec{R}^{0} - 2\vec{n}\left(\vec{R}^{0} \cdot \vec{n}\right), \tag{2.144}$$

where $\vec{n}$ is the unit vector normal to the underlying surface interface.

As we switch to the far-field zone (2.143) with respect to the backscattered field $(\vec{x}_0 = -r \cdot \vec{R}^{0}, r \to \infty)$, we can express the point source field as follows:

$$\begin{gathered} \vec{\mathcal{E}}(\vec{x}|\vec{x}_0, \vec{p}\,) \sim \Omega(k_0 r)\vec{\mathcal{E}}(\vec{x}|\vec{R}^{0}, \vec{p}\,), \\ \vec{\mathcal{H}}(\vec{x}|\vec{x}_0, \vec{p}\,) \sim \Omega(k_0 r)\vec{\mathcal{H}}(\vec{x}|\vec{R}^{0}, \vec{p}\,), \end{gathered} \tag{2.145}$$

where $\Omega(k_0 r) = \dfrac{e^{jk_0 r}}{4\pi k_0 r}$.

The field $\vec{\mathcal{E}}\left(\vec{x}|\vec{R}^{0}, \vec{p}\right)$, $\vec{\mathcal{H}}\left(\vec{x}|\vec{R}^{0}, \vec{p}\right)$ is excited by the plane wave

$$\begin{gathered} \vec{\mathcal{E}}_0\left(\vec{x}\middle|\vec{R}^{0}, \vec{p}\right) = k_0^2\omega\sqrt{\frac{\mu_0}{\varepsilon_0}}\vec{p}^{\,T} \exp\left(jk_0\left(\vec{R}^{0} \cdot \vec{x}\right)\right), \\ \vec{\mathcal{H}}_0\left(\vec{x}\middle|\vec{R}^{0}, \vec{p}\right) = -k_0^2\omega\vec{p}^{\,\perp} \exp\left(jk_0\left(\vec{R}^{0} \cdot \vec{x}\right)\right), \end{gathered} \tag{2.146}$$

where $\vec{p}^{\,\perp} = \vec{R}^{0} \times \vec{p}$, $\vec{p}^{\,T} = \vec{p} - \vec{R}^{0}\left(\vec{R}^{0} \cdot \vec{p}\right)$.

As we take into account the phase difference between the two irradiation waves incident on the object's surface, we can express the electric field component at point $\vec{x}$ of surface S as follows [19]:

$$\begin{aligned} \vec{\mathcal{E}}\left(\vec{x}\middle|\vec{R}^{0}, \vec{p}\right) &= k_0^2\omega\sqrt{\frac{\mu_0}{\varepsilon_0}}\Omega(k_0 r) \\ &\times\left[\vec{p}^{\,0} \exp\left(jk_0\left(\vec{R}^{0} \cdot \vec{x}\right)\right) + \vec{p}^{\,1} \exp\left(jk_0\left(\left(\vec{R}^{0} - \vec{R}^{1}\right) \cdot \vec{c} + \vec{R}^{1} \cdot \vec{x}\right)\right)\right], \end{aligned} \tag{2.147}$$

where $\vec{c} = \vec{x} - \dfrac{(\vec{x} \cdot \vec{n}) + h}{\left(\vec{R}^{1} \cdot \vec{n}\right)}\vec{R}^{1}$, $\vec{p}^{\,1}$ is the vector coefficient of reflection from the underlying surface, and h is the distance from the origin of the object's reference frame to the plane of interface D. Magnetic field component can be presented in the same manner.

Now, using the above expressions for the incident field ($\vec{\mathcal{E}}_0, \overline{\mathcal{H}}_0$) we can express the field backscattered by the object in the direction ($-\vec{R}^0$) in the following form:

$$\vec{p} \cdot \vec{E}\left(\vec{R}^0\right) = $$
$$-jk_0\Omega(k_0r) \int_S \left[\sqrt{\frac{\mu_0}{\varepsilon_0}} \left[\vec{p}^{\,0} \exp\left(jk_0\left(\vec{R}^0 \cdot \vec{x}\right)\right) + \vec{p}^{\,1} \exp\left(jk_0\left(\left(\vec{R}^0 - \vec{R}^1\right) \cdot \vec{c} + \vec{R}^1 \cdot \vec{x}\right)\right)\right] \cdot \vec{H}^{\perp}(\vec{x}) \right. \tag{2.148}$$
$$\left. + \left[\vec{p}^{\,0\perp} \exp\left(jk_0\left(\vec{R}^0 \cdot \vec{x}\right)\right) + \vec{p}^{\,1\perp} \exp\left(jk_0\left(\left(\vec{R}^0 - \vec{R}^1\right) \cdot \vec{c} + \vec{R}^1 \cdot \vec{x}\right)\right)\right] \cdot \vec{E}^{\perp}(\vec{x}) \right] \mathrm{d}S.$$

The surface S can be broken down into 'illuminated' and shadowed parts that depend on the two irradiation directions $\vec{R}^0$ and $\vec{R}^1$. Now, if we apply the PO approximation to the field components $\vec{E}^{\perp}$ and $\vec{H}^{\perp}$ for the two irradiation directions $\vec{R}^0$ and $\vec{R}^1$, and substitute the corresponding expressions into (2.148), we obtain the backscattered field in the form of four integrals [19] that represent the individual contribution of four distinct wave propagation routes into the backscattering. These four routes are: 'from transmitter – to object – to receiver', 'from transmitter – to ground – to object – to receiver', 'from transmitter – to ground – to object – to ground – to receiver', 'from transmitter – to object – to ground – to receiver'.

If the object's surface is non-perfect in terms of its conductivity, which for instance is due to the presence of a radar absorber layer, then the field components $\vec{E}^{\perp}$ and $\vec{H}^{\perp}$ can be found using the solution to the benchmark problem of EMW scattering from a layer of radar absorber backed by a perfectly conducting plate [19].

In particular, for the vector coefficient of wave reflection from such a structure $\vec{p}^{\,2}$ one can obtain the following expression:

$$\vec{p}^{\,2} = \vec{p}^{\,2T} - \vec{n}\frac{\left(\vec{p}^{\,2T} \cdot \vec{R}^0\right)}{\cos\theta}, \tag{2.149}$$

$$\vec{p}^{\,2T} = \frac{jg\cos\theta + 1}{jg\cos\theta - 1}\vec{p}^{\,0T} - \frac{2jg}{jg\cos\theta - 1}$$
$$\times \left[\vec{R}^{0T}\frac{\left(\vec{R}^{0T} \cdot \vec{p}^{\,0}\right)}{jg - \cos\theta} + \vec{R}^{0\perp}\frac{\left(\vec{R}^{0\perp} \cdot \vec{p}^{\,0}\right)}{\varepsilon'_1\mu'_1\left(jg - \dfrac{\cos^2\theta_1}{\cos\theta}\right)} \right], \tag{2.150}$$

where $g = \sqrt{\dfrac{\mu'_1}{\varepsilon'_1}}\cos\theta_1 tg(k_1\delta\cos\theta_1)$, $\cos^2\theta_1 = 1 - \dfrac{\sin^2\theta}{\varepsilon'_1\mu'_1}$, $k_1 = k_0\sqrt{\varepsilon'_1\mu'_1}$, ε'_1 and μ'_1 are the relative permittivity and permeability of the radar absorber material, and θ is the angle of the wave incidence upon the layer of the absorber.

Now we allow for an irregular underlying surface, which is described by a random function $z = \zeta(\xi, \eta)$ having a normal (Gaussian) distribution with probability density

$$\omega(\zeta) = \frac{1}{\sqrt{2\pi}\sigma}\exp(-\zeta^2/2\sigma^2).$$

Then, it was shown in [130] that one can handle the irregularity of the surface by introducing a scalar multiplier (characteristic function) by which one corrects the vector reflection coefficient $\vec{p}^{\,1}$

$$\overline{\mathrm{e}^{j2k_0R_3^0\varsigma}} = \exp\left(-2\left(k_0R_3^0\sigma\right)^2\right). \tag{2.151}$$

In addition, bearing in mind the basic asymptotic approximation, one may assume that the normal to irregular underlying surface is the same as the normal to flat one.

Thus formula (2.148) that would be 'corrected' by multiplying $\vec{p}^{\,1}$ by expression (2.151) allows us to compute the field backscattered from the ground object while taking into account the irregularities (roughness) of the underlying surface the object stands on.

As a next step towards the solution of our problem, we need to evaluate the degree of the object's 'contrast' against the ground in its immediate vicinity, and in particular against the ground 'shadowed' by the object. In order to do that, we want to compute the RCS of that particular portion of the underlying surface. We evaluate the latter RCS by multiplying the area of the object's footprint on the surface (expressed in terms of the object's radar shadow) by the value of the specific RCS of the ground, which depends on the kind of soil and can be obtained from reference tables [131].

The geometry of the object's surface was modeled in [19] as a set of adjoint segments of triaxial ellipsoids. Here, for the sake of definiteness, we limit ourselves to the model that does not include any edges. In this case, we can employ the PO method for computation of the object's backscattered field. In addition, integration over the surface involves breaking the surface down into a set of flat triangular facets. It is worth noting that the work [19] was dealing with a more complex model that accounted for the surface discontinuities.

The area of the radar shadow created by the object on the underlying surface can be evaluated using the following approach. Figure 2.102 shows an arbitrary ith facet irradiated by an EMW incident upon it in the direction $\vec{R}^{\,0}$. The area of the ith facet is defined by the coordinates of the triangle apices.

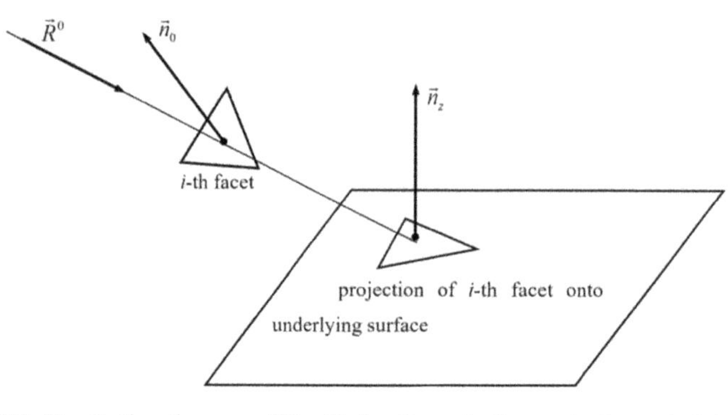

Figure 2.102. For finding the area of the ith facet's projection onto the underlying surface.

The ith facet is projected onto the underlying surface along the incidence direction $\vec{R}^0$ and the area of its projection can be computed as follows:

$$S_i^{\text{pr}} = S_i \frac{|\vec{R}^0 \cdot \vec{n}_0|}{|\vec{R}^0 \cdot \vec{n}_z|},$$

where $\vec{n}_0$ is the normal to the ith facet, and $\vec{n}_z$ is the normal to the underlying surface.

In the latter expression, the area of the ith facet is multiplied at first by the cosine of the angle between the ith facet's normal and the normal to plane perpendicular to direction of incidence $\vec{R}^0$. In this manner one finds the area of the ith facet's projection onto the plane perpendicular to the incidence direction $\vec{R}^0$. Next, the latter area is divided by the cosine of the angle between the normal to underlying surface and the plane perpendicular to the incidence direction $\vec{R}^0$. The sum of all such projections of irradiated facets yields the area of the object's radar footprint (radar shadow) on the ground in the object's immediate vicinity given that object was irradiated by radar from the direction $\vec{R}^0$.

In our computations, we employ a certain shortcut in order to generate a stable RCS value for the object, which we refer to as the non-coherent RCS [19]. Such a non-coherent RCS is a sum of particular RCS values pertaining to individual portions of the object's surface. As was shown in [19], the value of the non-coherent RCS of an object is very close to the value obtained by window averaging of an instantaneous object's RCS by irradiation directions or by irradiation signal frequencies.

Now, let us summarize the steps we take in order to compute the average (mean) RCS of a ground object. These steps are as follows:

1. Compute the individual RCSs of portions of the object's surface while these surface portions are considered to be perfectly conducting, the underlying surface being accounted for. These computations are carried out for several directions of EMW incidence upon the object's surface, and the average RCSs of such surface portions are computed.
2. Carry out the above computations given that these same portions of the object's surface are considered to be coated with a radar absorber; this round of computations outputs average RCSs for that condition.
3. Apply an exhaustive search to figure out the radar absorber application scheme for the object's surface that brings the mean (non-coherent) RCS of the object as close as possible to that of the object's radar footprint on the ground. In that search, we check the contribution of every particular portion of the object's surface to its non-coherent RCS given the two above conditions of present or absent radar absorber. The exhaustive search procedure results in a map by which the radar absorber is to be spread over the object's surface to bring the object's non-coherent RCS to a level maximally close to the mean RCS of the portion of underlying surface confined within the radar shadow of the object.

2.9.2 Results of computer simulation

Let us consider the simplified model of a tank shown in figure 2.103 as an example.

The model length (including the gun barrel) was 16.41 m, the hull length was 14.93 m, the hull width was 6.62 mm, and the tank height was 2.58 m. The model comprised segments of 13 triaxial ellipsoids.

All the results presented below correspond to a scenario in which this tank model (and the underlying surface beneath it) was irradiated by a radar probing signal obliquely, the elevation angle of the EMW incidence being 10° with respect to the ground. This particular scenario corresponded to radar observation of ground objects from helicopter platforms.

Examples of underlying surface types included the following:

- concrete, its dielectric constant being $\varepsilon = 6.5 + j1.5$, and its surface roughness having the root mean square $\sigma = 0.00015$ m [132].
- steppe (prairie) (a surface comprised of sandy loam, we assumed summer conditions, so it was covered with grass), its dielectric constant being $\varepsilon = 3 + j0.4$, and its surface roughness having the root mean square $\sigma = 0.05$ m.

Both underlying surfaces had relative permeability equal to unity. Actually any surface can be used, the only condition is that its dielectric constant, relative permeability, root mean square deviation of the surface irregularities, and specific RCS are known. The specific RCS of the ground can be taken from the reference literature where it has been measured experimentally [132–135], or it can be computed using corresponding models [129]. In our computations below, we used the data from [131] and, in particular, the specific RCS of concrete for its oblique irradiation at the elevation angle of 10° is 1×10^{-4} m^2 for horizontal polarization of the incident wave, and its specific RCS was 1×10^{-3} m^2 given vertical polarization of incident wave [131]; the specific RCS of a grass covered summer steppe is 0.025 m^2 [131] for both polarizations of incident wave.

Our objective here is to reduce the radar contrast of the tank model against the background of its immediate surroundings rather than bringing its RCS figure to a

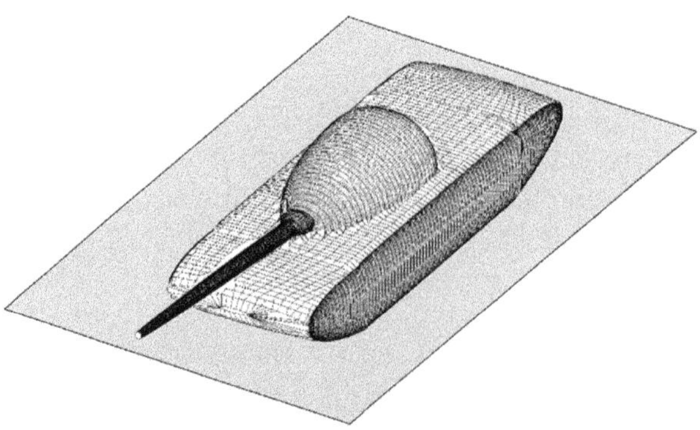

Figure 2.103. Simplified model of a tank.

minimum. To that end, we are going to apply the radar absorber selectively over the tank's surface to bring its radar returns as close as possible to the returns from the portion of ground situated in the radar shadow of the tank model for a certain direction of its irradiation by airborne radar.

The RCS of the above model was computed for the carrier frequency of radar signal equal to 10 GHz. Here, we considered only a monostatic radar layout. We also assumed a 1.3 mm thick radar absorber that had the following characteristics $\varepsilon'_1 = 20 + j0.1$ and $\mu'_1 = 1.35 + j0.8$. The radar absorber with these particular characteristics provides a 15 dB reduction in the RCS of a metallic plate for a normal incidence of EMW at the carrier frequency specified above upon the plate.

In our particular case, the object surface was broken down into 141 individual segments. We also computed the RCS of every surface segment for two conditions: the segment being coated with and being free from the above radar absorber. Now, by combining the coated and bare segments of the tank's surface we can change the mean RCS of the whole object in a controlled manner, so we can bring it to a desirable level. This desirable level will be defined by the mean RCS of the portion of the underlying surface embraced by the tank's radar shadow for a certain angular sector of its irradiation aspects.

In order to reduce computational burden as well as to reduce the quantity of radar absorbers needed to reach the objective, we actually tried applying the radar absorber to those segments of the tank surface that would have greater RCS values were they not coated with any absorber. In the case where we chose to apply the radar absorber to only the 25 surface segments (as opposed to all 141) providing the strongest radar returns, the RCS level of the first perfectly conducting segment to be disregarded in terms of its impact upon the total RCS figure of the tank would be $\approx 10^{-3}$–10^{-4} m^2 depending on the sector of irradiation aspects. Applying a radar absorber to such segments of the surface would not impact the total RCS of the tank in any significant way, yet it would increase the quantity of absorber used for camouflaging the tank dramatically, thus increasing the overall level of expense.

We selected the following sectors of the tank model radar irradiation aspect angles: head-on aspects from −5° to +5° and from −10° to +10°; side-on aspects from +40° to +50° and from +35° to +55°. The corresponding estimates of the RCS are presented in tables 2.5 and 2.6. All the values are given in meters squared.

The following notations are used in the tables: σ_g is the RCS of the tank's footprint in terms of its radar shadow; σ_{abs} is the RCS of the tank model if the tank's surface is fully coated with a radar absorber; σ_{pc} is the RCS of the tank if the tank's surface has no radar absorber at all; and σ_{opt} is the RCS of the tank model if the tank's surface has a radar absorber spread across it in the optimal manner explained above.

Analysis of the data presented in tables 2.5 and 2.6 shows the following:

1. The steppe as an underlying surface provides a significant level of background radar returns. Therefore, it is possible to come up with a radar absorber distribution pattern for the tank's surface which can bring the tank's radar contrast almost to an unit value. As a numerical measure for contrast, we used the ratio of the two RCSs, one is of the tank and the other is of the tank's footprint on the ground.

Table 2.5. Results of RCS computation with regard to the model of a tank for horizontal polarization of incident wave.

		Type of underlying surface	
Sector of irradiation aspects	RCS	Concrete	Steppe
−5° to +5°	σ_g	4.886E−03	1.2274
	σ_{abs}	0.5745	0.1927
	σ_{pc}	16.5733	5.869444
	σ_{opt}	0.5766	1.5089
−10° to +10°	σ_g	5.054E−03	1.2696
	σ_{abs}	0.4468	0.1573
	σ_{pc}	12.8057	4.6764
	σ_{opt}	0.4487	1.3265
+40° to +50°	σ_g	9.044E−03	2.2717
	σ_{abs}	0.1406	8.732E−02
	σ_{pc}	3.4512	2.7157
	σ_{opt}	0.1510	2.3607
+35° to +55°	σ_g	9.008E−03	2.2627
	σ_{abs}	0.1173	6.887E−02
	σ_{pc}	2.6824	2.0692
	σ_{opt}	0.1303	2.0692

Table 2.6. The results of RCS computation with regard to the model of a tank for vertical polarization of the incident wave.

		Type of underlying surface	
Sector of irradiation aspects	RCS	Concrete	Steppe
−5° to +5°	σ_g	4.886E−02	1.2274
	σ_{abs}	0.2326	0.1924
	σ_{pc}	7.3317	6.0018
	σ_{opt}	0.2345	1.5146
−10° to +10°	σ_g	5.054E−02	1.2696
	σ_{abs}	0.1838	0.1516
	σ_{pc}	5.8111	4.7938
	σ_{opt}	0.18545	1.3236

(*Continued*)

Table 2.6. (*Continued*)

		Type of underlying surface	
Sector of irradiation aspects	RCS	Concrete	Steppe
+40° to +50°	σ_g	9.044E−02	2.2717
	σ_{abs}	9.308E−02	8.673E−02
	σ_{pc}	3.3128	3.2195
	σ_{opt}	9.994E−02	2.8747
+35° to +55°	σ_g	9.008E−02	2.2627
	σ_{abs}	7.175E−02	6.596E−02
	σ_{pc}	2.5671	2.4874
	σ_{opt}	9.789E−02	2.2693

2. The concrete as an underlying surface provides an insignificant level of background radar returns. Therefore, coating the tank's entire surface with a radar absorber does not allow reducing the tank's contrast against the background as desired. At the same time, the application of a radar absorber in a certain pattern to only 25 segments of the tank's model surface out of a total of 141 segments is practically equivalent to the effect of coating the entire tank's surface with a radar absorber.

The optimal patterns of spreading the radar absorber over the tank model's surface for various types of underlying surface and various sectors of irradiation aspects are illustrated in figures 2.104–2.107. In these figures, we assume that radar absorber was in limited supply, so only 25 segments out of 141 of the tank's surface could be coated. The segments coated with radar absorber are shaded in light gray and outlined with a bold line.

Analysis of figures 2.104–2.107 taken together with the data of tables 2.5 and 2.6 reveals two remarkable tendencies.

1. In the case of the underlying surface providing an insignificant level of background returns (as in the case of concrete), the optimal pattern for

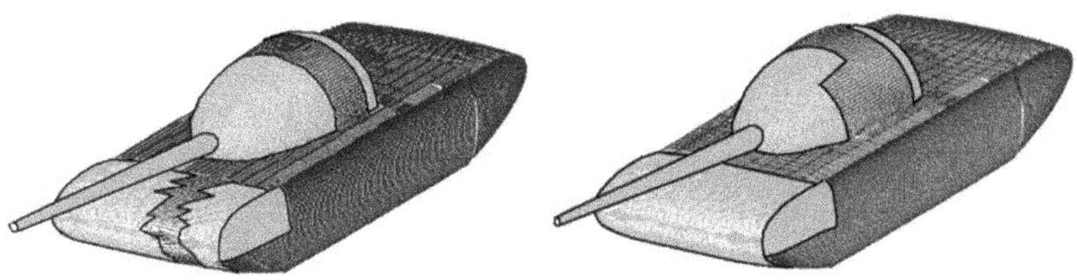

a) b)

Figure 2.104. The pattern for the optimal distribution of the radar absorber for the head-on sector of the irradiation aspects from −5° to +5° against a background of steppe (a) and concrete (b).

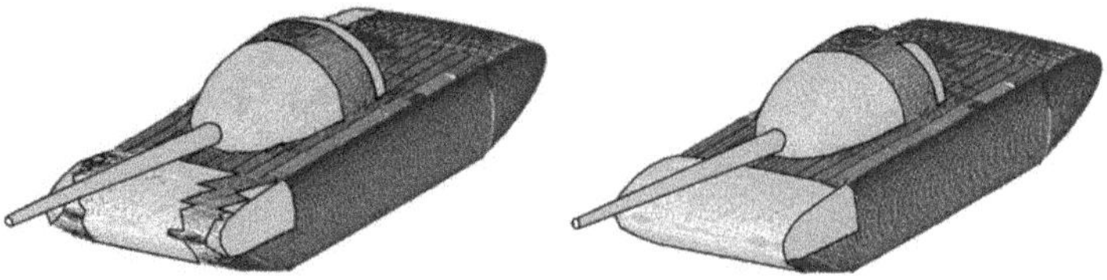

a) b)

Figure 2.105. The pattern for the optimal distribution of the radar absorber for the head-on sector of the irradiation aspects from $-10°$ to $+10°$ against a background of steppe (a) and concrete (b).

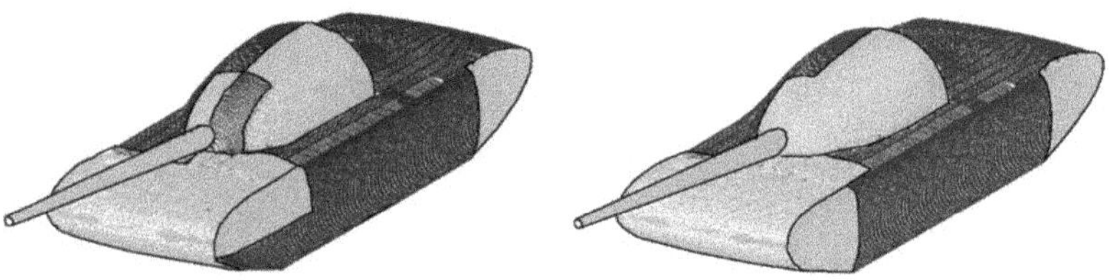

a) b)

Figure 2.106. The pattern for the optimal distribution of the radar absorber for the side-on sector of the irradiation aspects from $+40°$ to $+50°$ against a background of steppe (a) and concrete (b).

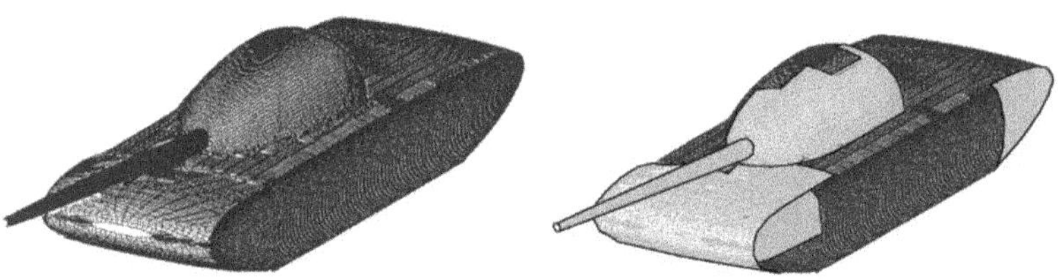

a) b)

Figure 2.107. The pattern for the optimal spread of the radar absorber for the side-on sector of the irradiation aspects from $+35°$ to $+55°$ against a background of steppe (a) and concrete (b).

applying the radar absorber to the tank's surface boils down to coating its 25 most salient segments (in terms of the RCS pertaining to a perfectly conducting model) with radar absorber. Using such an approach, one obtains the tank model with the RCS approaching that of a fully coated model.

2. In the case of the underlying surface providing a significant level of background returns (as in case of steppe), the optimal pattern may not use up the entire supply of radar absorber sufficient to coat all 25 segments of the tank's surface. In addition, the resulting optimal pattern may be quite elaborate. Moreover, there are cases (given for instance the side-on radar irradiation in the sector of aspects $+35°$ through $+55°$, figure 2.107(a)), where there is no need to apply any radar absorber to the tank's surface at all and where the

initial model with perfectly conducting surface is already pretty close to its footprint on the ground in terms of the RCS.

As a development of the second of the above two tendencies, we have considered the possibility of an additional reduction in the number of surface segments coated with radar absorber if the tank model is observed by radar against the background of steppe. Using the procedure described above aimed at finding the optimal pattern for radar absorber application, we also showed that one can obtain a decent reduction of radar contrast while applying as little radar absorber as possible.

Table 2.7 and figure 2.108 present the results of a computer simulation with regard to the RCS of the tank model observed by radar against a background of steppe for its side-on irradiation within the sector of aspects from +40° to +50° and given that the radar absorber was applied to only 7, 10, and 25 segments of the tank's surface (out of 141). Given this particular sector of irradiation aspects, the RCS of tank's footprint on the ground is $\sigma_g = 2.27$ m^2, the RCS of the fully coated tank model is $\sigma_{abs} = 0.087$ m^2, and the RCS of the tank model without any radar absorber coating applied is $\sigma_{pc} = 2.71$ m^2.

The computer simulation results presented in table 2.7 clearly demonstrate that a further decrease in the number of segments coated with radar absorber from 25 to 7 does not alter the tank model's RCS significantly, its variation not exceeding 2%. However, as seen from figure 2.108 the quantity of radar absorber that allows us to reduce the radar contrast against the background of the underlying surface with a significant level of radar returns can be decreased considerably.

Table 2.7. The results of the computer simulation involving a reduction in the quantity of radar absorber applied to the tank's model surface for the sector of its side-on irradiation aspects from +40° to +50° and horizontal polarization of the incident wave.

The number of surface segments coated with absorber	σ_{opt}, m^2	$\lvert \sigma_{opt} - \sigma_g \rvert$, m^2
7	2.405923	0.134241
10	2.380756	0.109074
25	2.360677	0.088995

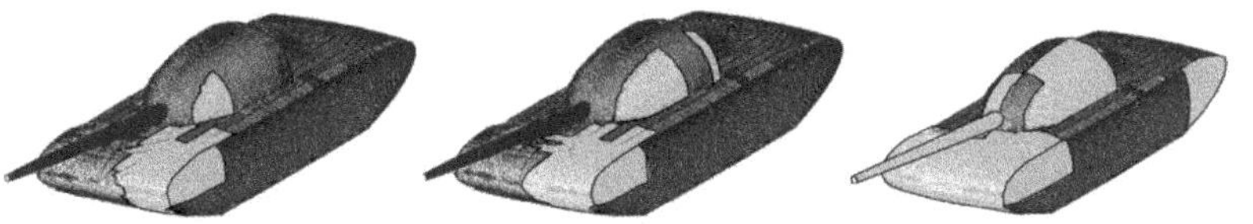

a) b) c)

Figure 2.108. The pattern for an optimal distribution of radar absorber over the surface of the tank model for its side-on irradiation with azimuth aspects from +40° to +50° and for various numbers of tank surface segments coated with radar absorber: 7 segments (a), 10 segments (b), and 25 segments (c).

It is worth noting that the simulation results presented in this section correspond to a simplified model of a tank, however, they indicate that the method for reducing the radar contrast of ground vehicles is operable. They also demonstrate that in certain situations, the application of small amounts of radar absorber distributed according to a certain pattern, can provide a good cloaking of ground vehicles against airborne radar surveillance systems.

References

[1] Sukharevsky O I, Muzychenko A V and Sazonov A Z 2004 Asymptotic solution of a problem of electromagnetic wave scattering by perfectly conducting cylindrical object embedded in a dielectric half-space *Telecommun. Radio Eng.* **61** 721–33

[2] Sukharevsky O I, Muzychenko A V and Sazonov A Z 2004 Asymptotic solution of a problem of electromagnetic wave scattering by impedance cylindrical object embedded in a dielectric half-space *Systemy obrobky Informatsii, Kharkiv Military University* **8** 71–7 (in Russian)

[3] Dmitriev V I and Zakharov E V 2008 *Integral Equation Method in Computational Electrodynamics* (Moscow: MAKS) (in Russian)

[4] Sirenko Yu K, Sukharevsky I V, Sukharevsky O I and Yashina N P 2000 *Fundamental and Applied Problems of Electromagnetic Wave Scattering* (Kharkiv: Krok) (in Russian)

[5] Nikolsky V V and Nikolskaya T I 1989 *Electrodynamics and Propagation of Radio Waves* (Moscow: Nauka) (in Russian)

[6] Tranter C J 1962 *Integral Transforms in Mathematical Physics* (York: Methuen)

[7] Gradshteyn I S and Ryzhik I M 2000 *Table of Integrals, Series and Products* 5th edn (New York: Academic)

[8] Fedoryuk M V 1987 *Asymptotics: Integrals and Series* (Moscow: Nauka) (in Russian)

[9] Sukharevsky O I, Zalevsky G S and Muzychenko A V 2001 Frequency and impulse responses of subsurface objects *Radiotekhnika* **6** 6–13 (in Russian)

[10] Ilinski A S and Galishnikova T N 1987 *The Numerical Methods in the Diffraction Problems* (Moscow: Moscow State University) (in Russian)

[11] Sukharevsky O I, Muzychenko A V and Sazonov A Z 2002 Estimation of geometrical parameters of perfectly conducting cylindrical object buried in dielectric half-space by its scattering characteristics *Int. Conf. on Mathematical Methods in Electromagnetic Theory (MMET) (Kiev)* pp 296–8

[12] Sukharevsky O I and Muzychenko A V 2000 Algorithm of imaging of 'illuminated' part of convex perfectly conducting surface *Radio Phys. Radio Astron.* **5** 311–9

[13] Bojarski N N 1967 *Three-dimensional Electromagnetic Short Pulse Inverse Scattering* (Syracuse, NY: Syracuse University Research Corporation)

[14] Lewis R M 1969 Physical optics inverse diffraction *IEEE Trans. Antennas Propag.* **17** 308–14

[15] Chaudhuri S K and Boerner W M 1976 A monostatic inverse scattering model based on polarization utilization *Appl. Phys.* **11** 337–50

[16] Chaudhuri S K and Boerner W M 1977 Polarization utilization in profile inversion of a perfectly conducting prolate spheroid *IEEE Trans. Antennas Propag.* **25** 505–11

[17] Bennet C L 1981 Time domain inverse scattering *IEEE Trans. Antennas Propag.* **29** 213–9

[18] Varganov M E *et al* 1985 *Radar Characteristics of Air Vehicles* ed L T Tuchkov (Moscow: Radio i svyaz) (in Russian)

[19] Sukharevsky O I, Vasylets V O, Nechitaylo S V and Orlenko V M 2014 *Electromagnetic Wave Scattering by Aerial and Ground Radar Objects* ed O I Sukharevsky (Boca Raton/London: CRC Press/Taylor and Francis) p 334

[20] Sukharevsky O I and Vasilets V A 1996 Impulse characteristics of smooth objects in bistatic case *J. Electromagn. Waves Appl.* **10** 1613–22

[21] Dines K A and Lytle R J 1979 Computerized geophysical tomography *Proc. IEEE* **67** 1065–73

[22] Kozakoff D J 2010 *Analysis of Radome-Enclosed Antennas* 2nd edn (Boston, MA: Artech House)

[23] Kaplun V A 1974 *UHF Antenna Radomes* (Moscow: Soviet Radio) (in Russian)

[24] Kaplun V A and Zelenkevich V M 1987 Introduction to computer-aided design of microwave antenna radomes *Radiotechnika i Electronika (USSR)* **42** 89–93 (in Russian)

[25] Zamyatin V I and Sukharevsky O I 1990 The backscattering of reflector antenna with dielectric envelope from side source waves *Antenny* (Moscow: Radio i svyaz) pp 78–87 (in Russian)

[26] Mikhailov G D, Kutischev S N and Kiryanov O E 1999 Radar cross section of reflector antenna with dielectric radome *Izvestiya vuzov. Radiophysika* **42** 879–85 (in Russian)

[27] Lansink Rotgerink J, van der Ven H, Voigt T, Jehamy E, Schick M and Schippers H 2016 Modelling of effects of nose radomes on radar antenna performance *10th European Conf. on Antennas and Propagation (EuCAP)* pp 1–5

[28] Shangji L, Huiyuan Z, Chang D and Lizhong S 2016 Electrical properties analysis of conical radome *2016 IEEE Int. Conf. on Ubiquitous Wireless Broadband (ICUWB)* pp 1–4

[29] Mrdakovic B Lj and Kolundzija B M 2016 A method for full wave analysis of electrically large transparent radomes *2016 IEEE Int. Symp. on Antennas and Propagation (APSURSI)* pp 1331–2

[30] Sheret T, Parini C and Allen B 2016 Efficient design of a radome for minimized transmission loss *IET Microw. Antennas Propag.* **10** 1662–6

[31] Rim J-W, Koh I-S and Choi K S 2016 IPO analysis of performance of arbitrary shaped radome *10th European Conf. on Antennas and Propagation (EuCAP)* pp 1–5

[32] Nair R U, Suprava M and Jha R M 2015 Graded dielectric inhomogeneous streamlined radome for airborne applications *Electron. Lett.* **51** 862–3

[33] Kim J H, Park Y B and Song S-C 2016 Electromagnetic analysis of a tangent-ogive dielectric radome with a metallic cap *2015 Int. Workshop on Antenna Technology (iWAT)* pp 367–8

[34] Dwivedi S and Mishra V 2013 Numerical analysis for directivity improvement of radome structure *2013 IEEE Applied Electromagnetics Conf. (AEMC)* pp 1–2

[35] Borovikov V A and Kinber B Y 1994 *Geometrical Theory of Diffraction* (London: IEE)

[36] Sukharevsky I V and Vashinski S E 1998 About the stationary phase points and caustic influence on lateral radiation of antenna systems with radomes *Int. Conf. on Mathematical Methods in Electromagnetic Theory* vol 2 *(Kharkiv)* pp 537–9

[37] Rosloniec S 2008 *Fundamental Numerical Methods for Electrical Engineering* (Berlin: Springer) p 284

[38] World Meteorological Organization 2003 *Guide to Practices for Meteorological Offices Serving Aviation* (Geneva: Secretariat of the World Meteorological Organization)

[39] Bazlova T A, Bocharnikov N V and Brylev G B 2002 *Automated Weather Radar Networks* (St Petersburg: GidrometeoIzdat)

[40] Vavriv D M, Volkov V A, Bormotov V N, Vynogradov V V, Kozhyn R V, Trush B V, Belikov A A and Semenyuta V Y 2002 Millimeter-wave radars for environmental studies *Radio Phys. Radio Astron.* **7** 121–38

[41] Vavriv D M, Bezvesilniy A A and Belikov A A 2012 Development of weather radars of mm wavelength range in the Radioastronomical Institute of National Academy of Sciences of Ukraine *First Ukrainian Conf. 'Electromagnetic Methods for Environmental Studies' (Kharkiv)* pp 77–82 (in Russian)

[42] Vavriv D M, Bezvesilniy O O, Volkov V A, Kravtsov A A and Bulakh E V 2015 Recent advances in millimeter-wave radars *Int. Conf. on Antenna Theory and Techniques (ICATT), Kharkiv, Ukraine* pp 19–24

[43] National Research Council (U. S.) Committee on Weather Radar Technology 2002 *Weather Radar Technology Beyond NEXRAD* (Washington, DC: National Academies Press)

[44] Atlas D 1990 *Radar in Meteorology: Battan Memorial and 40th Anniversary Radar Meteorology Conference* (Boston, MA: American Meteorological Society)

[45] Belov Y N, Voytovich O A, Makulina T A, Rudnev G A, Khlopov G I and Khomenko S I 2010 Software and hardware complex for investigation of meteorological radar echo *Telecommun. Radio Eng.* **69** 1517–27

[46] Boyarskii D A, Tikhonov V V, Kleeorin N I and Mirovskii V G 1994 Inclusion of scattering losses in the models of the effective permittivity of dielectric mixtures and applications to wet snow *J. Electromagn. Waves Appl.* **8** 1395–410

[47] King R W P and Smith G S 1981 *Antennas in Matter: Fundamentals, Theory, and Applications* (Cambridge, MA: MIT Press)

[48] Hallikainen M T, Ulaby F T and Abdelrazik M 1986 Dielectric properties of snow in the 3 to 37 GHz range *IEEE Trans. Antennas Propag.* **34** 1329–39

[49] Boyarskii D A, Mirovskii V G and Tikhonov V V 1994 Frequency depending model of effective permittivity of wet snow *Radiotekhnika i Electronika* **39** 1479–85 (in Russian)

[50] Shimba M, Sato T and Koike H 1988 The antenna pattern degradation by snow accretion on the reflector surface *IEEE Antennas and Propagat. Soc. Int. Symp.* vol 2 (Piscataway, NJ: IEEE) pp 871–4

[51] Ufimtsev P Y 2003 *Theory of Edge Diffraction in Electromagnetics* (Encino: Tech Science Press)

[52] Rahmat-Samii Y 1988 Reflector antennas *Antenna Handbook: Theory, Applications, and Design* ed Y T Lo and S W Lee (Princeton, NJ: Van Nostrand-Reinhold) pp 949–1072

[53] Ip H-P and Rahmat-Samii Y 1998 Analysis and characterization of multilayered reflector antennas: rain/snow accumulation and deployable membrane *IEEE Trans. Antennas Propag.* **46** 1593–1605

[54] Bakhrakh A D and Galimov G K 1981 *Dish Scanning Antennas* (Moscow: Nauka) (in Russian)

[55] Sukharevsky O I, Nechitaylo S V, Khlopov G I and Voitovych O A 2015 Influence of the snow cover on radiation characteristics of reflector antennas *J. Commun. Technol. Electron.* **60** 594–601

[56] Sukharevsky O I, Khlopov G I, Nechitaylo S V and Voitoich O A 2012 Method of pattern calculation for reflector antenna with surface partly covered by snow *Int. Conf. on Mathematical Methods in Electromagnetic Theory (MMET)* pp 308–11

[57] Sukharevsky O I, Nechitaylo S V, Khlopov G I and Voitovich O A 2013 Precipitations influence on reflector antenna radiation characteristics *Int. Conf. on Antenna Theory and Techniques (ICATT)* pp 447–9

[58] Brekhovskikh L M 1976 *Waves in Layered Media* (New York: Academic)

[59] Belov E N, Voitovich O A and Makulina T A 2009 Hardware and software complex for studying radar returns from meteoformations *Radiofizika I elektronika, Kharkiv* **14** 57–63 (in Russian)

[60] State Service for Hydrometeorology and Environmental Monitoring 1993 Manual on conducting observations and using information by non-automated radars MRL-1, MRL-2, and MRL-5 *Regulation* B2.04.320-91 GidroMeteoIzdat, St Petersburg (in Russian)

[61] State Service for Hydrometeorology and Environmental Monitoring 1974 Manual on conducting observations and using information by radars MRL-1 and MRL-2 *Regulation* B2.04.320-91 GidroMeteoIzdat, Leningrad (in Russian)

[62] Aizenberg G Z 1957 *Microwave Antennas* (Moscow: Svyazizdat) (in Russian)

[63] Finkelshtein M I, Lazarev E I and Chizhov A N 1984 *Aerial Radar Survey of Ice on Rivers, Lakes, and Storage Reservoirs* (Leningrad: Gidrometeoizdat) (in Russian)

[64] Sukharevsky O I, Nechitaylo S V, Khlopov G I and Voitovich O A 2015 The influence of asymmetric water layer on the radiation characteristics of reflector antenna *Int. Conf. on Antenna Theory and Techniques (ICATT) (Kharkiv, Ukraine)* pp 219–21

[65] Sukharevsky O I, Nechitaylo S V, Khlopov G I and Voitovych O A 2015 Radiation characteristics of single-dish antennas partially covered by a layer of water *Radioelectron. Commun. Syst.* **58** 61–8

[66] Ray P S 1972 Broadband complex refractive indices of ice and water *Appl. Opt.* **11** 1836–44

[67] Institute of Electronic Measurements Kvarz 1979 High frequency signal generators G4-78, G4-79, G4-80, G4-81, G4-82, G4-83 *Technical operation manual and logbook* 3.260.043 PS (in Russian)

[68] Kaunas Research Institute of Radio Measuring Equipment 1982 Selector amplifier U2-8 *Technical Operation Manual* (in Russian)

[69] Burenin N I, Popovsky V V and Lemansky A A 1990 Problems and methods for ensuring electromagnetic compatibility by means of antennas *Review, Antennas: Collection of Papers* **37** 107–23 (in Russian)

[70] Yampolsky V G and Frolov O P 1983 *Antennas and Electromagnetic Compatibility* (Moscow: Radio I sviaz)

[71] Nanos A A, Samaras T, Vafiadis E E and Sahalos J N 2003 On the safety evaluation of the space in the vicinity of an aperture antenna *IEEE Trans. Antennas Propag.* **51** 1706–15

[72] Knott E F 1985 A progression of high-frequency RCS prediction techniques *Proc. IEEE* **73** 252–64

[73] Sukharevsky O I, Nechitaylo S V and Sazonov A Z 2005 A method for computing the near-field field radiated by an antenna with non-plane edge *Kharkiv, ONDI ZS, Collection of Papers, #1* pp 174–83 (in Russian)

[74] Ivanov V K, Nechitaylo S V, Sazonov A Z, Sukharevsky O I and Ryabokon E A 2004 A method for computing the near-field of a reflector antenna *Institute of Problems in Modeling in Energy Industry, Collection of Papers, # 25* pp 48–55 (in Russian)

[75] Sukharevsky O I and Dobrodniak A F 1988 Three-dimensional problem of diffraction by an ideally conducting wedge with a radio-absorbing cylinder on the edge *Radiophys. Quantum Electron.* **31** 763–9

[76] Nechitaylo S V, Sukharevsky O I and Vasilets V A 2017 Near-field calculation for reflector antenna with shape of asymmetrical cutting from a paraboloid of revolution *Int. Conf. on Antenna Theory and Techniques (ICATT) (Kyiv, Ukraine)* pp 88–91

[77] Austin R 2011 *Unmanned Aircraft Systems: UAVS Design, Development and Deployment* (New York: Wiley)

[78] Yuricich J G and Radar A 2016 *Cross Section Estimations to Minimize Radar Echo in Unmanned Combat Air Vehicle Design* (Columbus, OH: The Ohio State University)

[79] Wilton D and Butler C 1976 Efficient numerical techniques for solving Pocklington's equation and their relationships to other methods *IEEE Trans. Antennas Propag.* **24** 83–6

[80] Sosa-Pedroza J, Barrera-Figueroa V and López-Bonilla J 2006 Pocklington equation method versus curved segments technique for the numerical study of circular antennas *Apeiron* **13** 260–73

[81] Barrera-Figueroa V, Sosa-Pedroza J and López-Bonilla J 2009 Pocklington equation via circuit theory *Apeiron* **16** 45–69

[82] Sosa-Pedroza J, Barrera-Figueroa V and Lypez-Bonilla J 2005 Pocklington equation and the method of moments *Proc. Pakistan Acad. Sci.* **42** 243–7

[83] Tretiakov Y and Pan G 2004 Malvar wavelet based Pocklington equation solutions to thin-wire antennas and scatterers *Prog. Electromagn. Res.* **47** 123–33

[84] Claeys X 2009 On the theoretical justification of Pocklington's equation *Math. Models Methods Appl. Sci.* **19** 1325–55

[85] Marasini C 2008 Efficient computation techniques for Galerkin MoM antenna design *PhD Thesis* Technische Universiteit Eindhoven p 262

[86] Katsenelenbaum B Z 2006 *High-frequency Electrodynamics* (Weinheim: Wiley)

[87] Tsang L, Kong J A and Ding K-H 2000 *Scattering of Electromagnetic Waves: Theories and Applications* (New York: Wiley)

[88] Rogers D F and Adams J A 1990 *Mathematical Elements for Computer Graphics* 2nd edn (New York: McGraw-Hill)

[89] Sukharevsky O I and Gorelyshev S A 1997 Passing the ultra-wideband signal in layer-uniform half-space with the parameters of the ground *J. Electromagn. Waves Appl.* **11** 1091–102

[90] Sukharevsky O I, Vasilets V A and Gorelyshev S A 2012 UWB pulse backscattering from objects located near uniform half-space *Ultrawideband Radar: Application and Design* ed J D Taylor (London: Taylor and Francis)

[91] Altair Feko https://www.altair.com/feko/

[92] Knott E F, Shaeffer J F and Tuley M T 2004 *Radar Cross Section* 2nd edn (Raleigh, NC: SciTech Publishing, Inc.)

[93] Shirman Y D, Gorshkov S, Leshenko S P, Orlenko V, Sedyshev S and Sukharevskiy O I 2002 *Computer Simulation of Aerial Target Radar Scattering Recognition, Detection and Tracking* ed Ya D Shirman (Boston, MA: Artech House)

[94] Lvova L A 2003 *Radar Visibility of Air Vehicles* (Snezhinsk: RFYaC—VNIITPh) (in Russian)

[95] Gibson W C 2008 *The Method of Moments in Electromagnetics* (London: Taylor and Francis)

[96] Zalevsky G S, Vasilets V A and Sukharevsky O I 2014 High resolution range profiles of cruise missiles in different wavelength ranges *Prikladnaya Radioelektronika* **13** 20–8 (in Russian)

[97] Maizels Y N and Torgovanov V A 1972 *Measurement of Scattering Characteristics of Radar Targets* ed M A Kolosov (Moscow: Soviet Radio) (in Russian)

[98] Sirman Ya D 2007 *Radioelectroni Systems: Design Basis and the Theory. Handbook* 2nd edn (Moscow: Radiotekhnika) (in Russian)

[99] MVG Microwave Vision Group *Solutions for Aerospace and Defense* http://www.microwavevision.com/panels/industries/aerospace-defense

[100] The Howland Company *Radar Cross section (RCS) Range Design & Evaluation* http://www.thehowlandcompany.com/radar_stealth/RCS-ranges.htm

[101] Damaskos, Inc. *Material Measurement Solution. Anechoic Chamber* http://www.damaskosinc.com/products.htm

[102] Kamensky E F, Leonov I G, Maksyuta D V, Nakonechny V S, Nechitaylo S V, Orekhov S V and Poberezhny A A 2000 Microwave anechoic chamber of centimeter wavelength range and methods of their main capability check *Collected Scientific Papers of Kharkiv Military University* **4** 72–8 (in Russian)

[103] Sukharevsky O I, Zalevsky G S, Ryapolov I Y and Vasilets V A 2020 Backscattering calculation of inflatable dielectric lifting-turning device for ground measurement of radar object scattering characteristics *IEEE Ukrainian Microwave Week (UkrMW)* pp 964–69

[104] Silver S 1949 *Microwave Antenna Theory and DesignMIT Radiation Laboratory Series, No 12* (New York: McGraw-Hill)

[105] Byzova N L, Ivanov V N and Garger E K 1989 *Turbulences in the Boundary Layer of Atmosphere* (Leningrad: Gidrometeoizdat) (in Russian)

[106] *Handbook on Radiometeorology* https://www.itu.int/dms_pub/itu-r/opb/hdb/R-HDB-26-2013-OAS-PDF-E.pdf

[107] Chernikov A A 1979 *Clear Sky Radar Returns* (Leningrad: Gidrometeoizdat) (in Russian)

[108] Martin W J and Shapiro A 2007 Discrimination of bird and insect radar echoes in clear air using high-resolution radars *J. Atmos. Ocean. Technol.* **24** 1215–30

[109] Fukao S and Hamazu K 2014 *Radar for Meteorological and Atmospheric Observations* (Tokyo: Springer Japan) p 561

[110] Yano J-I 2014 Basic convective element: bubble or plume? A historical review *Atmos. Chem. Phys.* **14** 7019–30

[111] Akhmetov D G 2009 *Vortex Rings* (Berlin: Springer) p 150

[112] Yusupaliev U, Shuteev S A, Vinke E E and Yusupaliev P U 2010 Vortex rings and plasma toroidal vortices in a homogeneous infinite medium. I. Maximum vortex path *Bull. Lebedev Phys. Inst* **37** 227–33

[113] Tzarouchis D C, Ylä-Oijala P and Sihvola A 2017 Resonant scattering characteristics of homogeneous dielectric sphere *IEEE Trans. Antennas Propag.* **65** 3184–91

[114] Tzarouchis D and Sihvola A 2018 Light scattering by a dielectric sphere: perspectives on the Mie resonances *Appl. Sci.* **8** 1–22

[115] Kucharski A A 2018 The FIT-MoM hybrid method for analysis of electromagnetic scattering by dielectric bodies of revolution *IEEE Trans. Antennas Propa.* **66** 1384–91

[116] Kong B and Sheng X 2018 Discontinuous Galerkin surface integral equation method for scattering from multiscale homogeneous objects *IEEE Trans. Antennas Propag.* **66** 1937–46

[117] Huang L, Hou Y, Zhang H, Zhou L and Yin W 2019 Time-domain discontinuous Galerkin PMCHW integral equation method with MOD scheme for simulating electromagnetic pulse responses of arbitrarily shaped dielectric objects *IEEE Trans. Electromagn. Compat.* **61** 1157–66

[118] Huang X and Sheng X 2019 A discontinuous Galerkin self-dual integral equation method for scattering from IBC objects *IEEE Trans. Antennas Propag.* **67** 4708–717

[119] Shariff K and Wray A 2002 Analysis of the radar reflectivity of aircraft vortex wakes *J. Fluid Mech.* **463** 121–61

[120] Bibby M M, Coldwell C M and Peterson A F 2013 A high order numerical investigation of electromagnetic scattering from a torus and a circular loop *IEEE Trans. Antennas Propag.* **61** 3656–61

[121] Vainshtein L A 1988 *Electromagnetic Waves* (Moscow: AST) (in Russian)

[122] Fedoryuk M V 1977 *Method of Steepest Descent* (Moscow: Nauka) (in Russian)

[123] Vulfson A N and Borodin O O 2016 System of convective thermals as a generalized ensemble of Brownian particles *Phys. Usp.* **59** 109–20

[124] Ufimtsev P Y 1996 Comments on: Diffraction principles and limitations for RCS reduction techniques *Proc. IEEE* **84** 1828–51

[125] Ponkin V F and Yarygin F P 2005 Reducing radar visibility of weapons and military equipment: problems and solutions *Voyennaya mysl* **8** 10–13 (in Russian)

[126] Alekseev A G, Shtager Y A and Kozyrev S V 2007 *Physical Bases of STEALTH Technology* (St Petersburg: VVM) (in Russian)

[127] Lagarkov A N and Pogosyan M A 2003 Fundamental and applied problems of stealth technologies *Vestnik RAN* **73** 779–87

[128] Nikolsky B A 2007 *Methods and Equipment of Electronic Protection of Aircraft* (Samara: Samara State University) (in Russian)

[129] Kulemin G P 2003 *Millimeter-wave Radar Targets and Clutter* (Boston, MA: Artech House)

[130] Semenov A A 1968 *Electromagnetic Wave Theory* (Moscow: Moscow State University) (in Russian)

[131] Verba V S, Neronsky L B, Osipov I G and Turuk V E 2010 *Space-based Ground-Survey Radar Systems* (Moscow: Radoitekhnika) (in Russian)

[132] Solunina V L, Pavelev V A and Khaminov D V 2009 *Scattering of Millimeter Range Electromagnetic Waves by Natural and Anthropogenic Objects* (Moscow: Bauman State Technical University) (in Russian)

[133] Ivanov V K, Kurekin A S, Yatsevich S E, Mozhaev A A and Gutnik A V 2005 Investigation of the backscattering of radio waves from small-sized objects in remote methods of earth sensing *Kharkiv, Systemy Obrobky Informatsii* **2** 161–5 (in Russian)

[134] Vnotchenko S L, Dostovalov M Yu, Yermakov R V, Musiyants T G and Sevalkina Ye P 2013 The main results of sounding the Earth and water surface by a multi-frequency radar complex of radars with a synthetic aperture 'Kompakt' *Vestnik SibGAU* **5** 35–8 (in Russian)

[135] Verba V S 2008 Aerial equipment of radar surveillance and targeting *Status and Prospects of Development* (Moscow: Radiotekhnika) (in Russian)

[136] Sukharevsky O I *et al* 2017 Scattering and radiation characteristics of antenna systems under nose dielectric radomes *Prog. Electromagn. Res.* B **76** 141–57

www.ingramcontent.com/pod-product-compliance
Ingram Content Group UK Ltd.
Pitfield, Milton Keynes, MK11 3LW, UK
UKHW051859100926
8556IPUK00005B/700

9 780750 339773